Advances in
APPLIED BIOLOGY

Advances in
APPLIED BIOLOGY

Vol. VI

Edited by

T. H. Coaker

*Department of Applied Biology, University of Cambridge,
Cambridge, England*

1981

ACADEMIC PRESS
London   New York   Toronto   Sydney   San Francisco
A Subsidiary of Harcourt Brace Jovanovich, Publishers

ACADEMIC PRESS INC. (LONDON) LTD.
24/28 Oval Road
London NW1

United States Edition published by
ACADEMIC PRESS INC.
111 Fifth Avenue
New York, New York 10003

574

A 652

British Library Cataloguing in Publication Data

Advances in applied biology—Vol. 6
1. Agriculture—Periodicals
630   S493

ISBN 0-12-040906-2
LCCN 76-1065

Text set in 11/12 Monophoto Baskerville by
Northumberland Press Ltd., Gateshead, Tyne and Wear
Printed in Great Britain by Fletcher and Son Ltd., Norwich

208635

Contributors

M. P. BROOKER, *Welsh Water Authority, Nash Area Authority, West Nash, Newport, Gwent, Wales.*

P. A. CARLING, *FBA Teesdale Unit, c/o Northumbrian Water Authority, Lartington Treatment Plant, Barnard Castle, UK.*

D. T. CRISP, *FBA Teesdale Unit, c/o Northumbrian Water Authority, Lartington Treatment Plant, Barnard Castle, UK.*

D. E. DAVIS, *777 Picacho Lane, Santa Barbara, California, USA.*

R. E. GOODIER, *Nature Conservancy Council, Edinburgh EH9 2AS, Scotland.*

W. B. JACKSON, *Environmental Studies Center, Bowling Green State University, Bowling Green, Ohio, USA.*

J. N. R. JEFFERS, *Institute of Terrestrial Ecology, Grange-over-Sands, Cumbria, LA11 6GU, England.*

R. R. B. LEAKEY, *Institute of Terrestrial Ecology, Bush Estate, Penicuik, Midlothian, EH26 0QB, Scotland.*

N. J. MILNER, *Welsh Water Authority, Area Laboratory, Highfield, Priestley Road, Caernarfon, Wales.*

H. A. ROBERTS, *National Vegetable Research Station, Wellesbourne, Warwick, UK.*

J. SCULLION, *UWIST Field Centre, Newbridge-on-Wye, Powys, Wales.*

Preface

The main purpose of this series is to draw together subject matter from currently important fields of Applied Biology to produce a synthesis for students, teachers and specialists from other fields. The widely diversified and multidisciplinary nature of Applied Biology emphasizes the importance of the series since interdependence of the various subjects it embraces cannot be ignored, especially if the fruits of research are ultimately to be applied for the benefit of man.

Contributions to the series are written by specialists in the field, some with a biological background, others with chemical or physical qualifications, each presenting his subject matter in his own style.

This volume, like preceding ones, contains papers which between them span the widely different fields of seed banks in soils and the adaptive biology of vegetatively regenerating weeds, the impact of impoundments and discharge on the ecology of rivers, biosphere and biogenetic reserves, and rat control. Each article seeks to fulfil four aims. They review the state of a subject generally, but select one or more aspect for particular critical discussion; they summarize what is known to be of practical importance and indicate where each author sees the greatest need for future work.

It is the editor's view that it would be difficult to mould contributions in a particular volume into a homogenous unit resembling a monograph. It is the aim, however, that volumes accumulate to provide the individual with the material he needs to synthesize his own views on Applied Biology as a whole.

July 1981   T. H. COAKER

Contents

Seed Banks in Soils

H. A. ROBERTS

National Vegetable Research Station,
Wellesbourne, Warwick, UK

I. Introduction

The term "seed bank" is a short and convenient one which has been widely adopted in recent years to denote the reserves of viable seeds present in the soil and on its surface. "Seeds" is used here in the broad sense to include fruits as well as true seeds, but not spores or propagules which are produced vegetatively.

Although Darwin (1859) took samples of pond mud and remarked upon the large numbers of seedlings that arose from seeds present in it, the first detailed study of seeds in the soil appears to be that of Putensen (1882) who examined their occurrence at three depths. Another early investigation was that by Peter (1893) who determined the seeds present in soil samples from forests. He showed that while only seeds of woody species occurred in ancient forests, those of species characteristic of arable fields or grassland could be found in the soil

1

beneath forests of more recent origin, and he concluded that many seeds could remain viable in the soil for more than 50 years. In the early years of this century, examinations both of seed populations in soils and of the techniques for quantifying them were made by workers in various countries (Mal'tsev, 1909; Snell, 1912; Wehsarg, 1912; Brenchley, 1918). Subsequently, detailed studies have been made in a range of vegetation types. Much of the information obtained about seed banks of grasslands has been reviewed by Major and Pyott (1966), who stressed the importance of considering the buried seeds as part of the flora. The investigations of seed banks of arable fields have been discussed by Kropáč (1966) and Roberts (1970). In comparison with the extensive literature on plant communities, however, that dealing with seed banks is relatively sparse, and it is only recently that much attention has been devoted to their role in "natural" habitats as distinct from those of agricultural importance.

The size of the seed bank, the relative contributions made to it by different species and the pattern of seed distribution reflect the seed production in the plant community and its surroundings not only in the preceding year but perhaps in many previous years. Seed banks are therefore of interest in the study of recent vegetational history, although the floristic representation is biased because of the considerable differences among species in seed output, dispersal and survival (Harper, 1977). The period for which seeds remain in or on the soil is also one of the phases which needs to be quantified in studies of the population dynamics of individual species (Sagar and Mortimer, 1976). For the applied biologist in particular, however, the aspect of greatest significance is the role of the seed bank in determining the future vegetation, especially after natural or deliberate perturbation.

In this article the techniques for studying seed populations in soils are considered first, since they have a bearing on the validity of the conclusions which can be drawn from investigations of seed banks. Its main purpose, however, is to review recent literature on seed banks in different vegetation types, with emphasis on the practical significance of the findings.

II. Estimation of Seed Banks

The most direct method of detecting the presence of viable seeds in soil is to observe the emergence of seedlings *in situ*. If the existing vegetation is removed and the underlying soil is disturbed, the seeds present may be released from enforced dormancy (Harper, 1957) and innate or induced dormancy overcome because of exposure to light

(Wesson and Wareing, 1967) or to enhanced fluctuations of temperature (Thompson *et al.*, 1977). This method of estimation has been used in studies of seeds beneath steppe and forest vegetation (Korneva, 1970; Petrov, 1977), of long-term seed survival in undisturbed ruderal sites (Ødum, 1978) and of the establishment of *Trifolium subterraneum* from seed reserves (Beale, 1974). If influx of fresh seeds is prevented, the technique is positive in that all seedlings must have arisen from viable seeds. It is likely, however, that the germination requirements of many of the seeds present will not be met during the period of observation. So that although this method can be useful in certain circumstances, and can provide a measure of the seeds which are "available" at the time of soil disturbance, it does not provide a full statement of the seed bank.

The technique much more generally employed is to take representative soil samples and to determine seed numbers either by placing the soil under conditions suitable for germination, or by using physical methods of separation from the soil particles based on differences in size and/or density.

A. SAMPLING

The most usual requirement is for an assessment of the numbers of viable seeds of each species present in the soil of a defined area such as a plot or a field. The results are normally expressed as mean numbers of seeds m^{-2} and related to a soil layer of specified depth. The problems in sampling fields for dormant seeds are similar to those encountered in work with other organisms such as nematodes (Southey, 1970). There are errors associated with heterogeneity, which may occur on more than one scale. In the absence of prior information on distribution and when an average assessment over an area is required the arrangement of sampling points in a "W" configuration, as recommended for sampling soil for chemical analysis (Jackson, 1958) or sampling clustered plant-disease distributions (Lin *et al.*, 1979), seems likely to be satisfactory. If the aim is to determine changes in the seed bank along a gradient, for example in topography or distance from seed-bearing plants, then a transect approach would be appropriate.

For determinations of the surface seed bank, the seeds can be retrieved by suction from a delineated sample area. Samples of soil to a specified depth can be obtained by removing blocks of soil of known dimensions or by using a sampling tool which extracts a soil core perhaps no more than 2·5 cm in diameter. Provided that the design of the corer minimizes compaction, the cores can be sectioned to obtain samples from different depths. Clay soils, however, are often difficult to sample by means of corers.

Studies of the sampling intensity necessary to estimate seed populations with prescribed degrees of accuracy in grassland (Rabotnov, 1958) and in arable soil (Smirnov and Kurdyukov, 1965) show that large numbers of samples are needed in order to obtain satisfactory results for individual species. Much depends on the seed density and distribution in the particular community under consideration, and most workers have compromised by taking what appeared to be a "reasonable" number of samples. As a general principle it seems wise to take a large number of small samples rather than a small number of large ones (Kropáč, 1966; Roberts, 1970; Dospekhov and Chekryzhov, 1972). Ideally a preliminary sampling study would be made, and even where this is not feasible it is desirable to obtain sufficient data to calculate whether increasing the sampling intensity would bring about a worthwhile gain in accuracy (Jones and Bunch, 1977).

One approach adopted in Japan has been to combine the sampling units to construct a species–soil volume curve in order to derive a minimum volume which will detect most species present. Minimum volumes found were 400 cm^3 in the early stages of a succession (Numata et al., 1964), 500–600 cm^3 in grassland (Hayashi and Numata, 1971) and 4000–6000 cm^3 in climax forest (Hayashi and Numata, 1968). Hayashi (1975) recommends that sufficient samples should be taken to make up twice the minimum volume.

Whipple (1978) tabulated the numbers and sizes of samples used in published studies of seed banks of non-agricultural communities in North America and concluded that because of the small surface areas they represented, the results should be viewed as estimates of the presence of species rather than of seed densities. Nevertheless, provided that the limitations are recognized, useful and valid results can be obtained, and if data are recorded separately for individual sampling units or replicated groups of them, estimates of error can be assigned to the mean values (Moore and Wein, 1977; Takahashi and Hayashi, 1978; Archibold, 1979).

The depth to which samples must be taken depends on the type of vegetation and the purpose of the investigation. In arable fields it may be sufficient to obtain an overall estimate of seed numbers in the cultivated soil layer, to a depth of perhaps 15 or 25 cm. Often, however, the occurrence of seeds in successive 5- or 10-cm layers is recorded (Kropáč, 1966; Fekete, 1975). Such data are necessary to evaluate the effects of different cultivation techniques on the numbers of seeds in the soil (Kazantseva and Tuganaev, 1972; Fay and Olson, 1978).

Some experiments have been done in which seeds of different kinds have been sown on the soil surface and their distribution in depth determined after various cultivations (Soriano *et al.*, 1968; Pawłowski and Malicki, 1969). In relatively undisturbed habitats, most seeds are usually found near the soil surface and for many purposes samples to a depth no greater than 5 cm may be adequate (van Altena and Minderhoud, 1972; Jones and Bunch, 1977). The depths at which concentrations of seeds of particular species occur may be significant in interpreting seed origin and vegetation history; sampling successive depth increments can provide data of this kind (Komendar *et al.*, 1973; Howard, 1974). Some detailed studies have been made in which both horizontal and vertical distributions of seeds in the soil have been determined (Želev, 1965; Kellman, 1978; Symonides, 1978).

The time of sampling in relation to the annual vegetation cycle must also be considered. In studies of the "survey" type, samples are often taken on a single occasion only, but although useful comparative data for different fields or localities can be obtained in this way, the results can be influenced by recent major inputs of seeds of particular species. Samples taken at the same time in successive years, however, can provide valid assessments of quantitative changes in the seed bank (Warington, 1958; Roberts and Dawkins, 1967; Barralis, 1972). To detect the changes that occur within the annual cycle, more frequent sampling is necessary. Recent studies of this kind have been made in a variety of plant communities including forest in Mexico (Guevara and Gómez-Pompa, 1972), sagebrush in Nevada (Young and Evans, 1975), psammophytes in Poland (Symonides, 1978) and for *Sporobolus airoides* in New Mexico (Knipe and Springfield, 1972).

B. SEEDLING EMERGENCE

In this method of determining the numbers of seeds in a sample, the soil is placed directly into a shallow container or spread in a thin layer on a suitable medium, kept moist, and the seedlings that emerge are identified and recorded. The aim is to ensure that as many as possible of the viable seeds present germinate and produce seedlings. If the soil contains a great deal of clay, its removal by washing through a fine sieve not only reduces the bulk but is likely to improve germination (Brenchley and Warington, 1930; Kropáč, 1966).

If only a single species is being studied, the samples can be kept under those conditions known to promote maximum germination of that particular species. Seeds of some species may require special procedures to overcome dormancy; Gratkowski (1964) obtained germination of *Ceanothus velutinus* by first heating and then chilling the soil. Usually, however, a range of species is involved which differ in their requirements for germination. If the samples are kept for only a short period, and particularly if the conditions remain fairly constant, a biased estimate of the seed bank will be obtained. This may itself be of interest in providing a measure of the "available" flora at a particular time (Barralis, 1972; Young and Evans, 1975). Results reported by Mott (1972) give a striking illustration of the effects of temperature. When soil samples from an arid annual pasture were kept under diurnal fluctuations of 30/25°C seedlings of monocotyledons predominated, whereas at 20/15°C those of dicotyledons were more numerous. Other workers have subjected soil samples to a range of treatments in order to assess the effects of management practices on the resulting flora (Strickler and Edgerton, 1976; Howard and Smith, 1979). The time of year at which the samples are taken can also have a marked effect; those taken after the seeds have been exposed to natural chilling during winter may yield appreciably more seedlings than samples taken in autumn (Raynal and Bazzaz, 1973; Leck and Graveline, 1979).

Germination and seedling emergence are favoured if the soil layer is shallow and if it is periodically disturbed. Fluctuating temperatures promote germination of seeds of many species, and exposure to periods of low temperature or desiccation may be given to overcome dormancy (Barralis, 1972; Hurle, 1974; Johnson, 1975). Many workers have noted that seedlings of some species appear rapidly while others continue to produce seedlings over a long period. The samples must therefore be kept for a considerable time if a valid estimate of all viable seeds present is to be obtained. Vega and Sierra (1970) kept samples of a tropical soil for three years and found that 83% of all seedlings appeared in the first year, 16% in the second and 1% in the third. This is consistent with earlier results for arable seed populations in temperate countries, suggesting that a period of two years would be a reasonable compromise.

Vanesse (1976) reported that preliminary drying improved seedling emergence from samples of forest soil but Feast and Roberts (1973) found that while drying promoted germination of *Chenopodium album*, the numbers of seedlings of other annual weeds were either not affected or were depressed. None of the simple treatments which they tried gave any consistent improvement over their standard procedure in which the samples were exposed to diurnally and seasonally fluctuating temperatures in an unheated glasshouse. Germination of many aquatic species is affected by whether the substrate is covered by water or not, and in studies of seed banks of wetlands it may be necessary to provide both these conditions (van der Valk and Davis, 1978; Leck and Graveline, 1979).

C. SEPARATION BY PHYSICAL METHODS

The second widely used method of enumerating seeds in soil samples utilizes differences in size or density to separate them from the soil components, although usually this is not entirely effective and hand-sorting is also required. Some published procedures are summarized in Table I.

1. *Sieving*

This entails washing the soil samples on sieves of appropriate mesh or pore size to reduce the volume of soil from which the seeds must be removed. In studies of individual species, sieving may be largely satisfactory by itself. With large seeds such as those of *Avena fatua* (Wilson, 1972; Wilson and Cussans, 1975) or *Ulex europaeus* (Ivens, 1978; Zabkiewicz and Gaskin, 1978), the soil can be reduced to a volume from which hand-sorting, directly or after airflow separation, can readily be achieved. Fay and Olson (1978) described a technique for seeds of *A. fatua* in which the soil is placed in nylon net bags which

TABLE I. Some procedures for separating seeds from soil

Authors	Steps in the procedure				
Hyde and Suckling (1953)	Sieving (wet)	→ Flotation (organic)	→ Airflow separation	→ Hand-sorting (dry)	→ Viability determination (germination)
Malone (1967)	Dispersion/ flotation (solution)	→ Sieving (wet)	→ Hand-sorting (dry)	→ Viability determination (tetrazolium)	
Jones and Bunch (1977)	Sieving (wet)	→ Airflow separation	→ Flotation (organic)	→ Hand-sorting (dry)	→ Viability determination (germination)
Roberts and Ricketts (1979)	Sieving (wet)	→ Flotation (water)	→ Flotation (solution)	→ Hand-sorting (wet)	→ Apparent viability determination (pressure)
Standifer (1980)	Soaking	→ Sieving (wet then dry)	→ Airflow separation (3-stage)	→ Hand-sorting (dry)	→ Viability determination (germination)

are mechanically agitated in water and the seeds then removed by hand. They quote the time required for the whole procedure as 5–26 minutes for a 3–5 kg soil sample and note that it would be applicable to other large seeds. If the seeds are very small, like those of *Orobanche* spp., sieving can remove most of the soil components (Krishna Murty and Chandwani, 1974). Ashworth (1976) described a technique for the quantitative detection of *Orobanche ramosa* in which the particle size range was first reduced to 125–495 μm by sieving, this fraction was centrifuged in calcium chloride solution (specific gravity 1·396), and the seeds were then counted directly on a filter. As few as five seeds per 500 g of soil could be reliably detected, and field infestations defined accurately.

The problem of isolating seeds is much more difficult when, as is usual, a range of species is present which differ in seed size. Sieving can then only concentrate the seeds in a fraction with a fairly large range of particle size, and although in a few soils this may permit hand-sorting (Hayashi and Numata, 1971) another step is usually needed before it becomes feasible.

2. *Flotation*

Viable seeds usually sink in water, although separation from mineral matter and some of the organic soil components can be achieved in a rotary tumbler with controlled water flow (Thorsen and Crabtree, 1977). More commonly, the soil containing the seeds is added to a liquid with a density greater than that of the seeds so that they can be skimmed off; devices which facilitate this have been described (Hayashi, 1975; Sinyukov, 1975).

Many different liquids have been used as flotation media. Solutions of mineral salts are often favoured, since they are cheap and present little health hazard. Among those used in recent studies are potassium carbonate (Hayashi and Numata, 1971), sodium carbonate (Hayashi, 1975; Rodríguez Bozán and Alvarez Rey, 1977), zinc chloride (Fekete, 1975; Hunyadi and Pathy, 1976) and calcium chloride (Barbour and Lange, 1967; Roberts and Ricketts, 1979). Soils with a high clay content may need treatment with a dispersant; sodium hexametaphosphate + sodium bicarbonate has been used, either with magnesium sulphate as the flotation medium (Malone, 1967) or prior to sieving (Carretero, 1977).

Other workers have preferred organic liquids of appropriate density, such as tetrachloromethane (Hyde and Suckling, 1953; Dechkov, 1975) or perchloroethylene (Jones and Evans, 1977). In Australia solvents such as perchloroethylene (S.G. 1·62), trichloroethylene (S.G. 1·46) and trichloroethane (S.G. 1·31) are used to separate soil from legume seeds recovered from the soil surface, the choice depending on the species (Loch and Butler, 1977). Flammability, toxicity and the need for a reasonably high boiling point to reduce losses must be taken into account, and adequate ventilation is needed during use (Jones and Bunch, 1977; Loch and Butler, 1977).

3. *Further separation*

Airflow separation has been used either as a complement to flotation on liquids (Hyde and Suckling, 1953; Jones and Bunch, 1977) or as a replacement for it. Tulikov (1976) described a method in which each of a series of size fractions obtained by sieving the soil sample is placed in a seed cleaner with vertical airflow, and claimed the technique to be reliable and twice as rapid as liquid flotation. Standifer (1980) also used a vertical airflow, first removing the light seed coats and bulky debris and then increasing the velocity of flow to separate the seeds from the heavier mineral particles.

With all techniques of physical separation the final step is normally hand-sorting in which the seeds are picked out individually from the residual organic matter, either directly or after drying, under a magnifier or binocular microscope. Because seed coats are often resistant to decay they may persist for long after the seeds have died, either as fragments or as entire structures. *Stellaria media, Chenopodium album* and *Portulaca oleracea* are species of which large numbers of empty seed coats are commonly found (Jensen, 1969; Carretero, 1977). Their presence greatly hinders the process of hand-sorting and imposes the need for a standard of acceptable seed damage. Roberts and Ricketts (1979) included flotation in water as one step of their procedure in order to reduce the numbers of empty seed coats.

4. *Determination of viability*

The numbers of dead seeds present may be significant in studies of their fate or of vegetation history, but usually concern is with the viable seeds and a method of determining viability is needed. The obvious method is to place the seeds, or a random sample of them, under conditions favourable for germination. This is most likely to be effective when a single species is involved, because optimum conditions can then be given. Pre-treatments can also be applied, such as pricking for *Avena fatua* (Wilson, 1972) or scarification for seeds of legumes; comparisons of germination with and without scarification provide useful information on the proportion of "hard" seeds (Jones and Evans, 1977).

Germination tests to assess viability are less satisfactory when a range of species is present since no single set of conditions will give a valid estimate for all of them. Other methods which have been used include direct examination of the embryo and testing with tetrazolium salts (Malone, 1967; Hayashi and Numata, 1971), but these are tedious and time-consuming when the seeds are small. One alternative is to record those seeds which appear to be intact and which resist gentle pressure, and to regard them as "apparently viable" (Zelenchuk, 1961; Hayashi *et al.*, 1978; Roberts and Ricketts, 1979). Carretero (1977) found that of the intact, apparently viable seeds of four species, 71–88% germinated when tested. For many purposes, assessment of the apparently viable seeds may be adequate.

D. COMPARISON OF TECHNIQUES

Most workers have employed the sieving/flotation technique or the seedling emergence technique, but not both. Separation by physical means would be expected to reveal many more seeds than seedling emergence because many will be non-viable, and comparisons have shown this to be so (Ramírez and Riveros, 1975; Williams and Egley, 1977). Although the percentage viability of legume seeds separated from the soil may be high (Charlton, 1977), for most species it is usually low. The value obtained depends on the criteria used in determining what to record as seeds and also on the nature of the test used to determine the viability. Mean values for viability which have been recorded include 15% (Wehsarg, 1912), 13·5% (Platon, 1955), 6–22% (Želev, 1965) and 12–36% (Carretero, 1977), while Kropáč (1966) considered that a figure of 20% could reasonably be assumed. Mean values can be misleading, however, since there is likely to be considerable variation from species to species.

In a detailed comparison with samples from 57 fields, Jensen (1969) found that separation gave an average of 135 000 seeds m^{-2} of which 27% germinated when tested after chilling and a further 11% were judged to be viable at the end of the tests. Seedling emergence gave an average estimate of 19 000 m^{-2}, only 38% of the viable seeds recovered by the separation technique. However, this difference was largely accounted for by the species with the highest seed numbers, *Juncus bufonius*, of which only 10% of the viable seeds emerged as seedlings in the glasshouse. Excluding this species, seedling emergence gave an estimate equivalent to 75% of the viable seeds recovered by separation.

The main advantages of the seedling emergence technique are that the effort required is spread over a period, each seedling represents a viable seed, and seedlings are usually easier to identify than seeds. The main disadvantages are the delay between sampling and obtaining the results and the fact that seed banks of some species may be underestimated. The shorter the period for which the samples are kept, the greater the risk of underestimation. Nevertheless, this method has been preferred for monitoring long-term experiments (Brenchley and Warington, 1930; Roberts, 1962; Barralis, 1972) and for studying seasonal changes in the seed bank (Guevara and Gómez-Pompa, 1972; Young and Evans, 1975). For some purposes it may not be necessary to attempt complete enumeration of the seeds present. If the aim is simply to determine whether species form seed banks which persist from year to year, or to detect gross changes in seed numbers within the year, a short germination period under standard conditions may prove adequate (Grime, 1979; Thompson and Grime, 1979).

The main virtue of techniques involving physical separation of seeds from the soil is that the results are available as soon as processing is complete; this is particularly desirable if the information is to be used predictively (Carretero, 1977). Unless viability is determined, however, these methods greatly overestimate the potential flora. Which technique is employed depends largely on the nature and purpose of the investigation; some workers have used a sequence involving seedling emergence followed by sieving and the tetrazolium test (Johnson, 1975; Moore and Wein, 1977). Results obtained by separation techniques are often quoted in terms of total seeds, but are sometimes adjusted for viability. Because of the differences in the methods used care is needed in making quantitative comparisons of results obtained by different investigators.

III. Seed Banks of Arable Soils

Because of the obvious economic importance of arable weeds and the fact that many of them reproduce only by seeds, considerable attention has been paid to the seed banks of cultivated soils. Published investigations can be considered under three headings. Firstly, there are those of a "survey" nature in which determinations have been made of the seeds present in the soil of fields in a particular locality or over a wider area. Secondly, there are those in which the seed bank has been used as an index to monitor improvements in agrotechnical measures or to compare the effects of different crop rotations, cultural techniques or weed control practices. Finally, there have been limited investigations of the quantitative relationships between the seed bank and the weed vegetation developing after cultivation.

A. SURVEYS

The main conclusion from the early work reviewed by Kropáč (1966) was that the numbers of seeds present in arable soils are usually very high. Data from various sources which were tabulated by Jensen (1969) show that the average number of viable seeds in the plough depth (0–15 or 0–25 cm) is usually greater than 4000 m^{-2} and in very weedy fields may be as high as 70000–80000 m^{-2}.

Recent studies confirm the generally high level of infestation by weed seeds. One of the most extensive series of investigations is that of Dechkov (1974, 1975, 1976, 1978), covering more than 800 fields in various regions of Bulgaria. Although the species composition of the seed bank varied according to soil type and district, certain species were particularly well represented. These included *Amaranthus* spp., which accounted for more than 50% of all seeds found in the western Danube Plain, and *Echinochloa crus-galli*. Others which occurred widely in appreciable numbers were *Digitaria sanguinalis*, *Setaria* spp. and *Sinapis arvensis*. Another extensive survey covering some 800 spring cereal fields was made in Finland (Paatela and Erviö, 1971) and showed a high general level of infestation. Although there were differences between districts and between soil types the ten most abundant species, which included *Chenopodium album*, *Spergula arvensis*, *Ranunculus* sp., *Galeopsis* sp., *Viola arvensis* and *Stellaria media*, accounted for 90% of all seeds found. Seed numbers, especially of *Chenopodium album* and *Stellaria media*, were highest following potato and root crops and least after leys. Raatikainen and Raatikainen (1972) discussed the same data from the point of view of the length of time for which the fields had been under arable cultivation. They concluded that seed numbers in general increased with the period of cultivation, while those of *Carex* spp., present before clearance, decreased with time.

A survey of 57 fields in Denmark by Jensen (1969) gave an average value of 50250 viable seeds m^{-2} in 0–20 cm, with a range from 600 to 496000 m^{-2}. The seven most common species, *Chenopodium album*, *Gnaphalium uliginosum*, *Juncus bufonius*, *Plantago major*, *Poa annua*, *Spergula arvensis* and *Stellaria media* accounted for 78% of all the seeds recorded. In Great Britain a survey of 58 vegetable fields gave a range of 1600–86000 viable seeds m^{-2} in 0–15 cm, with a median value of 10000 m^{-2} (Roberts and Stokes, 1966). The seven most common species, *Poa annua*, *Stellaria media*, *Urtica urens*, *Senecio vulgaris*, *Capsella bursa-pastoris*, *Chenopodium album* and *Veronica persica*, accounted for 80% of the seeds found. Samples from 32 cereal fields in the English Midlands gave a range of 1800–67000 viable seeds m^{-2} in 0–15 cm with a median of 5500 m^{-2}, and *Poa annua, Polygonum aviculare, Stellaria media, Chenopodium album* and *Aethusa cynapium* were the species most frequently represented (Lockett and Roberts, 1976).

Dvořák and Krejčíř (1974) examined fields of wheat, maize, sugar beet and lucerne in Czechoslovakia and showed that of the seeds found *Chenopodium album* accounted for 50%, *Stellaria media* for 14% and *Sinapis arvensis* for 10%. Other recent studies include those by Fekete (1975) and Sárkány (1975) of maize fields in Hungary, where *Echinochloa crus-galli* was the main species, and by Kazantseva and Tuganaev (1972) in the Tatar ASSR who recorded *Chenopodium album, Polygonum aviculare, P. convolvulus, Stachys annua* and *Amaranthus retroflexus* as major contributors to the seed bank. *A. retroflexus, Echinochloa crus-galli* and *Sinapis arvensis* were three of the main species represented in seed banks of cereal fields in Romania (Pocinoc, 1969) and in Spain the viable seeds in five fields ranged from 1000–21 000 m^{-2} with *Portulaca oleracea* the most abundant species (Carretero, 1977). A maize field in Argentina had 90 600 viable seeds m^{-2} in 0–20 cm, with *Portulaca oleracea* and *Chenopodium album* the main species (Leguizamón and Cruz, 1980) while two other arable fields had approximately 4000 m^{-2}, mainly *Amaranthus quitensis, Chenopodium album, Digitaria sanguinalis* and *Echinochloa* spp. (Leguizamón *et al.*, 1979).

These results are in accord with earlier investigations, summarized by Kropáč (1966), in showing that the main contributors to seed banks of arable soils are annual weeds. These often account for 95% or more of all seeds, and those of perennial weeds and crop species are usually only poorly represented. Moreover, there are often one or two species which have seed numbers much greater than the rest and a fairly small number of species normally accounts for about 80% of all seeds found, even when the fields sampled are distributed over a region. In cool temperate areas such as northern Europe, *Chenopodium album* and *Stellaria media* are among the species consistently found to be major contributors to arable seed banks. In areas with higher summer temperatures, *Amaranthus* spp. and *Echinochloa crus-galli* are among those of which high numbers of seeds are generally found.

Many of the studies of arable fields have included determinations of the numbers of seeds present at different depths in the soil (e.g. Kropáč, 1966; Dvořák and Krejčíř, 1974; Fekete, 1975). Where the soil is regularly worked to the same depth, the distribution of seeds tends to be fairly uniform throughout that depth with only a few seeds below it. However, much depends on the time of cultivation in relation to major recent seed inputs, on the depth of cultivation, and on whether or not the soil is wholly or partially inverted. If weeds have seeded and the land is then ploughed, the seeds may be concentrated in a zone at depth; Kazantseva and Tuganaev (1972) found a direct correlation between the vertical distribution of seeds and the depth of ploughing. A detailed study of the distribution of seeds of *Tribulus terrestris* showed that the numbers varied inversely and linearly with depth, but that soil type, texture and method of cultivation affected the gradient of distribution (Goeden and Ricker, 1973). The very small seeds of *Striga asiatica* were concentrated in the upper 30 cm of a sandy soil, but some were found even at a depth of 1·5 m and the percentage viability was higher in seeds which were deeply buried (Robinson and Kust, 1962).

The very limited data for tropical arable soils indicate that these too contain large numbers of viable weed seeds. Vega and Sierra (1970) sampled a lowland rice field in the Philippines and found 80 000 m^{-2} in 0–15 cm; the highest numbers were those of *Fimbristylis miliacea* and other Cyperaceae, which accounted for 94% of all the seeds found. A rice field in Java sampled in the dry season had more than 20 000 viable seeds m^{-2} in 0–10 cm, almost all Cyperaceae (Hayashi *et al.*, 1978). Kellman (1974b) recorded an average of 7600 viable seeds m^{-2} in 0–4·2 cm of arable fields in Belize, with *Amaranthus spinosus* a notable contributor. In further samples from 0–10 cm in a maize field there were 9800 m^{-2}, mainly concentrated near the surface (Kellman, 1978). A study of peat soil in Malaysia which had been cropped with pineapples for periods of 6 months to 10 years revealed only small numbers of viable seeds ranging from 70–630 m^{-2}, but the numbers of fern spores present were almost ten times greater than this (Wee, 1974).

B. EFFECTS OF CULTURAL PRACTICES

The seed bank reflects the history of the land in terms of cropping and the extent to which the cultural practices associated with the crops have been successful in controlling weeds and limiting their seed production. The weed vegetation present at any one time is only a partial representation of the potential weed flora (Rola, 1962), but assessment of the seed bank provides a means of determining long-term changes in the overall level of weed infestation and in the relative abundance of the different seed-producing species. Some examples of the reductions in numbers of seeds achieved in particular circumstances are shown in Table II.

Roberts (1962) showed that in the absence of significant production of fresh seeds, the numbers of viable seeds of arable weeds in a soil subject to frequent cultivation decreased exponentially. The rate of

TABLE II. Some reductions in size of seed banks of arable soils brought about by cultural treatments

Location	Cultural treatment	Period (years)	Reduction (%)	Authors
Bulgaria	Crop rotations; herbicides	6	65	Dechkov and Atanassov (1976)
Hungary	Maize; improved techniques; herbicides	10	75	Fekete (1975)
France	Sunflower, wheat, barley; herbicides	5	33	Barralis *et al.* (1978)
	Sugar beet, wheat, barley; herbicides	4	63	
England	Fallow; four cultivations per year	5	87	Roberts and Dawkins (1967)
	Fallow; two cultivations per year	5	83	
	Chemical fallow	5	71	
England	Fallow; seven cultivations per year	4	97	Roberts and Feast (1973b)
	Fallow; two cultivations per year	4	89	
	Chemical fallow	4	81	
USA	*Abutilon theophrasti*; lucerne	5	61	Lueschen and Andersen (1978)
	Abutilon theophrasti; fallow; two ploughings per year	5	93	
	Abutilon theophrasti; chemical fallow	5	73	
USA	*Brassica kaber*; bromegrass	4	40	Warnes and Andersen (1978)
	Brassica kaber; fallow; three ploughings per year	4	93	
	Brassica kaber; chemical fallow	4	45	
USA	Maize; herbicides	3	67	Schweizer and Zimdahl (1979)
	Maize, sugar beet, barley; herbicides	3	49	
USSR	Vegetable crops; herbicides, ploughing	3	52	Kolesnikov and Sidorov (1974)
USSR	Cotton; herbicides	4	83	Lozovatskaya (1968)

decline varied with the species and also with frequency of cultivation. When soil was cultivated monthly during the growing season the annual loss was almost 60% (Roberts and Feast, 1973b). In another experiment the seed bank in 0–23 cm declined at a rate of 36% per year with four cultivations annually and 22% per year with none at all (Roberts and Dawkins, 1967). The results for *Abutilon theophrasti* and *Brassica kaber* (Table II) suggest comparable loss rates. The seed population in a very shallow layer of cultivated soil would be expected to decline more rapidly (Roberts and Feast, 1972). In an experiment in Louisiana, continuous cropping for two years without ploughing eliminated viable seeds of *Eleusine indica* from 0–5 cm and reduced those of *Cyperus iria* by 76%, although there was much less effect on those at 15–20 cm (Standifer, 1980). Jan and Faivre-Dupaigre (1977) found that where effective weed control was achieved on direct-drilled plots, the numbers of viable seeds in 0–5 cm were reduced by 39% and 68% in two three-year experiments.

In practice, it is difficult to ensure that all seeding by weeds is prevented, and the addition of fresh seeds can easily counterbalance the natural losses from the seed bank (Roberts, 1962). Data quoted by Kropáč (1966) and Roberts (1970) show that fallowing, when properly carried out, can bring about appreciable reductions in the seed bank. With traditional methods of cropping, however, the evidence shows that only moderate reductions could be achieved over relatively long periods. Modern selective herbicides allow more complete weed control, so that the natural losses are cumulative and the seed bank can be substantially reduced within a few years (Table II).

Individual herbicides rarely kill all the weed species that may be present when applied at rates which are selective in a crop. If the same crop/herbicide combination is repeated, seed inputs from those species which survive may soon drastically alter the composition of the seed bank. Fekete (1975) recorded an increase in seeds of *Echinochloa crus-galli* where atrazine was used in maize, while Zuza (1973) showed that the use of triazine herbicides and 2,4-D in proso millet reduced seed numbers of dicotyledons but that 2,4-D in particular increased those of grasses. Results from an experiment in Germany in which various weed control measures had been applied to the same cereal plots for 12 years were reported by Hurle (1974). With no control at all, there were 43 800 viable seeds m^{-2} in 0–25 cm. The best treatment, DNOC, gave a value only 40% of this while those for MCPA, 2,4-D, calcium cyanamide and harrowing were 64, 56, 61 and 68% respectively. The effect of rotating these treatments was less than expected; the weak link, harrowing, allowed the greatest seed set and this reduced the benefits from previous treatments. Other investigations in which seed bank determinations have been used to assess the effects of herbicide treatments include those of Everest and Davis (1974), Dale and Chandler (1976, 1977) and Burnside (1978).

Estimates of seed numbers have also been made in experiments involving crop rotations and fertilizers (Watanabe and Ozaki, 1964; Dotzenko et al., 1969; Dechkov and Atanassov, 1976). These show, as might be expected, that changes in the numbers of seeds of the different weed species reflect opportunities for seed production under the various regimes. In an experiment in Oklahoma in which different fertilizers had been applied to winter wheat for 47 years, the lowest numbers of seeds were found on the unfertilized plots, and they increased as the fertilizer treatments became more complete (Banks *et al.*, 1976). As in earlier work in Britain (Brenchley and Warington, 1930), some species were favoured by particular fertilizer treatments.

Two experiments with raspberries also illustrate the way in which seeds can accumulate under particular regimes. Allott (1970) found that after eight years there were 1780 viable seeds m^{-2} in 0–10 cm where plots had been cultivated, but only 240 m^{-2} where a herbicide programme had been used. Clay and Davison (1976) established an experiment on land which had previously been under grass with 1300 viable seeds m^{-2} in 0–15 cm. On cultivated plots, annuals such as *Capsella bursa-pastoris*, *Stellaria media* and *Poa annua* were able to set seed between cultivations and after 8 years there were 14 700 seeds m^{-2}. In contrast, there was little change during this time in the numbers of seeds on bare plots which received herbicides.

C. RELATIONSHIPS WITH THE FLORA AFTER CULTIVATION

Relatively little attention has been devoted to the quantitative relationships between seed banks in arable soils and the populations of weed seedlings that appear after the ground has been disturbed. Von Hofsten (1947) recorded good correlation between the numbers of plants of *Sinapis arvensis* and the numbers of seeds in the soil at points on a grid over a single field. Chancellor (1965) noted that one very dense stand of *Matricaria matricarioides* accounted for only about 4% of the seeds in the soil, while Barralis (1965) found that in a crop of winter wheat the weeds that appeared represented 2·5% of the viable seeds present in 0–10 cm. Assessments made more recently (Table III) confirm that the seedlings account for only low percentages of the viable seeds present.

Variation in the percentage of viable seeds giving rise to seedlings can arise from several sources. The general suitability of the conditions for germination is obviously important; if it is very dry, for example, few, if any, seedlings may establish. The physical condition of the surface soil can also affect the numbers of seedlings, and twice as many may appear on a fine, firm seedbed as on a rough one (Roberts and Hewson, 1971). Soil disturbance acts selectively on the seed bank, since the prevailing conditions are likely to meet the germination requirements of some species but not others. Many species show distinct

Table III.   Numbers of seedlings of weed species emerging in crops in relation to the numbers of viable seeds of those species present in the soil

Location	Crop	Depth (cm)	Seedlings as % of viable seeds	Authors
Spain	Winter crops	10	3·4	Carretero (1977)
	Summer crops	10	1·7–10·4	
France	Winter wheat	10	1·9–13·3	Barralis (1972)
France	Winter and spring cereals; mean	10	5·3	Barralis and Salin (1973)
France	Winter wheat	10	0·6–8·9	Barralis and
	Spring barley	10	3·7–8·6	Chadoeuf (1976)
France	Winter wheat	10	3·0–10·8	Barralis et al.
	Spring barley	10	5·8–15·3	(1978)
England	Seedbeds throughout the year; median	10	2·0	Roberts and Ricketts (1979)
FR Germany	Cereals	25	1	Hurle (1974)
USA		20	2·4	Williams and Egley (1977)

patterns in the seasonal distribution of seedling emergence and the species composition of the plant population is likely to be influenced by the time of year at which the soil is disturbed. This is shown strikingly by the data of Carretero (1977). Comparisons of seed banks and resulting weed floras showed that the species could be divided into two groups, one of which produced no seedlings in summer crops, the other none in winter crops. The overall ratio between the numbers of viable seeds and the numbers of seedlings that appeared thus depended both on the time of cultivation and on the proportions of seeds of the two groups present in the soil.

Variation will also arise if species differ appreciably in the percentages of seeds which give rise to seedlings under optimum conditions, and there is evidence that this is so. Carretero (1977) showed that whereas the percentages were always low for *Chenopodium album* and several other species, they tended to be much higher for *Setaria adhaerens*. In one summer crop, about 22% of the viable seeds of this species in 0–10 cm produced seedlings. Roberts and Ricketts (1979) also recorded low percentages for *Chenopodium album*; usually less than 4% of the viable seeds in 0–10 cm produced seedlings even at the season of peak emergence. For *Papaver rhoeas* and *P. dubium* the percentages were even less, no more than 1% at the times of year most favourable for emergence. In contrast, the seedlings of *Polygonum aviculare* emerging after spring cultivation accounted for up to 15% of the viable seeds. Seed size evidently plays some part in these differences between species (Carretero, 1977). Although cultivation may bring about germination of a proportion of the seeds throughout the disturbed layer (Kropáč, 1966), only those within a certain distance of the surface will have sufficient reserves to enable them to reach it. For the large-seeded *Avena fatua*, percentages higher than any mentioned above have been recorded (Wilson and Cussans, 1975).

The age of the seeds may be an additional source of variation. Many species of arable weeds, such as *Senecio vulgaris*, *Stellaria media* and *Urtica urens*, produce seeds of which a variable but often appreciable proportion is capable of immediate germination under suitable conditions. If seeds of very recent origin form a significant part of the total bank, the percentages giving rise to seedlings might thus be relatively high. Using a "mark-and-recapture" technique Naylor (1972) found that about 60% of the seedlings in populations of *Alopecurus myosuroides* were derived from seeds less than one year old. The converse may be true if seeds have a high level of initial dormancy. In a detailed study of *Avena fatua* Wilson and Cussans (1975) showed that when the seeds were mainly those shed in the previous crop, 17–24% of those present in 0–20 cm in February produced seedlings in the spring barley crop. Where no fresh seeding had taken place, 38% of the seeds present gave rise to seedlings.

TABLE IV. Numbers of seedlings emerged per year in relation to the numbers of viable seeds in 0–23 cm at the start of the year (from Roberts and Dawkins, 1967)

Cultivations per year	Seedlings as % of viable seeds					
	Year 1	Year 2	Year 3	Year 4	Year 5	Year 6
4	10·7	6·1	9·4	8·3	8·7	9·6
2	7·8	4·7	6·7	6·1	9·0	7·2

Roberts and Dawkins (1967) made successive annual determinations of the naturally-occurring viable seeds in field plots and assessed the total numbers of seedlings appearing in each year, allowing no fresh seeding to take place. Under consistent regimes of soil disturbance there were no systematic trends in the percentages of seeds giving rise to seedlings (Table IV), and the chances of a seed producing a seedling remained the same over the six-year period. Seely (1976), however, has suggested that sometimes estimates of seeds in the soil may give an erroneous impression of the potential contribution to the flora. He cites an example in which a reduction in seed numbers of *Spergula arvensis* by 97% took place over 12 years, whereas 97% reduction in seedlings occurred after 6–7 years and after 12 years the seedling population was less than 0·5% of that originally present. Data from experiments in which seeds of annual weeds were mixed with soil which was then regularly disturbed also showed that the decline in seedling numbers from year to year was somewhat more rapid that that of viable seeds (Roberts and Feast, 1973a). Estimates of the chances of a seed producing a seedling in a variety of situations have been compiled and discussed by Sagar and Mortimer (1976), and related aspects considered by Roberts (1972).

A question which has aroused some interest in recent years is whether data on seed banks can be used predictively to forecast weed problems and help in specifying control measures. The knowledge that a partic-ular species is present as a major component of the seed bank can itself be useful. In Bulgaria, Dechkov (1974, 1975, 1976, 1978) mapped the main species present in his surveys and was able to indicate the areas in which particular herbicides would be needed to give effective weed control in the maize crop. Seed bank data obtained from year to year can also be valuable in monitoring the overall success of weed control programmes and in suggesting improvements.

Several attempts have been made to determine whether the weed flora in crops to be sown at different times of year can be predicted from the presence of seeds in the soil. Lhoste *et al.* (1969) set out paired 1 m² plots, recovered the top 5 cm of soil from one of each pair, recorded the seedlings that emerged in the glasshouse, and then compared the result with field emergence in June and November on the remaining plots. Although there was good correspondence for some species, there were others which emerged in the field but not in the glasshouse and vice versa. They concluded that the technique was promising, and might be improved by reproducing the appropriate climate in the glass-house. Naylor (1970) related seedling emergence from soil samples in the glasshouse to subsequent field emergence in a study of *Alopecurus myosuroides* and obtained reasonable correlation.

The possibility of using determinations of seeds in soil samples as an advisory tool has been examined in Texas by Palmer *et al.* (1970). They listed a number of advantages that the information could confer, but conceded the need for better sampling methods, a better way of determining viability, and for an assessment of the economics of the technique as a self-sustaining service. Determinations of the seed content of *Amaranthus* spp. have since been used as a basis for suggesting weed control methods in future cropping plans (R. D. Palmer, pers. comm.).

Prediction of weed infestations has also been examined in the USSR (Artyushin and Libershteĭn, 1976).

If the relative numbers of seeds of different species are determined, some estimate of the composition of the weed flora after cultivation at different times of year could be made from the known seedling emergence patterns in the region concerned. However, this would not allow for the inherent differences between species in the percentage of seeds likely to produce seedlings which have already been mentioned. In an attempt to include this factor, Roberts and Ricketts (1979) derived empirical curves by plotting the percentages of seeds in 0–10 cm producing seedlings against the data of cultivation. If sufficient data were available, it would be possible to predict seedling numbers at any time of year from the numbers of seeds and the expected percentage emergence obtained from the curves. As previously mentioned, lack of soil moisture limits germination on some occasions and not on others, so that the estimates of seedling numbers would be subject to considerable error. Provided that this factor does not operate selectively to a major extent, however, it might at least be possible to forecast the relative contributions of different species to the seedling flora, and this could be of practical value.

IV. Seed Banks of Grasslands

The reserves of seeds in soils beneath different meadow and pasture types have been the subject of investigations in Europe and North America (e.g. Chippindale and Milton, 1934; Dore and Raymond, 1942; Prince and Hodgdon, 1946; Champness and Morris, 1948; Milton, 1948; Foerster, 1956; Zelenchuk, 1961; Golubeva, 1962). Much of this work has been reviewed elsewhere (Major and Pyott, 1966; Rabotnov, 1969) and will not be considered in detail here. As might be expected, the data show wide variation in the numbers of viable seeds found. Examples from published accounts of work in Great Britain listed by Thompson (1978) range from approximately 400 m^{-2} in a permanent pasture to 70000 m^{-2} in a pasture which was formerly arable, and results from the USSR also show a wide range (Rabotnov, 1978).

As already emphasized, because of differences in the depth of sampling and technique of estimation, comparisons between the results of different workers must be made with care. The time of sampling in relation to seed production is particularly important, and the annual grassland in California studied by Bartolome (1979) provides a striking illustration of this (Fig. 1). At the start of the growing season there were more than 60000 seeds m^{-2} but the numbers declined rapidly

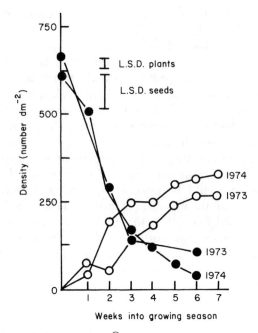

Fig. 1. Changes in total plant density (○) and numbers of germinable seeds in the soil (●) of a California annual grassland. The first soil sample was collected in late summer before the autumn rains (from Bartolome, 1979).

as germination took place after the autumn rains. Nevertheless, it appeared that many of the species carried over some viable seeds from year to year and these would be important for survival if drought were to occur during the growing season.

A. GRASSES

The species composition of grassland seed banks is of greater significance than the total number of seeds present, and some examples of the relative contributions made by four groups of species to the seed reserves of a range of grassland types are listed in Table V. Although seeds of grasses often form an appreciable percentage of the total, most workers have commented on the lack of quantitative correspondence between the representation of the different species in the sward and that in the seed bank. This reflects differences among species not only in seed production, but also in seed dormancy and survival. As Grime (1979) has pointed out, in habitats subject to summer drought viable

TABLE V. Contributions of different groups of species to the viable seed banks of some grassland soils

Location	Grassland type	% of total viable seeds				Authors
		Grasses	Legumes	Rushes + sedges	Others	
Wales	Lowland grazed, reseeded *Lolium perenne*	42	3	34	21	Jalloq (1975)
	Upland grazed, reseeded *Lolium perenne*	34	7	4	55	
Wales	Acidic pasture	62	0	1	37	King (1976)
Netherlands	Pastures; mean for 73 sites	62	0·5	11	26	van Altena and Minderhoud (1972)
Germany	Permanent grass; mean for 12 sites	46	3	7	44	Foerster (1956)
S. Bohemia	Meadows, peat soils; mean % for 8 sites	27	5	13	55	Mika (1978)
USSR	Meadow-steppe; mean of mown and unmown	2	7	1	90	Golubeva (1962)
USSR	Meadows and woodland pastures; mean % for 10 sites	23	2	25	50	Pyatin (1970)
Quebec	Permanent pasture; mean % for 4 sites	22	0·6	40	37	Dore and Raymond (1942)
California	*Stipa pulchra*; mean of grazed and ungrazed	36	2	0	62	Major and Pyott (1966)
Argentina	Old pasture	8	8	25	59	Leguizamón and Cruz (1980)
Tasmania	*Poa gunnii* grassland	16	0	35	49	Howard (1974)
Surinam	*Axonopus compressus*; grazed, mean for 2 sites	3	0·1	41	56	Dirven (1966)

seeds of many grasses are present only during the dry season; they are capable of germination under a wide range of conditions and moisture supply is the main factor determining the time of germination. Except during the summer period, therefore, seeds of these species make little or no contribution to the seed bank. Practically all the seeds germinate in autumn and large numbers of seedlings appear on bare ground which has developed in the dry season. Not only is this true for annual grassland (Fig. 1), but some of the commonest grasses of meadows and pastures in Europe show this pattern (Grime, 1979; Thompson and Grime, 1979). They include not only annuals such as *Hordeum murinum, Bromus sterilis, B. mollis* and *Lolium multiflorum,* but also perennials. Among them are desirable species which are frequently sown, such as *Dactylis glomerata, Festuca ovina, F. pratensis* and *Lolium perenne.* Surveys consistently demonstrate the absence of appreciable numbers of viable seeds of these species from the buried seed population (Chippindale and Milton, 1934; Champness and Morris, 1948; Foerster, 1956; Thompson and Grime, 1979). Van Altena and Minderhoud (1972) found that although *Lolium perenne* was a major component of pastures in the Netherlands, in terms of viable seeds present it ranked only seventh among the grasses and the numbers were low. Only very few seeds or none at all were recorded for other common sward components such as *Dactylis glomerata, Hordeum secalinum, Festuca pratensis* and *F. rubra.* The fact that these species do not accumulate reserves of buried seeds is of great importance for pasture management since it means that sown species, and *Lolium perenne* in particular, tend to be replaced by indigenous grasses (Chancellor, 1978; Grime, 1979; Thompson and Grime, 1979).

Many of these indigenous grasses, in marked contrast to the species cited above, produce seeds which do not germinate synchronously and of which a proportion can survive for at least a year when buried. Investigations of pastures and meadows in northern Europe consistently show that *Agrostis stolonifera, A. tenuis, Poa annua* and *P. trivialis* tend to be major contributors to the seed bank even though they may be present to only a limited extent in the vegetation (Chippindale and Milton, 1934; Champness and Morris, 1948; Foerster, 1956; Delpech, 1969; Rabotnov, 1969; van Altena and Minderhoud, 1972; Sparke, 1979). These species are thus well-placed to exploit gaps arising in the sward through death of the sown species, trampling and poaching by livestock or other disturbances. McRill (1974) showed that *Agrostis* spp., *Poa* spp. and *Holcus lanatus* were major components of the seed flora of earthworm casts from grasslands in North Wales, and argued that by providing a break in a closed community the casts could act as foci for invasion. Jalloq (1975) examined the possible role of molehills in this connection and recorded a high percentage of grass seeds in molehill soils from reseeded *Lolium perenne* pastures (Table V). Seeds of *Agrostis* spp. were prevalent in soil from upland sites, and those of *Poa annua* in lowland soils; the sown species accounted for less than 3% of the seeds found. Douglas (1965) considered the seed bank in relation to sward renewal and concluded that because of the absence of viable seeds, desirable grasses already present would not contribute to the new sward. Where *Poa* spp., *Agrostis* spp. and *Holcus lanatus* occurred in the vegetation, however, the presence of viable seeds in the surface layer of soil would be expected.

The lack of correspondence in representation of grass species in the vegetation and in the seed bank was very evident in the California grassland studied by Major and Pyott (1966), where no seeds of the dominant species, the perennial *Stipa pulchra*, were found in the soil. Dirven (1966), however, found that in two heavily grazed *Axonopus compressus* pastures in Surinam almost all the grass seeds recorded were of this species, although they accounted for only 3% of the total seed bank. Grasslands in Japan studied by Hayashi and Numata (1971, 1975) illustrate the differing relationships between the composition of the vegetation and that of the seed reserves. In pasture dominated by *Zoysia japonica* and other species regenerating both vegetatively and by seed, there were high numbers of seeds in the soil and good correlation between the floristic composition of the sward and the seed bank. The dominants of tall-grass meadows, *Miscanthus sinensis* and *Arundinella hirta*, however, regenerate mainly vegetatively and few seeds of these were recorded in the soil.

B. LEGUMES

Leguminous species are of particular interest as components of grassland seed banks because of their forage value. Krylova (1979) has summarized the published data obtained in the USSR and concluded that even in meadows of the forest zone, where the seeds are mainly in the top 5 or 10 cm of soil, the numbers are less than those of the herbs, grasses and sedges. As with other meadow species, seed numbers tend to be greater in wet soils than in dry ones and in flooded meadows there were up to 520 viable seeds m^{-2}, equivalent to 5·7% of all the viable seeds present. In pastures or grassland under seasonal grazing, the commonest leguminous species was *Trifolium repens*, with up to 1700 seeds m^{-2}. This is true elsewhere in Europe (Foerster, 1956; van Altena and Minderhoud, 1972); Champness and Morris (1948) record *T. repens* as one of the species which had a markedly greater representation in the seed bank than in the composition of the sward.

In most grassland types, the numbers of seeds of legumes tend to be low compared with those of other groups of species (Table V). Appreciable seed banks of *Ulex* spp. can occur, however, where seeding has taken place over a long period. Numbers of viable seeds at five sites in New Zealand ranged from 130 to 20500 m^{-2}, mainly in the 0–6-cm layer (Zabkiewicz and Gaskin, 1978), while Ivens (1978) recorded 10000 m^{-2} in 0–15 cm from a 20-year-old stand of *U. europaeus* in New Zealand. Viable seeds have been found up to 26 years after clearing (Moss, 1959), and in an eight-year-old *Lolium perenne* pasture in Wales seeds in *Ulex* spp. accounted for 11% of the total seed bank (Jalloq, 1975).

The occurrence of legume seeds in the soil of pastures in New Zealand has been reviewed by Suckling and Charlton (1978), who quote the ranges of seed levels for five species in areas of differing soil and climate. In lowland pastures the mean values for viable seeds of *Trifolium repens* in 0–5 cm ranged from 6·5 to 16·5 kg ha^{-1} (915–2320 m^{-2}) while those for *T. dubium* sometimes exceeded 50 kg ha^{-1} (10000 m^{-2}). There were also high numbers of seeds of this species in unploughable hill pastures, whereas those of *T. repens* were generally low and the seed levels of the other species were related to the annual rainfall. Grazing management also had an effect in determining the numbers of viable seeds in the soil.

Charlton (1977) examined the occurrence of seeds of legumes in an area of hill pasture in New Zealand and found that *Trifolium dubium* and *T. repens* were the main species, with mean values of 2540 and 290 m^{-2} respectively in 0–8 cm. The viability of seeds recovered from the soil was almost 100%, although 82% of *T. dubium* and 96% of *T. repens* seeds were hard. It was concluded that buried seeds are of high value both in maintaining and increasing the legume content of hill swards. Work in Queensland has also shown that high numbers of viable legume seeds can accumulate under pasture (Jones and Jones, 1975). At the end of a four-year grazing trial comparing *Trifolium semipilosum* and *T. repens* there were 8400 and 4900 m^{-2} respectively in 0–5 cm, although seedling regeneration was greater in *T. repens*.

Legume species used in subtropical pastures may also develop appreciable seed banks. Studies on 10-year-old pastures in Queensland gave an average number of 4700 m^{-2} in 0–5 cm for *Lotononis bainesii*, even though this species had been only a minor component of the vegetation for some years (Jones and Evans, 1977). Most of these seeds were hard, suggesting that loss of hard-seededness is a slow process. There were also 1300–3700 viable seeds m^{-2} of *Trifolium repens*, of which 70% were hard. Of the 150–500 m^{-2} seeds of *Desmodium intortum* found, only 9% were hard and there were no seeds where this species had been absent for some years in the heavily grazed areas. These results show that in mixed-legume pastures large reserves of viable seeds can accumulate which are not directly related to the current vegetation.

The role of seed reserves in the maintenance of pastures in Queensland based on the tropical legume *Macroptilium atropurpureum* has been studied in some detail. Data summarized by Tothill and Jones (1977) show that mean numbers of viable seeds of this species grown in association with different grasses ranged from 20 to 520 m^{-2} in 0–5 cm. Seed numbers were lowest under heavy grazing (Jones, 1979), and it is suggested that lenient grazing in the first year is desirable in order to establish a reserve of seeds in case, from necessity, the pasture has to be overgrazed at some future time (Jones and Jones, 1978).

Although the presence of viable seeds of legumes can be beneficial for pasture maintenance, problems can arise if circumstances make it necessary to reseed with a different species or cultivar. When *Trifolium subterraneum* cv. Yarloop pastures in Australia were reseeded with other cultivars, reversion often occurred in the following year. Beale (1974) showed that seed levels of cv. Yarloop at sites sown 8–13 years previously ranged from 3 to 169 g m^{-2} in 0–7·5 cm. When seeding was prevented in permanent quadrats, numbers fell by 74% in 2 years, probably largely as a result of predation. He concluded that several years with virtually no seeding would be needed to reduce the numbers to an acceptable level, although since most were near the surface ploughing down to a depth from which emergence could not occur might be feasible.

Suckling and Charlton (1978) pointed out that while seed populations of legumes can be reduced by cultivation, at least four years is required before returning to pasture. They cited an example of contamination of a seed production crop of *Trifolium repens* by *T. dubium*; after four years of arable cropping there were still 150 viable seeds m^{-2} of *T. repens* and 1600 m^{-2} of *T. dubium*. This species is especially serious as a contaminant of *T. repens* since the seeds are of similar size. Even more serious is the possibility of contamination of seed crops by different genotypes or cultivars of the same species, and in New Zealand areas for the production of breeders' seed are now checked for buried seeds (Suckling and Charlton, 1978).

C. OTHER SPECIES

Species of rushes (Juncaceae) and sedges (Cyperaceae) are frequently recorded as major contributors to seed banks in a range of grassland types. Rabotnov (1969) commented on the abundance of seeds of *Juncus* spp., especially *J. bufonius*, and of *Carex* spp. in meadow soils of the USSR. Seeds of *J. bufonius* were also found in most pasture soils in the Netherlands, often in high numbers (van Altena and Minderhoud, 1972). Champness and Morris (1948) recorded only low seed populations in the better types of British grasslands, but high numbers were found where rushes and sedges occurred in the vegetation; in one upland field dominated by rushes 11 000 viable seeds m^{-2} were recorded, mainly *Juncus effusus*. In a marshy area in Wales more than 50 000 viable seeds m^{-2} of *Juncus* spp. were found in 0–23 cm (Anon., 1967), while 8000–9000 m^{-2} in 0–10 cm were recorded for a pasture in Romania (Simtea, 1971). Species of Cyperaceae have been reported as important contributors to seed banks in diverse temperate and tropical grasslands in Tasmania (Howard, 1974), Japan (Hayashi and Numata, 1975), Surinam (Dirven, 1966) and Belize (Kellman, 1978).

One of the most consistent features evident from the investigations of grassland seed banks is the presence of appreciable numbers of viable seeds of dicotyledonous species (Table V). Some of these are typical grassland plants; in northwestern Europe, for example, *Ranunculus repens*, *Galium saxatile* and *Cerastium holosteoides* are frequently recorded as contributors to the seed bank. In the Netherlands, *Sagina procumbens* occurred in half of the fields sampled and the average numbers were high (van Altena and Minderhoud, 1972). In moorland soils, large populations of viable seeds of *Calluna vulgaris* can accumulate (Chippindale and Milton, 1934; Miles, 1973). Where the sward has been established on land formerly under arable cultivation, large numbers of seeds of arable weeds may be found; these tend to occur at greater depths than those of species which have seeded after sward establishment, and their numbers decrease with the age of the grassland (Douglas, 1965). Meadows which are subject to seasonal flooding may receive appreciable quantities of seeds of arable and ruderal species which then become incorporated into the seed bank (Rabotnov, 1969; Mika, 1978).

From their examination of the seed banks of British grasslands, Champness and Morris (1948) concluded that "Many of the herbs of our grasslands may give a very high buried seed population and yet occur as no more than occasionals in the pasture". This appears to be generally true, although not all dicotyledons behave in this way. Seeds of some species which regenerate vigorously by vegetative means, such as *Cirsium arvense*, may be absent altogether or found only in low numbers. Another group of plants which can be recognized comprises short-lived species of which almost all the seeds produced in autumn germinate early in the following spring. Among them are Umbelliferae such as *Seseli libanotis* and *Pimpinella saxifraga* (Rabotnov, 1978; Thompson and Grime, 1979); seeds of species in this group would not be found in the buried seed bank.

The type of management of grassland would be expected to have an effect on the size and species composition of the seed bank, and some studies have been made in which different regimes have been compared. Zelenchuk (1968) examined two grazed and two mown natural meadows in the L'vov region of the USSR and found that there were higher numbers of seeds under mowing. Few viable seeds of grasses and legumes occurred in the mown meadows, whereas with grazing they constituted more than 50% of the total. Golubeva (1962) observed no substantial difference in the seed banks of mown and unmown areas of meadow steppe, but in studies on peaty soils the seed numbers of grasses were less with mowing while those of *Carex* and *Juncus* spp. were greater (Zelenchuk, 1961). In Canada it was found that viable seeds of grasses were most numerous in ungrazed fields and lowest under heavy grazing, while the reverse was true for seeds of forbs and shrubs (Johnston *et al.*, 1969). Major and Pyott (1966) concluded from their own work on a California grassland and from other published studies that buried viable seed content is no indicator of grazing or mowing treatment. This is perhaps not surprising, since the differences in size and floristic composition of seed banks which develop under different management regimes will reflect the extent to which they favour or restrict growth and seed production of the range of species present in the vegetation.

The role of stock in relation to seed banks of grasslands has received some attention. The distribution of seeds in manure has been examined (Dore and Raymond, 1942; Boeker, 1959), and studies of the content of viable seeds in the surface soils of cattle resting places have been made in Germany (Boeker, 1959) and in Japan (Sugawara and Iizumi, 1964).

V. Seed Banks of Aquatic Ecosystems

It was pointed out by van der Valk and Davis (1976) that except for the investigation of salt marshes by Milton (1939), the study of seed banks of wetland habitats has been almost completely neglected, although a knowledge of their floristic composition could lead to better understanding of successional patterns and also help in management. In a preliminary study, they determined the numbers of viable seeds present in different vegetation zones of eight prairie glacial marshes in Iowa. A major conclusion from this work was that there appeared to be very little floristic similarity in the seed banks from different marshes, even when the vegetation types sampled were the same. Those of different vegetation types within the same marsh, however, were more similar; this is because seeds of nearly all aquatic and marsh plants float or are wind-dispersed, and so in time could reach all zones within a marsh. They concluded that the type of vegetation present at any one time depends mainly on the water level, but that its floristic composition reflects that of the seed bank.

In a later detailed investigation of two Iowa marshes (van der Valk and Davis, 1978), the numbers of viable seeds (including turions) found beneath different vegetation types ranged from 21 440 to 42 620 in 0–5 cm. The species contributing to the seed banks were of three kinds: emersed species with seeds which germinate on exposed mud flats or in very shallow water; submersed and free-floating species whose dormant seeds or turions can survive on exposed mud flats for a year and then germinate when there is standing water; ephemeral species whose seeds only germinate on the mud flats when there is no standing water. Primarily because of the fluctuations in water level and the effects of muskrat damage, cyclic changes take place in the vegetation of prairie marshes and the seed bank is the key to understanding their vegetation dynamics and survival (Fig. 2).

The possibility of reconstructing the recent vegetational history of

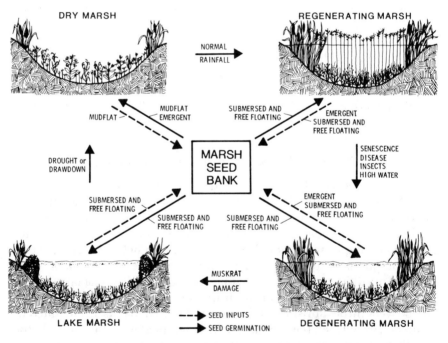

Fig. 2. Schematic presentation of the vegetation cycle and inputs and outputs of seeds from the seed bank during this cycle in a prairie glacial marsh in Iowa (from van der Valk and Davis, 1978; copyright 1978 by the Ecological Society of America).

a marsh from the seed bank was studied using soil samples from different depths in six vegetation types at Eagle Lake, Iowa (van der Valk and Davis, 1979). They concluded that this is possible to a greater or lesser extent, depending on the past and present communities, but that there are problems both of technique and interpretation. Some shallow-marsh vegetation types usually produce so few seeds that they cannot be detected, while others have dominant species whose seeds are difficult to germinate. Differential seed longevity must also be taken into account; seeds of mud-flat annuals and some emergent aquatics appeared to be long-lived, but those of many other aquatics remain viable for shorter periods.

Appreciable numbers of viable seeds were found in soils beneath several vegetation types of a fresh-water tidal marsh in New Jersey (Leck and Graveline, 1979), with mean values of 6400 to 32400 m^{-2} in 0–10 cm. Seven out of the ten species most frequently represented were annuals, and a striking feature was the correspondence between the floristic composition of the seed banks and that of the seedling populations observed. Comparison with published data for other ecosystems showed that in the tidal marsh there were relatively large numbers of seeds below the surface layer. Water movement may be partly responsible for the lower seed numbers near the surface, and the data suggest that viability of buried seeds in marsh soils may be prolonged.

Studies have been made of a very different aquatic habitat—the floating islands of Lake Rawa Pening in Central Java (Hayashi *et al.*, 1978). These islands are formed initially by massive growth of *Eichhornia crassipes* on the surface. Soil and peat rising upwards become trapped by the roots and provide a substrate for establishment of other species. Successional stages lead to a community dominated by the grasses *Leersia hexandra* and *Eragrostis amabilis*, which is then invaded by shrubs. Examination of soil samples taken from different successional stages showed the presence of large numbers of viable seeds, equivalent to 8780 to 12250 m^{-2} in 0–10 cm. The floristic composition of the seed banks did not correspond with that of the vegetational stages from which the samples were derived. Seeds of *Cyperus* spp. were abundant in soil from all the stages and also occurred in samples taken from the lake bottom. If the vegetation of the islands were to be destroyed, therefore, regeneration would occur from the seed reserves present.

Two studies of European marshland soils have recently been reported. Hunyadi and Pathy (1976) examined the seed banks of several marshes in western Hungary. Although high numbers of seeds of *Bulboschoenus maritimus* were found in some of them, the greatest numbers were those of annual weeds common in adjacent arable areas, such as *Echinochloa crus-galli*, *Amaranthus retroflexus* and *Stellaria media*. Mika (1978) studied the seed banks of peat bog soils under permanent grass in South Bohemia where the dominant species was *Carex davalliana*. Numbers of seeds ranged from 1500 to 9200 m^{-2} in 0–10 cm, and the floristic composition differed from that of the sward. The greatest numbers of seeds were of arable weeds, and their presence was apparently associated with water and insect distribution but not with wind dispersal. Except for some seeds of *Festuca rubra*, cultivated grasses were not represented; seeds of weedy grasses totalled 180–240 m^{-2} and those of legumes 40–360 m^{-2}.

The considerable populations of viable seeds which accumulate in marshy areas in northern and western Europe dominated by *Juncus* spp. have already been mentioned. In contrast, Moore and Wein (1977) were unable to find any viable seeds in a *Sphagnum* bog community in New Brunswick with *Kalmia angustifolia*, *Ledum groenlandicum* and *Rhododendron canadense*, and only 320 m^{-2} in 0–10 cm in a bog with *Larix laricina*.

VI. Seed Banks of Forests and Woodlands

Studies of the litter and surface soil in North American forests have demonstrated the presence of viable seeds of species not represented in the vegetation and have indicated long-term survival of seeds of some species of previous successional stages (Oosting and Humphreys, 1940; Olmsted and Curtis, 1947; Livingston and Allessio, 1968). The numbers of viable seeds in the forest soils, especially in the oldest stands, were low compared with those of other vegetation types. Several other studies have since been made of the seed banks of different forest types in North America (Table VI).

A consistent feature of the coniferous forests examined was the absence of viable seeds of the dominant tree species, attributable to low seed inputs and rapid loss from the surface seed bank (Kellman, 1974a; Whipple, 1978). The total numbers of viable seeds present were low compared with those in other communities and in the series of stands examined by Moore and Wein (1977) were lower than those in deciduous-dominated forest. The values quoted in Table VI for Oregon (Strickler and Edgerton, 1976) include spores of the fern *Cystopteris fragilis* which gave rise to plantlets; if these are excluded, values ranging from approximately 300–1000 seeds m^{-2} are obtained. The highest numbers were in the litter layer and most were of species such as *Epilobium watsonii* which produce abundant wind-borne seeds,

TABLE VI. Numbers of viable seeds in North American forest soils

Location	Forest dominants	Age (years)	Depth (cm)	Seeds (m^{-2})	Authors
British Columbia	*Tsuga heterophylla* *Pseudotsuga menziesii*	100	0–10	1016	Kellman (1970)
British Columbia	*Thuja plicata* *Tsuga heterophylla* *Pseudotsuga menziesii*	103	0–10	206	Kellman (1974a)
Northwest Territories	*Picea mariana* *Picea glauca* *Pinus banksiana*	180 100 100	0–10 0–10 0–10	0 0 0	Johnson (1975)
New Brunswick	*Betula papyrifera* *Fagus grandiflora*	40+	0–12	3400	Moore and Wein (1977)
	Acer rubrum *Fagus grandiflora*	40+	0–12	1950	
	Acer rubrum *Abies balsamea*	40+	0–16	1230	
	Picea mariana *Pinus strobus*	40+	0–14	580	
	Pinus mariana	40+	0–16	370	
Saskatchewan	*Picea glauca* *Populus tremuloides* *Betula papyrifera*	c.40	0–10	399	Archibold (1979)
Oregon	*Abies grandis* *Pinus contorta*	130	litter +4	421	Strickler and Edgerton (1976)
	Abies grandis *Picea engelmannii* *Larix occidentalis*	150	litter +4	1863	
	Abies grandis	175	litter +4	3447	
Colorado	Subalpine, mesic *Picea engelmannii* *Abies lasiocarpa*	325	0–5	53	Whipple (1978)
	Subalpine, dry *Pinus contorta*	325	0–5	3	

although the soil beneath contained some viable seeds which had probably been present for long periods. Kellman (1974a) compared the seed banks of a forest and the secondary community arising after clear-felling. The two were comparable in floristic diversity, but whereas the forest included both primary and secondary species, the appreciably larger seed bank of the secondary community was made up mainly of *Senecio sylvaticus* and other species reproducing locally.

Subalpine forest in Colorado yielded very low numbers of viable seeds (Whipple, 1978). With a single exception at each of the two sites examined, the tree and understory species present in the forest were not represented in the seed bank. Very low numbers of seeds were also recorded from subarctic forest stands of varying ages (Johnson, 1975). In some of them no viable seeds at all were found (Table VI); in others there were no germinable seeds but small numbers of those extracted, of *Empetrum nigrum* and *Betula* spp. in particular, proved to be viable when subjected to tetrazolium tests.

The occurrence of viable tree seeds in five northern hardwood stands in Pennsylvania dominated by *Prunus serotina* with *Acer saccharum* and/or *A. rubrum* was studied by Marquis (1975). In spring, when the previous season's seedfall was still present, the average number was 370 m^{-2}, mainly *Prunus pensylvanica*, *P. serotina* and *Betula* spp. Three groups of species could be distinguished. Seeds of some species germinated mainly in the first spring (*Acer saccharum*, *Tsuga canadensis*, *Fagus grandiflora*), others had some seeds which regularly survived for 2–5 years (e.g., *Fraxinus americanus*, *Prunus serotina*, *Betula* spp.), while high numbers of *Prunus pensylvanica* seeds remained viable for 30 years or more after this species had died out of the overstory. A detailed study of *P. pensylvanica*, a successional species appearing after disturbance of northern hardwood forests, was made by Marks (1974). He demonstrated the presence of appreciable populations of viable seeds of this species in forest soils well after it had disappeared from the vegetation, and discussed the "buried seed strategy" in relation to disturbance and the stability of the forest ecosystem. Evidence for long-term seed survival in a few other species has also been obtained. Gratkowski (1964) found 37 viable seeds m^{-2} of *Ceanothus velutinus* var. *laevigatus* in the soil of old *Pseudotsuga menziesii* forest, with none in the litter or duff layers and no plants present. Del Tredici (1977) concluded that seeds of *Comptonia peregrina* could persist for 70 years or more beneath *Pinus strobus* forest in Connecticut.

In the USSR, a study of a 100-year-old southern taiga (*Picea abies*/ *Vaccinium myrtillus*) forest revealed numbers of viable seeds ranging from 1200–5000 m^{-2} (Karpov, 1960). There was little correspondence between the floristic composition of the ground flora and that of the seed bank, and the species represented as seeds were largely those of early successional stages or of clearings. Similar results were reported by Petrov (1977) from a study of five forest types in the Moscow region in which the seedlings emerging on cleared plots were recorded. In old, comparatively undisturbed *Picea* or *Picea*/*Pinus* forests, seeds were present of herbs and shrubs which were absent or almost absent from the vegetation; those of *Carex pallescens* and *Luzula* sp. were especially numerous. Although seeds of a few forest species such as *Oxalis acetosella*, *Ajuga reptans* and *Moehringia trinervia* were found in significant numbers, there were few or none of most of the characteristic ground-layer species. If felling took place, these species could not regenerate from seeds in the soil and most of the seedlings would be those of non-forest species. In secondary *Betula* and *Populus tremuloides* woodlands 20–30 or 30–40 years old, the numbers of viable seeds were about three times those of the primary forest and the range of species was greater.

In a 45-year-old *Picea abies* forest in the Belgian Ardenne there were more than 21 000 viable seeds m⁻² in the 2–5 cm layer (Vanesse, 1976). The highest numbers were those of *Calluna vulgaris* and the seed bank provided a partial representation of the flora prior to afforestation. A study of cleared woodlands in Romania (Simtea, 1971) showed that the numbers of seeds decreased from the *Quercus* zone (300 m) through the *Fagus* zone (1000 m) to *Picea* (1400 m). There was a direct relationship between the extent of the seed bank and the degree of vegetation cover two years after clearing.

Only two studies appear to have been made of the seed banks of British woodlands. Thompson and Grime (1979) determined the germinable seeds present in the surface soil of a semi-natural wood in northern England dominated by *Quercus petraea* and noted a remarkable lack of correspondence between the vegetation and the seeds present. No seeds were found of some of the most frequent species in the ground flora, whereas of the five species with the highest seed numbers *Juncus effusus* was not recorded in the vegetation at all and *Digitalis purpurea*, *Milium effusum* and *Poa annua* were all infrequent. A detailed study of five coppice woods in southeast England which had been neglected for about 30 years was made by Brown and Oosterhuis (1981). They found an average of 1840 viable seeds m⁻² in 0–5 cm, with *Juncus effusus*, *Betula* spp. and *Rubus fruticosus* the most frequent. Soil samples from the 5–15 cm layer revealed seed numbers which were on average only 20% lower than those from 0–5 cm. Of the species present in the vegetation of the densely shaded areas from which the samples were taken, only *R. fruticosus* was represented in the seed bank. Most of the seeds found were those of species which are light-demanding or which will tolerate only partial shade, and which occupy the ground immediately after coppicing. They concluded that most of the seeds had been present since the last coppicing, and that the seed bank plays a major part in floristic recovery.

Data on the seed banks of coniferous woodlands in Japan were obtained in studies of a series of stages in a secondary succession

(Numata *et al.*, 1964; Hayashi and Numata, 1964, 1968). As found in investigations of North American old-field successions, seeds of species of earlier stages were present and the dominant tree species were poorly represented in the seed bank. There was a progressive decrease in seed numbers from a young stand of *Pinus thunbergii* through mature *P. thunbergii* and overmature *P. densiflora* to climax *Shiia sieboldii*. Another investigation was reported by Nakagoshi and Suzuki (1977) who examined the viable seeds beneath seven forest types, mainly with *Pinus densiflora* as the dominant species, on Miyajima Island in southwestern Japan. They were particularly concerned to establish the floristic correlation between the vegetation and the seed bank and they divided the species into four groups. The first comprised "differential" species which were present in some communities but not others, usually both in the vegetation and the seed bank. The second included the main species present in the stands; the highest seed numbers were those of *Pieris japonica* with a mean of 320 m^{-2} in 0–10 cm, followed by *Cleyera japonica* (80 m^{-2}), *Ilex peduncularis* (33 m^{-2}) and *Pinus densiflora* (32 m^{-2}). The species in this group are important in maintaining the communities. The third group comprised species of which seeds were found but which were not present in the vegetation; they were mainly endozoochores, and may be pioneers when the forest is destroyed by fire or felling. The final group was made up of species present in the forest vegetation but not detected in the seed banks.

Some data are available on seed banks beneath Australian woodlands. A study of viable seeds in the soil of different woodland and scrub communities in Victoria revealed a range from 670 m^{-2} in 0–10 cm for *Leptospermum myrsinoides* heath to 35 840 m^{-2} for savanna woodland (Carroll and Ashton, 1965). Seeds of dominant trees, such as *Nothofagus cunninghamii* and *Eucalyptus* spp., were few and most of the seeds found were those of herbaceous species. Similar results were obtained for a subalpine woodland dominated by *Eucalyptus pauciflora* var. *alpina* in Victoria (Howard and Ashton, 1967). The dominant species and several others common in the forest stand were not represented in the seed bank, which included many herbaceous species absent from the vegetation. Of the 1820 viable seeds m^{-2} in 0–18 cm, 83% were in the top 2·5 cm. Howard (1974) examined the seed bank beneath closed *Nothofagus cunninghamii* forest, grassland dominated by *Poa gunnii* and an ecotonal stage with a tree stratum of *Acacia melanoxylon* in Tasmania. Few seeds of *N. cunninghamii* were found, but those of *A. melanoxylon* occurred in the forest soil even though this species was not present in the forest vegetation. The total bank of viable seeds was least in the forest soil and herbaceous species were the main contributors, although largely absent from the vegetation. Howard and Smith (1979) studied the viable seeds beneath some forest and scrub communities in part of Sydney Harbour National Park. Significant numbers of seeds of weedy species were found even in apparently undisturbed vegetation, and would have to be taken into account in the management of the area.

The limited investigations of seed banks beneath tropical forests indicate the prevalence of seeds of species characteristic of secondary successions. Samples from mature forest in Nigeria (Keay, 1960) revealed seeds of only two economic timber trees. Most of the tree seeds recorded were those of typical forest regrowth species not present in the vegetation, while those of climbers, shrubs and herbs were also present. An examination of viable seeds beneath virgin forest in Sabah, Malaysia showed that 62% of those present were of secondary species (Liew, 1973). Guevara and Gómez-Pompa (1972) studied the viable seeds in soil beneath primary lowland tropical selva in Mexico and concluded that secondary species were the most important floristic element. Repeated sampling showed that seeds of some species were present throughout the year, indicating that dormancy is an important factor; no primary species showed this characteristic. They concluded that primary forest would not be restored if all trees in adjacent areas were destroyed, and that the seed bank is important in determining the direction of succession. In secondary vegetation of recent origin, the numbers of viable seeds were considerably higher than those beneath primary forest.

The results of these investigations indicate that most of the shade-tolerant true forest species do not produce seeds which enter the buried seed bank. This is also generally true for the dominant trees of mature forest. The greatest numbers of seeds found in soil beneath closed forest or woodland are those of species characteristic of more open habitats such as clearings or the early stages of secondary successions. In secondary forests seeds of these species which have become incorporated into the soil may persist in quantity for long periods, perhaps 40 years or more. The presence of seeds of secondary species in primary forest, however, suggests that immigration must occur. Kellman (1974a) showed that small quantities of seeds of secondary species were able to infiltrate for considerable distances, while Nakagoshi and Suzuki (1977) considered that bird dispersal was an important factor in the forest communities which they examined.

VII. Some General Considerations

The investigations which have been reported lead to the conclusion that the presence of appreciable reserves of viable seeds in the soil is a feature of a wide range of plant communities. The main exceptions appear to be mature temperate and tropical forests, where seed banks tend to be replaced by banks of persistent seedlings (Grime, 1979). The very limited evidence also suggests that seed numbers are likely to be low in the vegetation of arctic regions. Johnson (1975) found scarcely any viable seeds in the soil of subarctic forests, and concluded that there is a trend towards a decline in the size of the seed bank with increasing latitude.

As Thompson (1978) has pointed out, there are also general tendencies for the size of the seed bank to decline with increasing altitude and in the later stages of plant successions. He suggests that this variation, both on a local and on a continental scale, might be accounted for in terms of disturbance (anything which causes the destruction of all or part of the vegetation) and stress (mechanisms limiting the rate of production of biomass of all or part of the vegetation). A decline in seed numbers would be expected with decreasing intensity of disturbance and with increasing levels of stress, and evidence to support this is adduced from published studies of the seed banks of British grasslands and of North American herbaceous and forest vegetation. He concludes that the existing data fit reasonably well into an explanatory scheme based on gradients of stress and disturbance, but that more data are needed to test the hypothesis.

The species composition of a seed bank reflects the differing strategies of past and present components of the vegetation, and great diversity is apparent. At one extreme are species which produce large numbers of seeds, many of which are capable of remaining viable for long periods when buried; these are often the major contributors to seed banks. At the other are species in which regeneration is entirely or mainly clonal, or that produce seeds which all germinate rapidly, retain viability for only a short period, or are subject to severe predation. These species either do not occur in seed banks or are represented for only a limited part of the year by seeds present at or near the soil surface.

A useful classification (Fig. 3) which is applicable to individual species has been proposed by Thompson and Grime (1979) on the basis of a study of seed banks in a range of habitats. The main distinction is drawn between transient seed banks (Types I and II) in which no viable seeds remain for longer than a year, and persistent seed banks

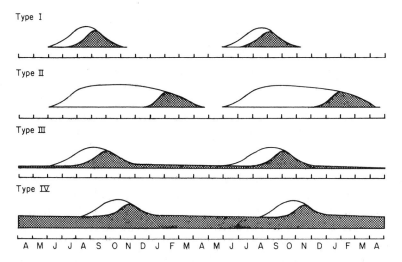

Fɪɢ. 3. Diagrammatic representation of four types of seed bank. Shaded areas: seeds capable of germinating immediately after removal to suitable laboratory conditions. Unshaded areas: seeds viable but not capable of immediate germination. Type I: annual and perennial grasses of dry or disturbed habitats. Type II: annual and perennial herbs colonizing vegetation gaps in early spring. Type III: species mainly germinating in the autumn but maintaining a small persistent seed bank. Type IV: annual and perennial herbs and shrubs with large persistent seed banks (from Thompson and Grime, 1979).

(Types III and IV) in which there is a carry-over of some viable seeds from year to year. The species which form transient seed banks are adapted to exploit the gaps created by seasonally-predictable damage and mortality in the vegetation, while the persistent seed bank (Type IV) confers the potential for regeneration in circumstances where disturbance of the vegetation is temporally and/or spatially unpredictable. Species with Type III seed banks show intermediate characteristics. The functional significance of the different types and the relationships with the seed morphology and germination physiology of the species concerned are discussed by Grime (1979) and Thompson and Grime (1979). In each of the ten plant communities which Thompson and Grime examined there were some species which had persistent seed banks, and there were usually others of which viable seeds were present for only part of the year.

The question of the fate of seeds while at the soil surface and after they have become part of the buried seed bank has been considered by Sagar and Mortimer (1976), who presented life tables for some of the few plant species in which the gains and losses from the seed bank have been quantified. Recent studies have provided further data for particular species and communities (e.g. Keeley, 1977; Nelson and Chew, 1977; Watkinson, 1978; Turkington *et al.*, 1979). One investigation of particular interest is that by van der Vegte (1978) of the population dynamics of *Stellaria media* on a sandy soil in the Netherlands. He found that there were two distinct populations, with the plants growing side by side, which differed in life-cycle strategy. In one of them almost all the seeds germinated in autumn and the seed bank was replaced annually; in the other, there was a persistent and phenotypically diverse seed reserve. As he points out, it seems a reasonable assumption that population differences of this kind will be detected in other species if more attention is paid to individual plants.

One aspect which is of particular significance in relation to the species which form persistent seed banks is the pattern of loss of viable seeds from the reserves and the way in which this may vary under different conditions. As already mentioned, there is evidence that in soil subjected to a consistent regime of disturbance the numbers of viable seeds of many arable weeds decrease exponentially from year to year (Roberts, 1962; Roberts and Dawkins, 1967). Further data have been obtained from experiments in which seeds have been mixed with soil and the numbers of viable seeds or of emerging seedlings recorded during successive years (Roberts, 1964; Roberts and Feast, 1973a; Watanabe and Hirokawa, 1975).

TABLE VII. Seedling emergence in successive years and survival of seeds freshly harvested in late summer or autumn, mixed with 7·5 cm of soil and cultivated three times per year. Means of three experiments begun in different years

Species	Initial[1]	Seedlings emerged as % of seeds sown					Viable seeds remaining as % of seeds sown
		Year[2] 1	Year 2	Year 3	Year 4	Year 5	
Hordeum murinum	87·8	0	0	0	0	0	0
Anthriscus sylvestris	0	77·3	1·2	0	0	0	0
Lamium amplexicaule	19·4	43·2	8·0	2·2	0·5	0·2	4·6
Rumex crispus	3·0	37·8	15·0	6·1	2·0	0·6	7·5
Atriplex patula	0	5·6	12·3	9·7	4·7	3·9	14·2
Aethusa cynapium	0	4·6	10·0	12·4	10·8	6·2	21·0

[1] The period from sowing until 31 December.
[2] Calendar years.

Some results from experiments of this kind are summarized in Table VII. Those for *Hordeum murinum* and *Anthriscus sylvestris* are representative of species with transient seed banks, germination occurring either in autumn soon after sowing (Type I of Thompson and Grime, 1979) or in early spring (Type II), with few if any viable seeds remaining after that. The other four species form persistent seed banks (Types III and IV). For *Lamium amplexicaule* and *Rumex crispus* the decrease in seedling numbers was approximately exponential once the initial emergence had taken place. In *Atriplex patula* maximum emergence did not occur until the second year, and in *Aethusa cynapium* not until the third; this appears to be a result of innate dormancy imposed by the seed coat. The depletion of the seed bank is most rapid when the seeds are present in a shallow layer of soil and when there is frequent disturbance (Roberts and Feast, 1972, 1973a); under these conditions the role of enforced dormancy is minimized.

The scheme proposed by Thompson and Grime (1979) and shown in Fig. 3 takes account of the presence of innate dormancy, but implies that once this has been overcome the seeds will germinate if transferred to suitable conditions. This may well be true, but there is evidence for some species that changes in the dormancy status of buried seeds can occur as a result of the temperature conditions which the seeds experience. Induced dormancy may develop which then requires a period of exposure to temperatures within a particular range before germination can take place. In these species, therefore, seeds held under enforced dormancy undergo cyclic physiological changes in response to seasonal temperature variations. These not only serve to regulate the time of year at which emergence of seedlings can occur (Montegut, 1975), but may well increase longevity since germination is possible during only part of the year (Fig. 4). The general question of dormancy in relation to seed survival in the soil has been discussed by Roberts (1972).

The pattern of decline for persistent seed banks in undisturbed surface soil also appears to be exponential (Roberts and Dawkins, 1967; Roberts and Feast, 1973a; Watanabe and Hirokawa, 1975), and the data of Forbes (1963) and Thurston (1966) for survival of seeds of *Avena fatua* and *A. ludoviciana* under leys also conform to this pattern. There are few other studies in which seed survival under natural conditions has been determined for long enough to characterize the form of the decline curve. Bowes and Thomas (1978) constructed a curve for seed survival of *Euphorbia esula* in 0–2·5 cm, using data from different sites, which showed a linear rather than an exponential decrease of 13% per year over an 8-year period.

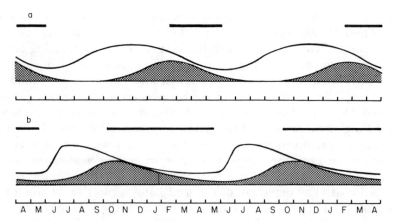

Fig. 4. Diagrammatic representation of persistent seed banks of two species of arable weeds. Shaded areas: seeds capable of germinating immediately after removal to suitable laboratory conditions. Unshaded areas: seeds viable but not capable of immediate germination. Horizontal bars: period of seedling emergence in the field. a: *Polygonum aviculare* (data of Courtney, 1968). Innate and induced dormancy overcome by low temperatures; induced dormancy re-imposed in late spring. b: *Veronica hederifolia* (data of Roberts and Lockett, 1978). Innate and induced dormancy overcome by high temperatures; induced dormancy re-imposed during winter and spring.

Seeds distributed in soil layers close to the surface are subject to predation and to wide fluctuations in temperature and moisture. These factors combined with inherent variability in the seed population, especially when seeds of different ages are involved, are presumably responsible for the exponential decline observed in species with persistent seed banks. The situation is different for seeds which are buried at depth, where conditions are much more uniform (Harper, 1957). Here it is probable that external factors are relatively less important and longevity depends much more on the inherent properties of the seed. Certain species appear to be particularly well able to survive in this situation; *Hyoscyamus niger* and *Verbascum thapsus* are two species for which there is good evidence of long-term seed survival in undisturbed sites (Ødum, 1978). In the Duvel experiment (Toole and Brown, 1946), seeds of *Solanum nigrum* buried at depths of 20 cm or more had approximately 80% viability after 39 years. In contrast, a naturally-occurring population of seeds of this species distributed in 0–23 cm of undisturbed soil declined exponentially at a rate equivalent to a loss of about 23% per year (Roberts and Dawkins, 1967).

In experiments in which seeds are buried at depth one might expect some initial loss, varying very much with species and with the time of year at which the experiment is begun, because of germination before dormancy is enforced. There might then be a period of little change in the numbers of viable seeds remaining, with finally a fall to zero as the seeds die. Data obtained by Burnside *et al.* (1977) for seeds of *Sorghum bicolor* buried at a depth of 22 cm show these three phases very clearly. There was an initial loss of about 60% of the seeds in the first year, little change in numbers for the next seven years, and then a gradual decrease in the final five years of the experiment, reflecting the variability within the seed sample.

In respect of a persistent, buried seed bank three situations can thus be recognized. In soil which is periodically disturbed, dormancy will be enforced for only part of the time and mechanisms of innate and induced dormancy will play a large part in determining persistence. Where the seeds are distributed in undisturbed soil fairly close to the soil surface, perhaps to a depth of 20 cm or so, the role of enforced dormancy becomes more important. Finally, in seeds buried at depth under comparatively uniform conditions the intrinsic capacity for survival in the imbibed state becomes dominant.

VIII. CONCLUDING REMARKS

The presence of viable seeds in the soil is clearly relevant to the study of plant communities: as Major and Pyott (1966) observed "plants occurring in this form are part of the flora, which helps to determine the community, even though they are not readily evident". In the past ecologists have tended to ignore the seed bank, perhaps partly deterred by the difficulties in assessing it. Recently, however, there has been an upsurge of interest in seed banks, related to developments in the study of life-cycle strategies of plant species, plant demography and vegetation dynamics (Sagar and Mortimer, 1976; Harper, 1977; Rabotnov, 1978; Grime, 1979; Miles, 1979; Thompson and Grime, 1979).

From the practical point of view the viable seeds in the soil may be regarded as posing problems or as conferring benefits, depending on the situation. In arable cultivation, the presence of usually large numbers of seeds means that there is a continuing need for weed control; here the objective is to maintain the seed bank at the lowest feasible level in order to minimize interference with crop production. As already mentioned, the advent of selective chemical methods of weed control for most crops has enabled this to be achieved more readily than before. Repeated use of particular herbicides, however, by allowing seed production in some species and not others can markedly alter the composition of the seed bank and hence that of the weed vegetation appearing in future crops. The increasing incidence of selection of biotypes resistant to herbicides, particularly the *s*-triazines (Stryckers, 1979), is especially significant in this respect. The period for which viable seeds remain in the soil is one of the factors determining when resistance is likely to appear (Gressel, 1978).

Seed bank studies can make a contribution to improving the efficiency of crop production in various ways. A knowledge of the extent to which viable seeds are likely to persist in the soil has obvious relevance in assessing the consequences of changes in cropping and cultivations. Quantitative studies of the fate of seeds of weed species under different conditions (Wilson and Cussans, 1975; Mortimer, 1979) are particularly important in providing a basis for sound advice on control measures. Although viable seeds of crop species normally form only a small proportion of the total seed bank of arable soils, their persistence can be troublesome. Crops with which problems can arise include cereals, where disease carry-over is a prime concern (Hughes, 1974), and sugar beet, in which the annual bolting types can persist from crop to crop as seeds in the soil (Longden, 1974). The persistence of potato seeds is also being examined (Lawson and Wiseman, 1979).

The use of seed-bank data in monitoring the success of long-term weed control programmes is an aspect which could attract greater attention as the concept of weed management gains wider recognition; species which regenerate mainly or entirely by vegetative means would, of course, require separate consideration. Altieri and Whitcomb (1979) have suggested that the weed flora might be manipulated, by soil disturbance at particular times of year, to encourage those species whose presence can greatly increase the populations of beneficial insects. Knowledge of the species composition of arable seed banks can be used predictively in a general way to give guidance on the choice of herbicides (Dechkov, 1978), but how far relationships between seed numbers and seedling densities can be quantified and used for short-term forecasting of weed floras remains problematical. For this, and indeed for other purposes, techniques of estimation of viable seeds more rapid than those at present available would be desirable; however, the prospects for improvement do not seem very hopeful.

In grasslands, persistent seed banks of the less-desirable grasses contribute to sward deterioration and can pose problems when it is renewed. At the same time reserves of seeds of leguminous species are valuable in the maintenance of pastures, especially in regions subject to drought. In forests and woodlands the presence of viable seeds is a major factor determining the direction of succession after destruction of the trees by fire or felling.

In these and other communites, the seed bank has a role in maintaining floristic diversity and must be taken into account in management. Baskin and Baskin (1978) cite results which demonstrate the function of a persistent seed bank in enhancing the genetic diversity and stability of a plant population at a particular site, and point out that this is especially significant for narrowly endemic species which occur in small populations. The wider genetic implications of seed banks have been discussed by Levin and Wilson (1978). One direct use of the seed bank is in the restoration of disturbed land; Beauchamp *et al.* (1975) found that the soil at potential strip-mining sites in Wyoming contained sufficient seeds in 0–5 cm to revegetate the areas, although most were those of species of secondary successions. This aspect has been discussed by Johnson and Bradshaw (1979).

The main concern of this article has been to review recent work on seed banks in a range of vegetation types. The seed bank, however, is in a state of dynamic flux; seeds enter it and leave it, and while remaining in it undergo physiological changes which can affect their response to the present and future environment. There has been a trend from mainly descriptive studies of the occurrence of seeds in the soil to those in which attempts are made to quantify seed inputs and losses and to determine the mechanisms by which they are brought about. Continuation of these approaches will not only increase understanding of how plant communities are regulated but is likely also to bring practical benefits.

References

Allott, D. J. (1970). Non-cultivation a success in 10-year raspberry soil trial. *Grower* **73**, 35–36.

Altieri, M. A. and Whitcomb, W. H. (1979). The potential use of weeds in the manipulation of beneficial insects. *HortScience* **14**, 12–18.

Anon. (1967). Hill land research. *Rep. Welsh Pl. Breed. Stn for 1966*, 97.

Archibold, O. W. (1979). Buried viable propagules as a factor in post-fire regeneration in northern Saskatchewan. *Can. J. Bot.* **57**, 54–58.

Artyushin, A. M. and Libershtein, I. I. (1976). The role of the State Agrochemical Service in increasing the efficacy of using herbicides. *Khim. Sel'. Khoz.* **14** (2), 7–9.

Ashworth, L. J. Jr (1976). Quantitative detection of seed of branched broomrape in California tomato soils. *Pl. Dis. Reptr.* **60**, 380–383.

Banks, P. A., Santelmann, P. W. and Tucker, B. B. (1976). Influence of long-term soil fertility treatments on weed species in winter wheat. *Agron. J.* **68**, 825–827.

Barbour, M. G. and Lange, R. T. (1967). Seed populations in some natural Australian topsoils. *Ecology* **48**, 153–155.

Barralis, G. (1965). Aspect écologique des mauvaises herbes dans les cultures annuelles. *C.r. 3ᵉ Conf. Com. franç. mauv. Herbes (COLUMA), Paris*, 5–25.

Barralis, G. (1972). Evolution comparative de la flore adventice avec ou sans désherbage chimique. *Weed Res.* **12**, 115–127.

Barralis, G. and Chadoeuf, R. (1976). Evolution qualitative et quantitative d'un peuplement adventice sous l'effet de dix années de traitement. *C.r. Vᵉᵐᵉ Colloque int. sur la Biologie et l'Ecologie des Mauvaises Herbes, Dijon*, 179–186.

Barralis, G. and Salin, D. (1973). Relations entre flore potentielle et flore réelle dans quelques types de sol de Côte d'Or. *C.r. IVᵉᵐᵉ Colloque int. sur la Biologie et l'Ecologie des Mauvaises Herbes, Marseille*, 94–101.

Barralis, G., Chadoeuf, R., Compoint, J. P., Gasquez, J. and Lonchamp, J. P. (1978). Etude de quelques aspects de la dynamique d'une agrophytocènose. *Proc. Symp. Mediterraneo de Herbicidas, Madrid*, 85–98.

Bartolome, J. W. (1979). Germination and seedling establishment in California annual grassland. *J. Ecol.* **67**, 273–281.

Baskin, J. M. and Baskin, C. C. (1978). The seed bank in a population of an endemic plant species and its ecological significance. *Biol. Conserv.* **14**, 125–130.

Beale, P. E. (1974). Regeneration of *Trifolium subterraneum* cv. Yarloop from seed reserves in soil on Kangaroo Island. *J. Aust. Inst. agric. Sci.* **40**, 78–80.

Beauchamp, H., Lang, R. and May, M. (1975). Topsoil as a seed source for reseeding strip mine soils. *Res. J. Wyo agric. Exp. Stn* **90**, pp. 8.

Boeker, P. (1959). Samenauflauf aus Mist und Erde von Triebwegen und Ruheplätzen. *Z. Acker- u. PflBau* **108**, 77–92.

Bowes, G. G. and Thomas, A. G. (1978). Longevity of leafy spurge seeds in the soil following various control programmes. *J. Range Mgmt* **31**, 137–140.

Brenchley, W. E. (1918). Buried weed seeds. *J. agric. Sci., Camb.* **9**, 1–31.

Brenchley, W. E. and Warington, K. (1930). The weed seed population of arable soil. I. Numerical estimation of viable seeds and observations on their natural dormancy. *J. Ecol.* **18**, 235–272.

Brown, A. H. F. and Oosterhuis, L. (1981). The role of buried seeds in coppicewoods. *Biol. Conserv.* **21**, 19–38.

Burnside, O. C. (1978). Mechanical, cultural and chemical control of weeds in a sorghum-soybean (*Sorghum bicolor*)–(*Glycine max*) rotation. *Weed Sci.* **26**, 362–369.

Burnside, O. C., Wicks, G. A. and Fenster, C. R. (1977). Longevity of shattercane seed in soil across Nebraska. *Weed Res.* **17**, 139–143.

Carretero, J. L. (1977). Estimacion del contenido de semillas de malas hierbas de un suelo agricola como prediccion de su flora adventicia. *An. Inst. Bot. Cavanilles* **34**, 267–278.

Carroll, E. J. and Ashton, D. H. (1965). Seed storage in soils of several Victorian plant communities. *Vict. Nat.* **82**, 102–110.

Champness, S. S. and Morris, K. (1948). The population of buried viable seeds in relation to contrasting pasture and soil types. *J. Ecol.* **36**, 149–173.

Chancellor, R. J. (1965). Weed seeds in the soil. *Rep. Weed Res. Org. for 1960–64*, 15–19.

Chancellor, R. J. (1978). Grass seeds beneath pastures. In "Changes in Sward Composition and Productivity". *Occasional Symposium of the British Grassland Society* No. 16, 147–150.

Charlton, J. F. L. (1977). Establishment of pasture legumes in North Island hill country. I. Buried seed populations. *N.Z. J. exp. Agric.* **5**, 211–214.

Chippindale, H. G. and Milton, W. E. J. (1934). On the viable seeds present in the soil beneath pastures. *J. Ecol.* **22**, 508–531.

Clay, D. V. and Davison, J. G. (1976). The effect of 8 years of different soil-management treatments in raspberries on weed seed and worm populations, soil microflora and on chemical and physical properties of the soil. *Proc. 1976 Br. Crop Protection Conf.— Weeds* **1**, 249–258.

Courtney, A. D. (1968). Seed dormancy and field emergence in *Polygonum aviculare* L. *J. appl. Ecol.* **5**, 675–684.

Dale, J. E. and Chandler, J. M. (1976). Impact of herbicides and crop rotation on weed plant succession. *Proc. sth. Weed Sci. Soc.* **29**, 167.

Dale, J. E. and Chandler, J. M. (1977). Weed seed populations in a cotton and soybean rotation treated with phenylurea herbicides. *Proc. sth. Weed Sci. Soc.* **30**, 96.

Darwin, C. (1859). "On the Origin of Species by means of Natural Selection". London, Watts. (Reprint of first edition, 1950.)

Dechkov, Z. (1974). Weed seed in some soils. *Rast. Nauki* **11** (9), 115–122.

Dechkov, Z. (1975). Weed seeds in the soil of some regions in the central part of the Danubian plain. *Rast. Nauki* **12** (2), 148–157.

Dechkov, Z. (1976). Weed seeds in soil of some regions of the western Danubian plain. *Rast. Nauki* **13** (8), 111–118.

Dechkov, Z. (1978). Efficient prognostication of the weed infestation in the Plovdiv district. *Rast. Nauki* **15** (7), 79–87.

Dechkov, Z. and Atanassov, P. (1976). Potential weediness of soil as influenced by the cropping sequence, *Rast. Nauki* **13** (10), 134–142.

Delpech, R. (1969). Essai d'estimation du stock de semences viables dans l'horizon superficiel du sol d'une vieille prairie pâturée de l'Auxois (Côte d'Or). *C.r. IIIème Colloque sur la Biologie des Mauvaises Herbes, Grignon*, **2**, 80–89.

Del Tredici, P. (1977). The buried seeds of *Comptonia peregrina*, the Sweet Fern. *Bull. Torrey bot. Club* **104**, 270–275.

Dirven, J. G. P. (1966). Zaden in de 0–5 cm zodelaag van beschuitgrasland. *Surin. Landb.* **14** (3), 90–94.

Dore, W. G. and Raymond, L. C. (1942). Pasture studies. XXIV. Viable seeds in pasture soil and manure. *Scient. Agric.* **23**, 67–79.

Dospekhov, B. A. and Chekryzhov, A. D. (1972). Counting weed seeds in the soil by the method of small samples. *Izv. timiryaz. sel'.-khoz. Akad.* **1972** (2), 213–215.

Dotzenko, A. D., Ozkan, M. and Storer, K. R. (1969). Influence of crop sequence, nitrogen fertilizer and herbicides on weed seed populations in sugar beet fields. *Agron. J.* **61**, 34–37.

Douglas, G. (1965). The weed flora of chemically-renewed lowland swards. *J. Br. Grassl. Soc.* **20**, 91–100.

Dvořák, J. and Krejčíř, J. (1974). Příspěvek ke studiu obsahu semen plevelů v ornici. *Sb. vys. Sk. zeměd. les. Fac. Brne, A* **22**, 453–461.

Everest, J. W. and Davis, D. E. (1974). Effects of repeated herbicide applications on weed-seed in soil. *Proc. sth. Weed Sci. Soc.* **27**, 337.

Fay, P. K. and Olson, W. A. (1978). Technique for separating weed seed from soil. *Weed Sci.* **26**, 530–533.

Feast, P. M. and Roberts, H. A. (1973). Note on the estimation of viable weed seeds in soil samples. *Weed Res.* **13**, 110–113.

Fekete, R. (1975). Comparative weed-investigations in traditionally-cultivated and chemically-treated wheat and maize crops. IV. Study of the weed-seed contents of the soils of maize crops. *Acta biol., Szeged* **21**, 9–20.

Foerster, E. (1956). Ein Beitrag zur Kenntnis der Selbstverjüngung von Dauerweiden. *Z. Acker- u. PflBau* **100**, 273–301.

Forbes, N. (1963). The survival of wild oat seeds under a long ley. *Expl Husb.* No. 9, 10–13.

Goeden, R. D. and Ricker, D. W. (1973). A soil profile analysis for puncturevine fruit and seed. *Weed Sci.* **21**, 504–507.

Golubeva, I. V. (1962). Some data on the number of viable seeds found in soils under meadow-steppe vegetation. *Byull. mosk. Obshch. Ispyt. Prir., Otdel Biol.* **67** (5), 76–89.

Gratkowski, H. J. (1964). Dormant, viable *Ceanothus* seeds in soil under old-growth Douglas-fir forests. *Res. Progr. Rep. west. Weed Control Conf. 1964*, 44–45.

Gressel, J. (1978). Factors affecting the selection of herbicide resistant biotypes of weeds. *Outl. on Agric.* **9**, 283–287.

Grime, J. P. (1979). "Plant Strategies and Vegetation Processes" Wiley, Chichester.

Guevara, S. S. and Gómez-Pompa, A. (1972). Seeds from surface soils in a tropical region of Veracruz, Mexico. *J. Arnold Arbor.* **53**, 312–335.

Harper, J. L. (1957). The ecological significance of dormancy and its importance in weed control. *Proc. 4th int. Conf. Pl. Protection, Hamburg*, 415–420.

Harper, J. L. (1977). "Population Biology of Plants". Academic Press, London and New York.

Hayashi, I. (1975). The special method of inventory of buried seed population of weeds. *Workshop on Research Methodology in Weed Science, Bandung* **1**, paper no. 4.

Hayashi, I. and Numata, M. (1964). Ecological studies on the buried seed population in the soil from the viewpoint the plant succession. III. A mature stand of *Pinus thunbergii. Physiol. Ecol.* **12**, 185–190.

Hayashi, I. and Numata, M. (1968). Ecological studies on the buried seed population of the soil as related to plant succession. V. From overmature pine stand to climax *Shiia* stand. *Ecological Studies of Biotic Communities in the National Park for Nature Study,* No. 2, 1–7.

Hayashi, I. and Numata, M. (1971). Viable buried-seed population in the *Miscanthus*- and *Zoysia*-type grasslands in Japan—ecological studies on the buried-seed population in the soil related to plant succession. VI. *Jap. J. Ecol.* **20**, 243–252.

Hayashi, I. and Numata, M. (1975). Viable buried seed population in grasslands in Japan. *In* "JIBP Synthesis Volume 13. Ecological Studies in Japanese Grasslands" (Numata, M., ed.), pp. 58–69. University of Tokyo Press.

Hayashi, I., Pancho, J. V. and Sastroutomo, S. S. (1978). Preliminary report on the buried seeds of floating islands and bottom of Lake Rawa Pening, Central Java. *Jap. J. Ecol.* **28**, 325–333.

Howard, T. M. (1974). *Nothofagus cunninghamii* ecotonal stages. Buried viable seed in North West Tasmania. *Proc. R. Soc. Vict.* **86** (2), 137–142.

Howard, T. M. and Ashton, D. H. (1967). Studies of soil seed in snow gum woodland (*E. pauciflora* Sieb. ex Spreng. var. *alpina* (Benth.) Ewart). *Vict. Nat.* **84**, 331–335.

Howard, T. M. and Smith, L. W. (1979). Weed seed storage in soil from native vegetation areas. *Proc. 7th Asian-Pacific Weed Sci. Soc. Conf.*, 371–376.

Hughes, R. G. (1974). Cereals as weeds. *Proc. 12th Br. Weed Control Conf.* **3**, 1023–1029.

Hunyadi, K. and Pathy, Z. (1976). Keszthely környéki rétláp talajok gyommagfertőzottsége. *Növényvédelem* **12**, 391–396.

Hurle, K. (1974). Effect of long-term weed control measures on viable weed seeds in the soil. *Proc. 12th Br. Weed Control Conf.* **3**, 1145–1152.

Hyde, E. O. C. and Suckling, F. E. T. (1953). Dormant seeds of clovers and other legumes in agricultural soils. *N.Z. Jl Sci. Technol. A* **34**, 375–385.

Ivens, G. W. (1978). Some aspects of seed ecology of gorse (*Ulex europaeus*). *Proc. 31st N.Z. Weed and Pest Control Conf.*, 53–57.

Jackson, M. L. (1958). "Soil Chemical Analysis". Constable, London.

Jalloq, M. C. (1975). The invasion of molehills by weeds as a possible factor in the degeneration of reseeded pasture. I. The buried viable seed population of molehills from four reseeded pastures in West Wales. *J. appl. Ecol.* **12**, 643–657.

Jan, P. and Faivre-Dupaigre, R. (1977). Incidence des façons culturales sur la flore adventice. *Proc. EWRS Symposium on Different Methods of Weed Control and their Integration, Uppsala* **1**, 57–64.

Jensen, H. A. (1969). Content of buried seeds in arable soil in Denmark and its relation to the weed population. *Dansk bot. Ark.* **27** (2), 1–56.

Johnson, E. A. (1975). Buried seed populations in the subarctic forest east of Great Slave Lake, Northwest Territories. *Can. J. Bot.* **53**, 2933–2941.

Johnson, M. S. and Bradshaw, A. D. (1979). Ecological principles for the restoration of disturbed and degraded land. *Appl. Biol.* **4**, 141–200.

Johnston, A., Smoliak, S. and Stringer, P. W. (1969). Viable seed populations in Alberta prairie topsoils. *Can. J. Pl. Sci.* **49**, 75–82.

Jones, R. J. and Jones, R. M. (1975). Animal and pasture production from Kenya white clover and white clover based pastures. *Rep. CSIRO Div. Trop. Agron. for 1974–1975*, 6–7.

Jones, R. J. and Jones, R. M. (1978). The ecology of Siratro-based pastures. *In* "Plant Relations in Pastures" (Wilson, J. R., ed.), pp. 353–367. CSIRO, Melbourne.

Jones, R. M. (1979). Effect of stocking rate and grazing frequency on a Siratro (*Macroptilium atropurpureum*)/*Setaria anceps* cv. Nandi pasture. *Aust. J. exp. Agric. Anim. Husb.* **19**, 318–324.

Jones, R. M. and Bunch, G. A. (1977). Sampling and measuring the legume seed content of pasture and cattle faeces. *CSIRO Trop. Agron. tech. Memorandum* No. 7, pp. 9.

Jones, R. M. and Evans, T. R. (1977). Soil seed levels of *Lotononis bainesii*, *Desmodium intortum* and *Trifolium repens* in subtropical pastures. *J. Aust. Inst. agric. Sci.* **43**, 164–166.

Karpov, V. G. (1960). On the species composition of the viable seed supply in the soil of spruce-*Vaccinium myrtillus* vegetation. *Trudy mosk. Obshch. ispyt. Prir.* **3**, 131–140.

Kazantseva, A. S. and Tuganaev, V. V. (1972). The species composition and distribution of weed seeds in soils of field plant communities in the Tatar ASSR. *Biol. Nauki* **15** (11), 72–74.

Keay, R. W. J. (1960). Seeds in forest soil. *Niger. For. Inf. Bull.* (*N.S.*) No. 4, 1–12.

Keeley, J. E. (1977). Seed production, seed populations in soil, and seedling production after fire for two congeneric pairs of sprouting and non-sprouting chaparral shrubs. *Ecology* **58**, 820–829.

Kellman, M. C. (1970). The viable seed content of some forest soil in coastal British Columbia. *Can. J. Bot.* **48**, 1383–1385.

Kellman, M. C. (1974a). Preliminary seed budgets for two plant communities in coastal British Columbia. *J. Biogeogr.* **1**, 123–133.

Kellman, M. C. (1974b). The viable weed seed content of some tropical agricultural soils. *J. appl. Ecol.* **11**, 669–677.

Kellman, M. C. (1978). Microdistribution of viable weed seed in two tropical soils. *J. Biogeogr.* **5**, 291–300.

King, T. J. (1976). The viable seed contents of ant-hill and pasture soil. *New Phytol.* **77**, 143–147.

Knipe, O. D. and Springfield, H. W. (1972). Germinable alkali sacaton seed content of soils of the Rio Puerco Basin, West Central New Mexico. *Ecology* **53**, 965–968.

Kolesnikov, V. A. and Sidorov, V. I. (1974). The influence of continual herbicide use combined with mouldboard and mouldboardless ploughing on the weed infestation of soil under vegetable crops. *Khim. sel'. Khoz.* **12** (3), 216–218.

Komendar, V. I., Dubanich, M. V. and Boïko, F. M. (1973). On the quantities of viable seeds in soils of some meadow plant associations in the lowlands and foothills of the Transcarpathia. *Ukr. bot. Zh.* **30** (2), 212–219.

Korneva, I. G. (1970). Seed reserves in soil under *Agropyron/Artemisia* steppes and grass/forb meadow steppes in the Talass Ala-Too mountain range. *In* "Vegetation of Hill Pastures in Kirgizia and its Improvement", pp. 66–69. Ilim, Frunze. (from *Herbage Abstr.* **41**, 2185).

Krishna Murty, G. V. G. and Chandwani, G. H. (1974). A method for isolation and estimation of germinated and ungerminated *Orobanche* seeds from soil. *Indian Phytopath.* **27**, 447–448.

Kropáč, Z. (1966). Estimation of weed seeds in arable soil. *Pedobiologia* **6**, 105–128.

Krylova, N. P. (1979). Seed propagation of legumes in natural meadows of the U.S.S.R.—Review. *Agro-Ecosystems* **5**, 1–22.

Lawson, H. M. and Wiseman, J. S. (1979). Weed control in crop rotations. *Rep. scott. hort. Res. Inst. for 1978*, 37.

Leck, M. A. and Graveline, K. J. (1979). The seed bank of a freshwater tidal marsh. *Amer. J. Bot.* **66**, 1006–1015.

Leguizamón, E. S. and Cruz, P. A. (1980). Población de semillas en perfil arable de suelos sometidos a distinto manejo. *Rev. Inst. Cienc. Agron. Univ. nac. Córdoba.* (In press.)

Leguizamón, E. S., Cruz, P. A., Guiamet, J. J. and Casano, L. (1979). Diagnostico de la población de semillas de malezas en suelos agricolas del distrito Pujato (Prov. Santa Fe). *Resúmenes de la VII Reunión Argentina de Ecologia, Mendoza.*

Levin, D. A. and Wilson, J. B. (1978). The genetic implications of ecological adaptations in plants. *In* "Structure and Functioning of Plant Populations" (Freysen, A. H. J. and Woldendorp, J. W., eds), pp. 75–100. North Holland, Amsterdam, Oxford, New York.

Lhoste, J., Raffray, B., Quintin, D. and Gianelloni, H. (1969). Premiers résultats expérimentaux sur la prévision des levées de plantes adventices dans les champs. *C.r. 3ᵉᵐᵉ Colloque sur la Biologie des Mauvaises Herbes, Grignon* **4**, 38–44.

Liew, T. C. (1973). Occurrence of seeds in virgin forest top soil with particular reference to secondary species in Sabah. *Malay. Forester* **36**, 185–193.

Lin, C. S., Poushinsky, G. and Mauer, M. (1979). An examination of five sampling methods under random and clustered disease distributions using simulation. *Can. J. Pl. Sci.* **59**, 121–130.

Livingston, R. B. and Allessio, M. L. (1968). Buried viable seed in successional field and forest stands, Harvard Forest, Massachusetts. *Bull. Torrey bot. Club* **95**, 58–69.

Loch, D. S. and Butler, J. E. (1977). Effects of heavy solvents in seed viability of *Stylosanthes guyanensis. J. Aust. Inst. agric. Sci.* **43**, 77–79.

Lockett, P. M. and Roberts, H. A. (1976). Weed seed populations. *Rep. natn. Veg. Res. Stn for 1975*, 114.

Longden, P. C. (1974). Sugar beet as a weed. *Proc. 12th Br. Weed Control Conf.* **1**, 301–308.

Lozovatskaya, M. A. (1968). The effect of diuron and monuron on the weediness and yield of cotton. *Khim. sel'. Khoz.* **6**, 686–687.

Lueschen, W. E. and Andersen, R. N. (1978). Effect of tillage and cropping systems on velvetleaf longevity. *Proc. N. centr. Weed Control Conf.* **33**, 110.

Major, J. and Pyott, W. T. (1966). Buried viable seeds in two California bunchgrass sites and their bearing on the definition of a flora. *Vegetatio* **13**, 253–282.

Malone, C. R. (1967). A rapid method for enumeration of viable seeds in soil. *Weeds* **15**, 381–382.

Mal'tsev, A. (1909). Die Unkräuter auf den Feldern im Petersburger Gouvernement. *Trudy Byuro prikl. Bot.* **2**, 81–170.

Marks, P. L. (1974). The role of pin cherry (*Prunus pensylvanica* L.) in the maintenance of stability in northern hardwood ecosystems. *Ecol. Monogr.* **44**, 73–88.

Marquis, D. A. (1975). Seed storage and germination under northern hardwood forests. *Can. J. For. Res.* **5**, 478–484.

McRill, M. (1974). The ingestion of weed seed by earthworms. *Proc. 12th Br. Weed Control Conf.* **2**, 519–524.

Mika, V. (1978). Der Vorrat an keimfähigen Samen in Südböhmischen Niedermoorböden. *Z. Acker- u. PflBau* **146**, 222–234.

Miles, J. (1973). Natural recolonization of experimentally bared soil in Callunetum in north-east Scotland. *J. Ecol.* **61**, 399–412.

Miles, J. (1979). "Vegetation Dynamics". Chapman and Hall, London.

Milton, W. E. J. (1939). The occurrence of buried viable seeds in soils at different elevations and on a salt marsh. *J. Ecol.* **27**, 149–159.

Milton, W. E. J. (1948). The buried viable seed content of upland soils in Montgomeryshire. *Emp. J. exp. Agric.* **16**, 163–177.

Montegut, J. (1975). Ecologie de la germination des mauvaises herbes. *In* "La Germination des Semences" (Chaussat, R. and le Deunff, Y., eds), pp. 191–217. Gauthier-Villars, Paris.

Moore, J. M. and Wein, R. W. (1977). Viable seed populations by soil depth and potential site recolonization after disturbance. *Can. J. Bot.* **55**, 2408–2412.

Mortimer, A. M. (1979). The influence of cultural measures and herbicide practices on the fates of weed seeds. *Proc. EWRS Symposium on the Influence of Different Factors on the Development and Control of Weeds, Mainz*, 135–143.

Moss, G. R. (1959). The gorse seed problem. *Proc. 12th N.Z. Weed Control Conf.*, 59–64.

Mott, J. J. (1972). Germination studies on some annual species from an arid region of Western Australia. *J. Ecol.* **60**, 293–304.

Nakagoshi, N. and Suzuki, H. (1977). Ecological studies on the buried viable seed population in soil of the forest communities in Miyajima Island, Southwestern Japan. *Hikobia* **8**, 180–192.

Naylor, R. E. L. (1970). The prediction of blackgrass infestations. *Weed Res.* **10**, 296–299.

Naylor, R. E. L. (1972). Aspects of the population dynamics of the weed *Alopecurus myosuroides* Huds. in winter cereal crops. *J. appl. Ecol.* **9**, 127–139.

Nelson, J. F. and Chew, R. M. (1977). Factors affecting seed reserves in the soil of a Mojave desert ecosystem, Rock Valley, Nye County, Nevada. *Am. midl. Nat.* **97**, 300–320.

Numata, M., Aoki, K. and Hayashi, I. (1964). Ecological studies on the buried-seed population in the soil as related to plant succession. II—Particularly on the pioneer stage dominated by *Ambrosia elatior. Jap. J. Ecol.* **14**, 224–227.

Numata, M., Hayashi, I., Komura, T. and Ôki, K. (1964). Ecological studies on the buried-seed population in the soil as related to plant succession. I. *Jap. J. Ecol.* **14**, 207–215.

Ødum, S. (1978). "Dormant Seeds in Danish Ruderal Soils". Royal Veterinary and Agricultural University, Hørsholm Arboretum, Hørsholm.

Olmsted, N. and Curtis, J. D. (1947). Seeds of the forest floor. *Lcology* **28**, 49–52.

Oosting, H. T. and Humphreys, M. E. (1940). Buried viable seeds in a successional series of old field and forest soils. *Bull. Torrey bot. Club* **67**, 253–273.

Paatela, J. and Erviö, L.-R. (1971). Weed seeds in cultivated soils in Finland. *Annls agric. fenn.* **10**, 144–152.

Palmer, R. D., Cordero, J. M. and Merkle, M. G. (1970). A new look at weed ecology in Texas. *Proc. sth. Weed Sci. Soc.* **23**, 347–352.

Pawlowski, F. and Malicki, L. (1969). Wplyw rodzaju orki na pionowe rozmieszczenie nasion chwastów w glebie wytworzonej z lessów. *Annls Univ. Mariae Curie-Skłodowska, Sect. E.* **23**, 161–174.

Peter, A. (1893). Culturversuche mit "ruhenden" Samen. *Nachr. Ges. Wiss. Göttingen* **17**, 673–691.

Petrov, V. V. (1977). Reserve of viable plant seeds in the uppermost soil layer beneath the canopies of coniferous and small-leaved forests. *Moscow Univ. biol. Sci. Bull.* **32** (4), 33–40.

Platon, M. (1955). Rezerva de seminte din sol şi puterea lor de germinare la gospodăria didactică Ezăreni-Iaşi. *Stud. Cerc. ştiint. Acad. R.P.R. Fil. Iaşi, Ser. 2* **6**, 181–195.

Pocinoc, V. (1969). Contribuţii la studiul efectului lucrărilor agrotehnice asupra gradului de imburuienare al solului. *Comun. Bot., Bucurest* **8**, 67–78.

Prince, F. S. and Hodgdon, A. R. (1946). Viable seeds in old pasture soils. *Tech. Bull N.H. agric. Exp. Stn,* No. 89.

Putensen, H. (1882). Untersuchungen über die im Ackerboden enthaltenen Unkrautsämereien. *Hannov. land- u. forstwirtsch. Vereinsbl.* **21**, 514–524.

Pyatin, A. M. (1970). On the content of viable seeds in the soils of meadow and woodland pastures, used without system. *Byull. mosk. Obshch. ispȳt. Prir., Otdel Biol.* **75** (3), 85–95.

Raatikainen, M. and Raatikainen, T. (1972). Weed colonization of cultivated fields in Finland. *Annls agric. fenn.* **11**, 100–110.

Rabotnov, T. A. (1958). On methods of studying the content of viable seeds in meadow soils. *Bot. Zh. SSSR* **43**, 1572–1581.

Rabotnov, T. A. (1969). Plant regeneration from seed in meadows of the USSR. *Herbage Abstr.* **39**, 269–277.

Rabotnov, T. A. (1978). On coenopopulations of plants reproducing by seeds. *In* "Structure and Functioning of Plant Populations" (Freysen, A. H. J. and Woldendorp, J. W., eds), pp. 1–26. North Holland, Amsterdam.

Ramírez G, C. and Riveros G, M. (1975). Contenido de semillas en el suelo y regeneración de la cubierta vegetal en una pradera de la Provincia de Valdivia, Chile. *Phyton, Argentina* **33**, 81–96.

Raynal, J. D. and Bazzaz, F. A. (1973). Establishment of early successional plant populations on forest and prairie soils. *Ecology,* **54**, 1335–1341.

Roberts, E. H. (1972). Dormancy: a factor affecting seed survival in the soil. *In* "Viability of Seeds" (Roberts, E. H., ed.), pp. 321–359. Chapman and Hall, London.

Roberts, H. A. (1962). Studies on the weeds of vegetable crops. II. Effect of six years of cropping on the weed seeds in the soil. *J. Ecol.* **50**, 803–813.

Roberts, H. A. (1964). Emergence and longevity in cultivated soil of seeds of some annual weeds. *Weed Res.* **4**, 296–307.

Roberts, H. A. (1970). Viable weed seeds in cultivated soils. *Rep. natn. Veg. Res. Stn for 1969*, 25–38.

Roberts, H. A. and Dawkins, P. A. (1967). Effect of cultivation on the numbers of viable weed seeds in soil. *Weed Res.* **7**, 290–301.

Roberts, H. A. and Feast, P. M. (1972). Fate of seeds of some annual weeds in different depths of cultivated and undisturbed soil. *Weed Res.* **12**, 316–324.

Roberts, H. A. and Feast, P. M. (1973a). Emergence and longevity of seeds of annual weeds in cultivated and undisturbed soil. *J. appl. Ecol.* **10**, 133–143.

Roberts, H. A. and Feast, P. M. (1973b). Changes in the numbers of viable weed seeds in soil under different regimes. *Weed Res.* **13**, 298–303.

Roberts, H. A. and Hewson, R. T. (1971). Herbicide performance and soil surface conditions. *Weed Res.* **11**, 69–73.

Roberts, H. A. and Lockett, P. M. (1978). Seed dormancy and periodicity of seedling emergence in *Veronica hederifolia* L. *Weed Res.* **18**, 41–48.

Roberts, H. A. and Ricketts, M. E. (1979). Quantitative relationships between the weed flora after cultivation and the seed population in the soil. *Weed Res.* **19**, 269–275.

Roberts, H. A. and Stokes, F. G. (1966). Studies on the weeds of vegetable crops. VI. Seed populations of soil under commercial cropping. *J. appl. Ecol.* **3**, 181–190.

Robinson, E. L. and Kust, C. A. (1962). Distribution of witchweed seeds in the soil. *Weeds* **10**, 335.

Rodríguez Bozán, J. I. and Alvarez Rey, H. (1977). Métodos para el conteo de semillas de malas hierbas en el suelo. *Centro Agricola* **4** (2), 79–89.

Rola, J. (1962). Badania nad dynamika zbiorowisk chwastów segetalnych w płodozmianie. *Roczn. Nauk roln. A* **85**, 515–553.

Sagar, G. R. and Mortimer, A. M. (1976). An approach to the study of the population dynamics of plants with special reference to weeds. *Appl. Biol.* **1**, 1–47.

Sárkány, L. (1975). Észak-zala kukorica kultúráinak gyomosodása és a talajok gyommagtartalom vizsgálata. *Növényvédelem* **11**, 304–308.

Schweizer, E. E. and Zimdahl, R. L. (1979). Changes in the number of weed seeds in irrigated soil under two management systems. *Proc. west. Soc. Weed Sci.* **32**, 74.

Seely, C. I. (1976). Weed seed in soil. *Proc. 26th Conf. Wash. State Weed Assoc.* 25–27.

Simtea, N. (1971). Cercetări asupra diseminării şi a rezervei de seminţe din sol şi rolul lor la înierbarea terenurilor recent defrişate. *Comun. Bot., Bucurest* **12**, 373–388.

Sinyukov, V. P. (1975). A new instrument for determining the potential weediness of soil. *Izv. timiryaz. sel'.-khoz. Akad.* **1975** (1), 227–229.

Smirnov, B. M. and Kurdyukov, Y. F. (1965). The determination of the weediness of fields. *Vest. sel'.-khoz. Nauki, Mosk.* **10** (2), 125–128.

Snell, K. (1912). Über das Vorkommen von keimfähigen Unkräutsamen im Boden. *Landw. Jbr* **43**, 323–347.

Soriano, A., Zeiger, E., Servy, E. and Suero, A. (1968). The effect of cultivation on the vertical distribution of seeds in the soil. *J. appl. Ecol.* **5**, 253–257.

Southey, J. F. (1970). Laboratory methods for work with plant and soil nematodes (5th edn). *Tech Bull. Minist. Agric., Fish., Fd*, No. 2.

Sparke, C. J. (1979). Seed populations in the soil in hill land infested with bracken (*Pteridium aquilinum*). *Proc. EWRS Symposium on the Influence of Different Factors on the Development and Control of Weeds, Mainz*, 265–274.

Standifer, L. C. (1980). A technique for estimating weed seed populations in cultivated soil. *Weed Sci.* **28**, 134–138.

Strickler, G. S. and Edgerton, P. J. (1976). Emergent seedlings from coniferous litter and soil in eastern Oregon. *Ecology* **57**, 801–807.

Stryckers, J. M. T. (1979). Veränderungen in der (Un)krautflora durch Herbizidan-wendung. *Proc. EWRS Symposium on the Influence of Different Factors on the Development and Control of Weeds, Mainz*, 25–38.

Suckling, F. E. T. and Charlton, J. F. L. (1978). A review of the significance of buried legume seeds with particular reference to New Zealand agriculture. *N.Z. Jl exp. Agric.* **6**, 211–215.

Sugawara, K. and Iizumi, S. (1964). Studies on the buried seed populations in the surface soils of *Zoysia* type grassland. *Sci. Reps Res. Inst. Tohoku Univ. (D)* **15**, 87–90.

Symonides, E. (1978). Numbers, distribution and specific composition of diaspores in the soils of the plant association Spergulo-Corynephoretum. *Ekol. Pol.* **26**, 111–122.

Takahashi, Y. and Hayashi, I. (1978). An experimental study on the development of herbaceous communities in Sugadaira, Central Japan. *Jap. J. Ecol.* **28**, 215–230.

Thompson, K. (1978). The occurrence of buried viable seeds in relation to environmental gradients. *J. Biogeogr.* **5**, 425–430.

Thompson, K. and Grime, J. P. (1979). Seasonal variation in the seed banks of herbaceous species in ten contrasting habitats. *J. Ecol.* **67**, 893–921.

Thompson, K., Grime, J. P. and Mason, G. (1977). Seed germination in response to diurnal fluctuations of temperature. *Nature, London.* **267**, 147–149.

Thorsen, J. A. and Crabtree, G. (1977). Washing equipment for separating weed seed from soil. *Weed Sci.* **25**, 41–42.

Thurston, J. M. (1966). Survival of seeds of wild oats (*Avena fatua* L. and *Avena ludoviciana* Dur.) and charlock (*Sinapis arvensis* L.) in soil under leys. *Weed Res.* **6**, 67–80.

Toole, E. H. and Brown, E. (1946). Final results of the Duvel buried seed experiment. *J. agric. Res.* **72**, 201–210.

Tothill, J. C. and Jones, R. M. (1977). Stability in sown and oversown Siratro pastures. *Trop. Grasslands* **11**, 55–65.

Tulikov, A. M. (1976). The parachute method for separating weed seeds from the mineral fraction of the soil. *Izv. timiryaz. sel'.-khoz. Akad.* **1976** (2), 39–46.

Turkington, R., Cahn, M. A., Vardy, A. and Harper, J. L. (1979). The growth distribution and neighbour relationships of *Trifolium repens* in a permanent pasture. III. The establishment and growth of *Trifolium repens* in natural and perturbed sites. *J. Ecol.* **67**, 231–243.

Van Altena, S. C. and Minderhoud, J. W. (1972). Keimfähige Samen von Gräsern und Kräutern in der Narbenschicht der Niederländischen Weiden. *Z. Acker- u. PflBau* **136**, 95–109.

Van der Valk, A. G. and Davis, C. B. (1976). The seed banks of prairie glacial marshes. *Can. J. Bot.* **54**, 1832–1838.

Van der Valk, A. G. and Davis, C. B. (1978). The role of seed banks in the vegetation dynamics of prairie glacial marshes. *Ecology* **59**, 322–335.

Van der Valk, A. G. and Davis, C. B. (1979). A reconstruction of the recent vegetational history of a prairie marsh, Eagle Lake, Iowa, from its seed bank. *Aquatic Bot.* **6**, 29–51.

Van der Vegte, F. W. (1978). Population differentiation and germination ecology in *Stellaria media* (L.) Vill. *Oecologia (Berl.)* **37**, 231–245.

Vanesse, R. (1976). Un exemple d'analyse du stock grainier dans un sol forestier. *C.r. Vème Colloque int. sur l'Ecologie et la Biologie des Mauvaises Herbes, Dijon*, 141–147.

Vega, M. R. and Sierra, J. N. (1970). Population of weed seeds in a lowland rice field. *Philipp. Agricst* **54**, 1–7.

Von Hofsten, C. G. (1947). Åkersenapens förekomst som ogräsbestånd och som fröförråd i jorden. *Växtodling* **2**, 175–182.

Warington, K. (1958). Changes in the weed flora on Broadbalk permanent wheat field during the period 1930–55. *J. Ecol.* **46**, 101–113.

Warnes, D. D. and Andersen, R. N. (1978). Dissipation of wild mustard seed under various cultural practices. *Proc. N. centr. Weed Control Conf.* **33**, 110.

Watanabe, Y. and Hirokawa, F. (1975). Ecological studies on the germination and emergence of annual weeds. 3. Changes in emergence and viable seeds in cultivated and uncultivated soil. *Weed Res., Japan* **19**, 14–19.

Watanabe, Y. and Ozaki, K. (1964). Studies on rotation systems. 4. The effect of crop sequence on the weed seed population of soil. *Res. Bull. Hokkaido natn. agric. Exp. Stn* **83**, 53–63.

Watkinson, A. R. (1978). The demography of a sand dune annual, *Vulpia fasciculata*. II. The dynamics of seed populations. *J. Ecol.* **66**, 35–44.

Wee, Y. C. (1974). Viable seeds and spores of weed species in peat soils under pineapple cultivation. *Weed Res.* **14**, 193–196.

Wehsarg, O. (1912). Das Unkraut im Ackerboden—Ergebnisse der Untersuchung von Ackerboden aus verschiedenen Teilen Deutschlands auf Unkrautsamen. *Arb. dt. LandwGes.* **226**, pp. 87.

Wesson, G. and Wareing, P. F. (1967). Light requirements of buried seeds. *Nature, Lond.* **213**, 600–601.

Whipple, S. A. (1978). The relationship of buried, germinating seeds to vegetation in an old-growth Colorado subalpine forest. *Can. J. Bot.* **56**, 1505–1509.

Williams, R. D. and Egley, G. H. (1977). Comparison of standing vegetation with the soil seed population. *Abstr. 1977 Mtg Weed Sci. Soc. Amer.*, p. 79.

Wilson, B. J. (1972). Studies of the fate of *Avena fatua* seeds on stubble as influenced by autumn treatment. *Proc. 11th Br. Weed Control Conf.* **1**, 242–247.

Wilson, B. J. and Cussans, G. W. (1975). A study of the population dynamics of *Avena fatua* L. as influenced by straw burning, seed shedding and cultivations. *Weed Res.* **15**, 249–258.

Young, J. A. and Evans, R. A. (1975). Germinability of seed reserves in a big sagebrush community. *Weed Sci.* **23**, 358–364.

Zabkiewicz, J. A. and Gaskin, R. E. (1978). Effect of fire on gorse seeds. *Proc. 31st N.Z. Weed and Pest Control Conf.*, 47–52.

Zelenchuk, T. K. (1961). The content of viable seeds in meadow peaty soils of the L'vov region. *Byull. mosk. Obshch. Ispyt. Prir., Otdel Biol.* **66** (3), 77–92.

Zelenchuk, T. K. (1968). Specific composition and quantity of viable seeds in the soil and on its surface in the meadow vegetation. *Bot. Zh. SSSR* **53**, 1755–1765.

Želev, A. (1965). Study on weed seed distribution in the soil. *Priroda* **14** (5), 61–62.

Zuza, V. S. (1973). The effect of *s*-triazines and 2,4-D on the potential weed seed bank of the soil. *Khim. sel'. Khoz.* **11**, 449–451.

Adaptive Biology of Vegetatively Regenerating Weeds

R. R. B. LEAKEY

Institute of Terrestrial Ecology, Bush Estate, Penicuik,
Midlothian, EH26 0QB, Scotland

I. INTRODUCTION

A. DEFINITIONS, GENERALIZATIONS AND LIMITATIONS

Vegetative regeneration is a term embracing lateral spread by creeping shoots, roots etc. and the production of new plantlets from dispersed, separated or fragmented parts of plants. It is often inextricably linked in many species with aspects of perennation, including the production of specialized storage organs.

Individuals of any plant species can be labelled weeds, if they grow where they are not wanted. However, some species because of the frequency with which they make unwanted appearances become recognized as weed species. These weed species are the subject of this review.

Although not exclusive to weed species, vegetative regeneration is a characteristic of the "Ideal Weed" (Baker, 1965), and is a common feature of many of the world's worst weeds (Holm *et al.*, 1977). However, not all weeds regenerate vegetatively, although it is a characteristic of many perennial weeds, some of which combine different methods of vegetative regeneration with sexual reproduction.

57

The growth strategies adopted by plants have been reviewed by Grime (1979), who identifies plant species as: (a) competitors; (b) stress tolerators; (c) ruderals; or (d) as any of the various intermediates. Most of the common perennial weeds (e.g. *Agropyron repens, Ranunculus repens, Cirsium arvense, Agrostis stolonifera*, etc.), which are well known for their ability to multiply asexually, fall into the category of perennial competitive ruderals (C-R strategists). In this instance, selection is maximizing the capture of resources in conditions where disturbance, even if fairly regular, does not usually occur during the growing season, or at least before the production of vegetative propagules. Lateral spread, however, increases the effectiveness of competition for light, water and mineral nutrients while providing a high plant density, in which plants are often interconnected, minimizing their risks of elimination. In contrast, regeneration by separation and fragmentation favours wider dispersal and the colonization of new locations.

The success of a weed species can be judged by the abundance of individuals in existence during a growing season; an assessment combining fecundity with the occupation of a wide range of habitats over a large area of the world's surface (Baker, 1975). In achieving this, many species exploit sexual reproduction and vegetative regeneration, their life cycles becoming integrated with those of the crops with which they are competing and becoming, at the same time, very difficult to control.

What are the specific adaptations to plant growth which allow weed species to exploit vegetative regeneration so successfully? This review, after first identifying the morphology of the range of organs which confer the ability to regenerate vegetatively, attempts to: (a) describe what is known of their physiology, and (b) highlight the characteristics which have ensured the success of some species as weeds. The restriction of many examples to a few common weeds, notably *Agropyron repens*, is unfortunately a sad reflection of the inadequacy with which other, perhaps less important, species have been studied.

B. METHODS OF VEGETATIVE REGENERATION AND THE DEVELOPMENT OF SPECIALIZED ORGANS

In Britain, 28% of dicotyledonous species have some form of vegetative regeneration; rhizomes and stolons being the commonest adaptations (Clegg, 1978).

The ability to regenerate vegetatively arises from modifications to stems, roots and inflorescences, and can take a wide range of forms. Some species exploit one of these different methods of regeneration, while others combine two or more.

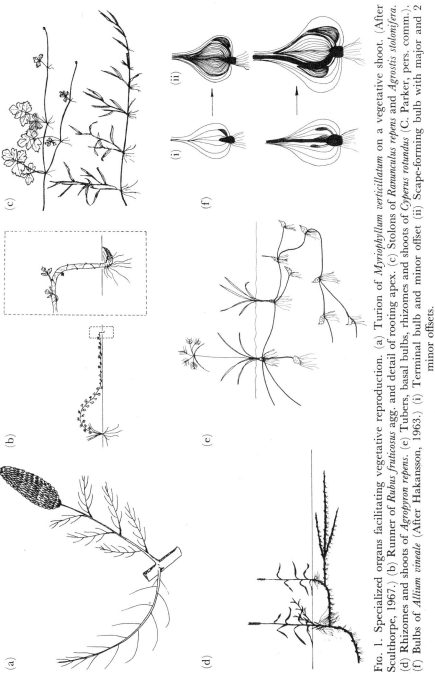

FIG. 1. Specialized organs facilitating vegetative reproduction. (a) Turion of *Myriophyllum verticillatum* on a vegetative shoot. (After Sculthorpe, 1967.) (b) Runner of *Rubus fruticosus* agg. and detail of rooting apex. (c) Stolons of *Ranunculus repens* and *Agrostis stolonifera*. (d) Rhizomes and shoots of *Agropyron repens*. (e) Tubers, basal bulbs, rhizomes and shoots of *Cyperus rotundus* (C. Parker, pers. comm.). (f) Bulbs of *Allium vineale* (After Hakansson, 1963.) (i) Terminal bulb and minor offset (ii) Scape-forming bulb with major and 2 minor offsets.

1. *Rooting of detached shoots*

Many plant species can vegetatively regenerate from shoot fragments if the environmental conditions are favourable and the fragment contains meristematic tissues, usually a bud. The aquatic environment, in particular, is very suitable for this form of vegetative regeneration and many species easily multiply from plant fragments, detached from the plant by either damage, abscission or decay (Sculthorpe, 1967). Indeed, there is a tendency among many hydrophytes towards the replacement of sexual reproduction by vegetative regeneration, thereby evading the hazards of elevating aerial flowers. The dwarf vegetative shoot, or turion, is a common adaptation of aquatic plants which is thought to be a replacement for an inforescence. Turions can be morphologically distinct from other sections of stem, as in *Myriophyllum verticillatum* (Fig. 1a) in which they are shed by abscission; or more or less unspecialized stem fragments released by decay or physical damage.

2. *Creeping stems*

Five basic types of creeping stem can be recognized, although to some extent they represent a continuum which is not easily partitioned; except into those occurring above and below ground.

Layers. These are the simplest form of creeping stems. They have no specific control of growth to keep them horizontal, but when shoots come in contact with soil they develop roots, usually at their nodes. Tillers of *Poa annua*, for example, layer when they are forced, by sheer weight of numbers, to travel horizontally along the soil surface for a short distance before becoming erect (Wells, 1974). Layering also occurs where a shoot becomes unable to sustain erect growth, and becomes decumbent under its own weight.

Runners. The term "runner" is not applied in this review to rosette-type dicotyledonous stolons. Instead, runners are considered to be plagiotropic shoots which are not in contact with soil along their entire length, and in which rooting is usually restricted to particular zones. They, therefore, have a more complex control of rooting than layers as, for example, in *Rubus fruticosus* agg. and *Calystegia sepium*. Both of these species root apically when their shoots droop down to ground level at the end of the growing season. Rooted runners develop as new erect plants the following spring. In *Rubus* spp., prolific rooting occurs immediately behind the cane apex, (Heslop-Harrison, 1959; Barnola, 1971; Amor, 1973) following a period of positive geotropism during which cane apices are inserted into the soil (Fig. 1b).

What triggers this rooting and reversal of shoot polarity? Darkening is known to be important (Heslop-Harrison, 1959), but whether short days, the positive geotropic orientation of the tip and the plants overall physiological condition are also important is questionable.

Stolons. Contrary to other definitions, which often include longevity (Clapham *et al.*, 1962), stolons are considered, in this review, to be horizontally-growing stems, produced from basal nodes of parent plants, which run along the soil surface (Fig. 1c). The main factor differentiating stolons from layers and runners is their controlled plagiotropism.

Plantlets usually develop at stolon nodes, occurring at almost every node in many species, e.g. *Ranunculus repens* and *Agrostis stolonifera*, although branches occasionally form instead. Stolon initiation is generally considered to be induced by long-days. Those of *Ranunculus repens* tend to appear from May to August (Harper, 1957) with sometimes only 2–3 plantlets produced per parent plant in grassland communities (Sarukhán, 1974). By contrast, in more open conditions, where competition for mineral nutrients is likely to be less intense, many more plantlets can be produced.

Rhizomes. Rhizomes are horizontally-growing stems, often rich in stored carbohydrates which, in contrast to stolons, are subterranean.

The term rhizome should not be applied to the vertically-growing underground stems connecting shoots to various subterranean organs, a more appropriate term for which is caudex.

Rhizomes are characteristically of two types (Hartmann and Kester, 1968):

(a) pachymorphic, which are short, thick, often highly branched and determinate;

(b) leptomorphic, which are long, slender, often unbranched and indeterminate (Fig. 1d).

Most of the important rhizomatous weeds (e.g. *Agropyron repens*, *Sorghum halepense*, *Cynodon dactylon*) are leptomorphic, producing rhizomes with many nodes; they have considerable regenerative capacities. Soil penetration is aided in *Poa pratensis* by circumnutation—the spiral movement of actively growing apices (Fisher, 1964)—and in *Agropyron repens* by sharp apical leaf scales. Pachymorphic rhizomes, on the other hand, have low powers of regeneration being mainly perennating/ storage organs, usually producing as in *Ranunculus acris* only one vegetative plant per year (Sarukhán and Harper, 1973).

Rhizomes of *A. repens* are initiated from basal buds on the primary shoot in long days and high light intensities, following a well-defined pattern, in which the four most basal of the seven axillary buds usually form rhizomes, while the remaining three develop as aerial tillers (Palmer, 1958).

Stem tubers. Stem tubers are swollen portions of underground stems which are especially adapted as perennating organs. Tubers of *Cyperus rotundus*, for example, form in meristematic regions near the tips of rhizomes (Fig. 1e). With accumulation of starch in parenchyma, inside and outside the endodermis, the tubers swell, their lateral meristems either becoming inactive or developing as new rhizome branches from basal buds (Holm *et al.*, 1977). In this way, tubers are produced in chains with 5–7 nodes between each pair (C. Parker, pers. comm.). The first "tubers" produced are often enclosed in fully developed leaves and are sometimes called "basal bulbs", although strictly they are neither bulbs, corms or tubers. Production is rapid with new "basal bulbs" and tubers sometimes being formed within three weeks of planting (Musik and Cruzado, 1953).

Tubers of *Equisetum arvense* differ from those of *C. rotundus* in that they are swollen, starch-filled internodes, and increase in size with greater depth (Williams, 1979).

3. *Modified shoot bases*

Bulbs. Bulbs are storage organs consisting of a collection of swollen leaf bases or scales, at or slightly below ground level, enclosing the elements of a shoot system with very short internodes. (Fig. 1f). While they are basically therefore perennating organs, some of their axillary buds develop as daughter bulbs called minor and major offsets. These are capable of regeneration but with different patterns of dormancy. The scales can be of two types: (a) membranous and (b) true, the former having a protective function and forming leaves; while the latter are simply storage organs. In *Allium vineale*, regeneration is almost exclusively by vegetative means although offsets are only rarely produced before plants are two years old (Richens, 1947; Håkansson, 1963), when the minimum bulb size to produce offsets is directly related to depth (Lazenby, 1960). Major offsets are uncommon unless an inflorescence (scape) has been produced.

Oxalis latifolia, similarly relies almost entirely on vegetative regeneration, again from offsets which, in this instance, develop at the tip of rhizomes originating from the axils of bulb scales. The bulb itself has a tap root with storage and contractile functions, the latter ensuring that

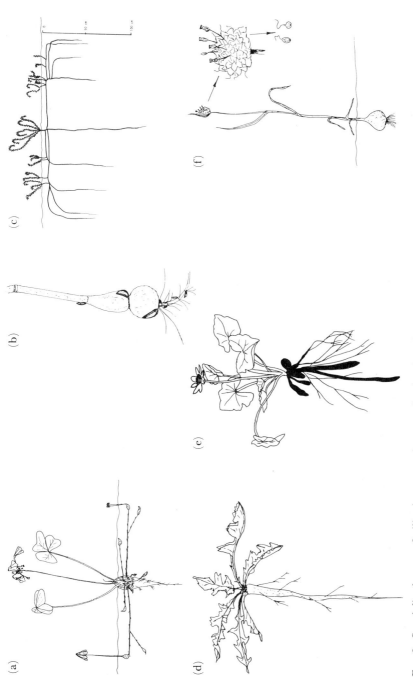

Fig. 2. Specialized organs facilitating vegetative reproduction (continued). (a) Bulb and offsets of *Oxalis latifolia*. (After Jackson (1960) and Chawdhry (1974).) (b) Corms of *Arrhenatherum elatius*. (c) Creeping root system of *Convolvulus arvensis*, after 13 weeks. (After Frazier, 1943.) (d) Taproot of *Taraxacum officinale*. (e) Root tubers and shoot of *Ranunculus ficaria*. (After Marsden-Jones, 1935.) (f) Scape forming plant of *Allium vineale* with detail of spathe and bulbils. (After Håkansson, 1963.)

bulbs are buried at the completion of their annual growth cycles (Chawdhry, 1974). Numbers of rhizomes formed can equal the numbers of true scales produced the previous season; 10–14 are commonly found (Jackson, 1960), with up to 28 on some occasions (Chawdhry, 1974). They often branch to produce secondary rhizomes which form "secondary" offsets. In this way, up to 60 offsets can be produced per plant (Fig. 2a). Rhizome length is variable, an almost sessile form occurring in Cornwall (Young, 1958). Daylength and night temperature have no obvious effects on offset production (Jackson, 1960).

Corms. Corms are swollen stems with dormant buds in the axils of scale-like leaf remains and, like bulbs, occur at the base of the shoot. The corms of *Ranunculus bulbosus* serve as perennating organs usually regenerating by self-replacement in the same position, and only occasionally are one or two daughter corms formed.

The swollen internodes of *Arrhenatherum elatius* have greater regenerative capacity, forming a chain, some of which may occur above the soil surface (Fig. 2b). These corms are easily separated by physical disturbance.

4. Root suckers

Three of the seven types of regeneration from roots (Hudson, 1955) are relevant to weed species, irrespective of the type of root system:

(a) Suckers growing from uninjured roots at a distance from the parent plant.

(b) Suckers developing on undisturbed roots after damage to, or removal of, shoots.

(c) Suckers growing from disturbed root fragments.

Suckers from creeping roots. Creeping root systems which can utilize all three methods of suckering are found in *Cirsium arvense* (Sagar and Rawson, 1964; Chancellor, 1970; Moore, 1975) in which daughter plants separate from their parents when connecting roots die and decompose, and *Convolvulus arvensis* (Frazier, 1943; Davidson, 1970) in which connecting roots are persistent (Fig. 2c).

Creeping root systems consist of a series of vertical and horizontal roots, formed by the positive geotropism of root apices after a period of diageotropic growth and the subsequent production of new roots where vertical roots arise, to continue the horizontal growth (Fig. 2c). Sucker shoots are also often initiated near the insertion of vertical roots (Frazier, 1943).

Primordia initiated in the dark, form in the pericyclic tissue near the protoxylem of roots of *Convolvulus arvensis* (Torrey, 1958) and subsequently develop into either lateral roots or sucker shoots (Bonnet and Torrey, 1966), red light being required to induce the latter (Bonnet, 1972). In *Cirsium arvense*, preformed buds appear to be initiated when nutrients are in short supply (McIntyre and Hunter, 1975).

Suckers from tap roots. Tap roots, large positively geotropic primary roots developed by secondary thickening as carbohydrate rich storage organs (Fig. 2d) only regenerate when defoliated, damaged or fragmented. Little is known about the processes of tap root development or their bud primordia initiation.

Suckers from root tubers. Root tubers which are swollen roots or root apices, with storage and perennating functions, are commonplace among aquatic plants and those inhabiting moist habitats, for example *Oenanthe crocata*, and *Ranunculus ficaria* (Fig. 2e).

5. *Agamospermy*

Agamospermous and apomictic species, 1·7% of British dicotyledonous plants (Clegg, 1978), flower and form seeds and embryos by non-sexual processes. *Taraxacum officinale* and some forms of *Allium vineale* are completely asexual (obligate), the former flowering and fruiting abundantly, with new variants being formed by somatic mutation (Valentine and Richards, 1967), while the latter produce bulbils instead of seeds (Fig. 2f). Other forms of *A. vineale* are facultative, like *Poa pratensis* and *Rubus fruticosus* agg., combining asexual propagation with sexual reproduction. In this instance, seeds and bulbils of *A. vineale* are both found in the same umbel (Richens, 1947: Håkansson, 1963; Davis and Heywood, 1963), each bulbil consisting of a cone-shaped, fleshy scale, the size of a wheat grain, with an associated meristem. Interestingly, agamospermy is associated with a wide range of other methods of vegetative regeneration involving a number of different dispersal mechanisms. For example, propagules of *T. officinale*, *R. fruticosus* and *A. vineale* are dispersed by wind, birds and man (in wheat samples) and regenerated from root fragments, runners and offset bulbs respectively.

II. ADAPTATIONS FOR VEGETATIVE REGENERATION

A. REGENERATION BY LATERAL SPREAD

Two strategies for the invasion of new territory by lateral spread have been identified by Clegg (1978). These vary in the way in which individuals make contact with other plants. In the "guerilla" strategy,

daughter plantlets develop on a few long creeping organs which pene-
trate surrounding vegetation and make many interspecific plant contacts, whereas the "phalanx" strategy establishes a solid, advancing
"front" of many daughter plantlets. In the latter type, contacts between plantlets are often intra-specific, or indeed intra-clonal. Among
British dicotyledons the "guerilla" species are often those with stolons
(*Rancunculus repens*) and runners (*Rubus fruticosus*) while the "phalanx"
species are those with rhizomes (*Tussilago farfara*) or creeping roots
(*Cirsium arvense*) (Clegg, 1978). There therefore appears to be a tendency
for different strategies of lateral spread to be related to aerial and subterranean adaptations in growth habit. Does this apply to monocotyledons? It may, but there are probably some notable exceptions;
Agropyron repens, for example, is probably a "guerilla" species, at least
during the early stages of colonization. There is, however, need for
more detailed studies of the strategies and adaptations found in weed
species, particularly the grasses, colonizing different ecological niches.
The same applies to the methods of exerting competition, because so
far this approach has only been applied to *Trifolium repens* (Turkington,
1979a, b, c; Turkington *et al.*, 1979), in which clones have been found
to differ in their aggressiveness, thereby suiting them either as colonizers
of new sites or as regular occupants of already colonized communities. The stoloniferous growth habit of clones of the latter type
allows them to: (a) escape from competition; (b) grow among species
with which they can co-habit; and (c) adapt to the aggressiveness of
the dominant species in their community.

1. *Production of creeping stems and roots*

As already mentioned, the simplest creeping stem is the decumbent
tiller, which produces roots by "layering" where nodes make contact
with soil. From these simple beginnings different species, often tax-
onomically unrelated, have developed different specializations and
adaptations.

In exploring their habitats stolons produce plantlets at fairly regular
intervals throughout the summer, although individual plantlets may
survive only a few months (Sarukhán, 1971). In *Ranunculus repens* it
seems that the production of stolons takes precedence over the perpetuation of individual plantlets, dry matter being invested so as to
maximize its spread (Ginzo and Lovell, 1973a). Thus, the mean length
of primary stolons has been found to be fairly constant at the expense
of: (a) numbers of primary stolons produced; (b) their dry weight per
unit length; and (c) the frequency of stolon branching even under
conditions of widely varying nutrient status (Ginzo and Lovell, 1973a).

Production of primary stolons is very sensitive to nutrient supply, for example, increasing the supply of nitrogen from 10 to 50 ppm increased numbers of stolons produced per plant from two to nine. How does the production of stolons compare with that of other organs of lateral spread? Single plants of *Agropyron repens* and *Ranunculus repens* can produce individual rhizomes and stolons achieving lengths of *c*. 2 m., but while total rhizome lengths per plant of 82·5 m per season have been recorded for *A. repens* (Tildesley, 1931), only 28 m with 480 nodes has been recorded for *R. repens* (Anon, 1958). Among aquatic stoloniferous plants, *Eichhornia crassipes* can double its plant numbers in open water in only two weeks. Twenty-five plants will therefore fill a hectare with 1 617 300 daughter plantlets in a single season (Penfound and Earle, 1948). Creeping roots while not as productive as this are more prolific than the terrestrial stolon producers, for the creeping roots of *Convolvulus arvensis* can achieve lengths of 1·5 m in 13 weeks (Frazier 1943) and *Cirsium arvense* spreads up to 5 m (Bakker, 1960) or even 12 m (Rogers, 1929) in a single season. Equivalent data for plants spreading by runners has not been found.

The importance of lateral spread to *R. repens* and other stoloniferous plants can also be inferred from the distribution of assimilates which usually move towards stolon apices. Ginzo and Lovell, (1973b) found that labelled assimilates were exported from rosette leaves of plantlets to developing stolons, and from the largest and oldest daughter plantlets to the younger and smaller plantlets on the same stolon, and in particular to the stolon apex. Similar patterns of assimilate transport occur in *Cynodon dactylon* (Forde, 1966a) and some non-weedy stoloniferous species, for example, *Saxifraga sarmentosa* (Qureshi and Spanner, 1971, 1973a, b) and in *Trifolium repens* (Harvey, 1970). However, while this is the normal pattern, it can be altered to supply assimilates to plantlets growing in unfavourable sites (Qureshi and Spanner, 1971; Ginzo and Lovell, 1973b). A similar switch in the flow of mineral nutrients, notably nitrogen, has been reported in *Cynodon dactylon* (Giddens *et al.*, 1962). It seems, therefore, that the production and maintenance of vegetative propagules, at least by stoloniferous species, is such that the parent plant produces only as many plantlets as it can support. This limitation occurring despite the potential of individual plantlets to be self supporting for assimilates and mineral nutrients. Propagule size is therefore relatively constant and thus is comparable with the known uniformity of seeds produced by plants growing in a wide range of contrasting environments (Harper *et al.*, 1970).

Surprisingly little is known about factors controlling the growth and development of rhizomes or creeping roots. Numbers of rhizomes formed by *A. repens*, unlike those of stolons in *R. repens*, appear to be greater in plants with small amounts of nutrients than in plants with a plentiful supply, the change being attributable to a slight alteration in the proportion of buds developing as rhizomes and tillers (McIntyre, 1965, 1967). Although this was not confirmed by Rogan and Smith (1975), it agrees with the results obtained when Leakey *et al.* (1977a) tested the effects of different amounts of artificial fertilizers. The lateral spread of rhizomes, therefore, unlike that of stolons, seems to be maximized when plants would benefit by spreading to more fertile sites.

Cyperus rotundus, a tuberizing rhizomatous species, unlike *A. repens*, restricts its lateral spread when nitrogen is scarce; in this instance, by the stimulation of tuberization. Thus, in these conditions, perpetuation of the individual takes preference over dissemination of the species (Chadwick and Obeid, 1963), a situation similar to that found in the corm-forming *Ranunculus bulbosus* (Sarukhán, 1971; Ginzo and Lovell, 1973a, b).

In contrast to the effects of limited supplies of nutrients, low light intensities favour the production of shoots. Thus, instead of spreading, *A. repens* consolidates its position by developing more tillers when severely shaded. These tillers, however, do not necessarily flower and set seed (Palmer and Sagar, 1963). Long days of fairly intense light (Palmer, 1958) and locations with limited supplies of nitrogen and fairly low average temperatures (McIntyre, 1967) seem optimal for rhizome production; these conditions favouring carbohydrate accumulations. The effect of these conditions on lateral spread was later confirmed with decapitated rhizomes (McIntyre, 1970) where rhizome initiation seems to be influenced by carbohydrate: nitrogen ratios and perhaps amounts of amino-nitrogen (Nigam and McIntyre, 1977). Similarly, in *Tussilago farfara* the proportion of annual dry matter production allocated to lateral spread of rhizomes varied from 4% to 23%, the proportion being greatest when the supply of nutrients was restricted (Ogden, 1974).

Unfortunately, studies of this nature have not been done on plants spreading by creeping roots or any of the less common forms of lateral spread, like the bulb-forming rhizomes of *Oxalis latifolia*.

In *Rubus* runners, however, the rooting of apices was greatest in shady areas less suitable for vigorous cane growth (Kirby, 1980). This seems to be related to the chances of runners reaching the ground rather than any differences in their ability to root.

2. *Plagiotropy*

The plagiotropic (prostrate) growth habit of stolons, rhizomes and creeping roots is an essential adaptation for their role in lateral spread, however, relatively little is known about the mechanisms, especially in roots. Light seems to be the commonest controlling factor, the effect presumably being mediated by growth regulators and/or phytochrome. Several types of response have been recorded:

(1) *Light-induced plagiotropism.*

(a) *Parent plant stimulation.* Rhizomes of *Agropyron repens*, like stolons of *Cynodon dactylon*, are strongly plagiotropic when the parent plant is illuminated; the degree of control lessening in shade, after defoliation, or with the onset of shoot senescence (Palmer, 1956, 1962). Shading *C. dactylon* stolons, however, had no effect on their orientation.

(b) *Apex stimulation.* Phytochrome pigments have been isolated from the apical zone of *Sorghum halepense* rhizomes, suggesting the photo-control of their orientation, although it is not known whether they respond negatively or positively (Duke and Williams, 1977). Negative phototropism has, however, been observed in intact rhizomes of *Aegopodium podograria*, unlike the tips of detached *A. repens* rhizomes (McIntyre, 1969), the former turning away from light when exposed for just 30 s to red light (Bennett-Clark and Ball, 1951). Rhizomes of *Polygonatum multiflorum* and *Urtica dioica* are thought to respond similarly to light penetrating the upper layers of soil (Clapham 1945). Similarly, the creeping growth of *Convolvulus arvensis* roots, is thought to be mediated by light perceptive phytochrome located in root tips (Tepfer and Bonnett, 1972). Red light induced negative phototropism whereas far-red and darkness encourage horizontal growth. By contrast, to the above examples, irradiation or rhizome apices of *Poa pratensis* stimulated erect growth, (Nyahoza *et al*; 1974).

The role of the parent plant in light-induced plagiotropism is poorly understood. In *Cynodon dactylon*, experiments done by Montaldi (1970) suggest that horizontal growth could be sustained by applying sucrose to the cut ends of severed rhizomes, but this experimental effect has not been successfully repeated (Chancellor, pers. comm.). In *Agropyron repens*, gibberellic acid seems to be the "message" from the parent plant for its application to cut ends of detached rhizomes prevents the upward curvature of the apices (Rogan and Smith, 1976). In contrast, the application of gibberellins to whole plants does not affect the orientation of their rhizomes (Palmer, pers. comm.).

(2) *Gravity-induced plagiotropism.*

Unlike rhizomes of *A. repens* (Palmer, 1962) rhizomes of *Aegopodium podograria*, when displaced from the horizontal, respond by curving back to their original position. It is not clear, however, if this is a true gravity response or whether the rhizomes are perhaps responding to concentrations of CO_2 in the upper soil (Bennett-Clark and Ball, 1951).

3. Apical dominance

Horizontally-growing shoots, whether naturally occurring stolons or trained branches on fruit trees (Wareing and Nasr, 1958, 1961), tend to lack apical dominance. In the former, erect aerial shoots are progressively larger towards the base of the stolon, the pattern only being broken by the occasional stolon which is also plagiotropic. In *Ranunculus repens*, frequency of branching directly reflects the availability of nutrients (Ginzo and Lovell, 1973a).

Rhizomes, unlike stolons, frequently have strong apical dominance despite their being horizontal; but there are a few exceptions, e.g. *Panicum repens* (Moreira, 1975). In *Agropyron repens*, apical dominance, which is strong, prevents 95% of axillary buds from being active (Johnson and Buchholtz, 1962). As a result, *A. repens* typically has a large reserve of dormant buds ranging from 620 m^{-2} in natural infestations (Fail, 1956) to 18800 m^{-2} in planted experimental plots (Leakey, 1974); a single plant producing 825 inactive buds annually (Haddad, 1968). The reserves of inactive buds differ from "seed banks" because they represent a restricted array of genotypes (Harper, 1977). If undisturbed, *A. repens* buds remain viable for 2–3 years (Sagar, 1960), while in *Pteridium aquilinum* they may survive 30–70 years. The buds sprout when rhizome systems are fragmented, the numbers becoming active depending on the size of the fragments. The few buds (5%) which are active on undamaged rhizomes of *A. repens* are usually those near apices, which form branch rhizomes or rhizome tillers either at anthesis in mid-summer or when the rhizome apices become erect in autumn (Sagar, 1961; Håkansson, 1967).

The strong apical dominance or rhizomes, in addition to forming a "bud bank", presumably aids lateral spread by conserving assimilates for the growth of primary apices. How is this strong apical dominance achieved? In *Agropyron repens*, some axillary buds are activated by both decapitation and separation from the parent plant (Leakey *et al.*, 1975; Rogan and Smith, 1976), the combination releasing all buds from dominance (Leakey *et al.*, 1978a). Both these influences were replaceable by a mixture of auxin and cytokinin for when applied in agar to the cut ends of rhizomes all buds remained inactive (Leakey *et al.*, 1975), some gibberellins also seem to contribute to the inhibitory effect of the parent plant factor (Rogan and Smith, 1976). Thus it seems that the rhizome apex itself and the parent plant are both actively involved in apical dominance. Is this a special adaptation of rhizomes? It seems likely that the parent plant effect is particularly strong in these subterranean stems which are strongly dependent on their parent shoots, but "root factors" also influence apical dominance in erect aerial shoots (Woolley and Wareing, 1972). Parent plant effects on apical dominance have also been noted in *Holcus lanatus* and *Trifolium repens* stolons (Watt and Haggar, 1980; Thomas, 1970).

The extensive study of apical dominance has yielded several theories (Phillips, 1969, 1975). When considering the phenomenon in rhizomes, McIntyre (1965, 1969, 1971, 1976, 1977) strongly emphasized the effects of nitrogen, carbohydrate and water, believing that competition for these is the basis of apical dominance. While accepting the role of nutrients, especially nitrogen, others have emphasized auxin-mediated inhibitors (Leaky *et al.*, 1975). However, the different approaches are probably not mutually exclusive for when large amounts of nitrogen are applied to *A. repens* the pattern of bud outgrowth (McIntyre, 1965) suggests that the establishment of dominance is prevented rather than being broken, this concurring with the pattern of sylleptic branching; while that following decapitation (Leakey, 1974) is characteristic of proleptic branching. The differentiation of branching processes into these two types is usually associated with trees (Tomlinson and Gill, 1973), but they also seem to occur in herbaceous plants and grasses. Support for the idea that these two processes are different in terms of their dominance mechanism comes from the effects of applying nitrogen at different stages during the process of dominance re-development in multi-node rhizome fragments, when nitrogen prevented slowly-growing buds from being inhibited but did not release any inactive buds from inhibition (Leakey *et al.*, 1978a).

In *Cyperus rotundus* the distal tuber of a chain supresses the shoot growth of the remaining tubers. This form of dominance, like that of the apical dominance in the shoots of herbaceous plants, is stronger when the chains of tubers are vertical than when horizontal (Musik and Cruzado, 1953). The production of a tuber chain is itself a peculiarity of the apical dominance of individual tubers, for the connecting "continuation rhizome" sprouts from the basal bud of the tuber (Fig. 1e). The other buds are only released from dominance by the fragmentation of the tuber chain.

4. *Assimilate and nutrient storage*

The storage of assimilates and mineral nutrients in regenerative organs, particularly those that are subterranean, has two purposes. Firstly, as

will be seen later, they allow regeneration following separation or fragmentation: and secondly they act as reserves, which by enabling the
early production of leaves and a rapid increase in height, confer considerable competitiveness advantages (Bradbury and Hofstra, 1976).
Thus, in the spring *Agropyron repens* mobilizes stored reserves, particularly nitrogen, from the rhizome network (Leaky *et al.*, 1977a); the consequent losses being gradually replenished later in the year.

Studies on assimilate transport to the rhizomes of *Poa pratensis* and
A. repens up to the 6–8 leaf stage, indicate that they are supplied from
a pool to which the main shoot and all its tillers contribute (Nyahoza
et al., 1973; Forde, 1966b; Fiveland *et al.*, 1972); after which time
reciprocal transport between parent plant and rhizome tillers of the
latter may occur (Rogan and Smith, 1974). This flexibility in terms of
assimilate transport, is likely to be of considerable importance for invasive and competitive growth, particularly perhaps in grazed or mown
plant communities.

In *Pteridium aquilinum*, which has a persistent rhizome system that
lacks an overwintering aerial shoot, carbohydrate reserves are stored in
special non-frond bearing rhizomes. These reserves supplying young
developing fronds until they start to photosynthesize for themselves
(Williams and Foley, 1976; Fletcher and Kirkwood, 1979), and are
subsequently replenished with firstly, photosynthates and, secondly, at
a later stage, by the re-mobilization of assimilates from senescing fronds.
In *Equisetum* species, tubers fulfil the storage function in a similar way,
with those at greater depths being bigger, presumably to allow the successful emergence of new shoots (Williams, 1979).

In *Cyperus rotundus*, assimilates are stored in tubers whose formation
is initiated at rhizome apices. The mechanism of initiation in this
species is not known, but in the potato (an important weed in some
circumstances (Lutman, 1977)), it is enhanced by short days and low
temperatures. Whereas cytokinins appear to be involved in the production of tubers in *Solanum tuberosum*, abscisic acid seems important in
Solanum andigena, substituting for leaf-produced inhibitors induced by
short-days (reviewed by Cutter, 1978). Gibberellins appear to delay
tuber formation in the potato. In *C. rotundus*, tuber production occurs
at almost any time during the summer and is commonest in autumn
and winter (Betria and Montaldi, 1975), which suggests that *C. rotundus*
may be responsive to factors similar to those influencing the potato.
"Basal bulbs" of *C. rotundus* appear to be initiated earlier in the year
than tubers and can be initiated in response to red light, this being
mediated by phytochrome (Betria and Montaldi, 1974).

Relatively little is known about assimilate storage in bulbs and corms of weed species, except that in *Ranunculus bulbosus* corms are the primary sink for assimilates, especially when in adverse conditions. Meanwhile in *Arrhenatherum elatius*, corm formation only occurs in part of the population, only 50% of those from an arable sample being sufficiently large to represent a weed threat (Ayres, 1977).

The role of stolons as storage organs has not been investigated, presumably because they are likely to be dependent on current photosynthesis to a greater extent.

5. *Subterranean life*

Many important weeds, including *Cyperus rotundus* "the world's worst weed" (Holm *et al.*, 1977), that successfully grow and regenerate vegetatively by lateral spread and the separation of plant parts, have developed specialized underground organs, for lateral spread. For this, morphological and physiological adaptations have evolved, affecting a range of plant structures. These adaptations seem to confer the following advantages:

(a) Protection from predation by herbivores.

(b) Enhanced exploitation of soils by competing, even at considerable distance from the aerial shoot, with neighbouring plants for water and nutrients.

(c) Protection from extreme temperature fluctuations, by the insulating effect of soils.

(d) Establishment of a substantial "bud-bank", attributable to the inhibition of axillary buds and undifferentiated bud primordia, in reserve against fragmentation and/or disturbance (see next section).

B. REGENERATION BY SEPARATION AND FRAGMENTATION

Plant parts can either be separated naturally by the decay of connecting tissues, by active shedding by abscission, or they can be fragmented by physical damage, caused by animals and man. Fragmentation often occurs at innately brittle points of attachment at the base of stolons, rhizomes, tubers etc., or at the crown of rootstocks. In both naturally shed organs and fragments, the processes of decay have to be controlled to prevent rot spreading too fast through tissues. In many species stem nodes seem to provide a barrier to uncontrolled decay.

1. *Natural separation*

Separation by decay and natural senescence. This is commonplace in many weed species, thus stolons of *Ranunculus repens*, rhizomes of *Tussilago farfara*, root suckers of *Cirsium arvense*, tubers of *Solanum tuberosum* and bulbs of *Oxalis latifolia* are released to establish themselves as independent units. In underground organs this necessitates a supply of stored reserves and sometimes a change in the balance of endogenous growth regulators to break bud dormancy. Furthermore, as in the potato tuber, efficient utilization of reserves is achieved by the development of dominance amongst sprouting buds, restricting growth to just a few, usually the most apical shoots (Goodwin, 1967).

In some stolon systems the rates of plantlet production and senescence are fairly rapid, thus the stolon complex advances forwards and recedes from behind. In this way, individual plantlets of *Ranunculus repens* sometimes only survive for a period of a few months (Sarukhán, 1971; Ginzo and Lovell, 1973a). Similarly, rhizomatous species form advancing and receding networks of rhizomes, as described for *Pteridium aquilinum* (Watt, 1947), individual rhizomes of this species remaining viable for long periods. Rhizomes of *Tussilago farfara* and *Agropyron repens* in contrast only survive for 1–3 years (Sagar, 1960; Myerscough and Whitehead, 1967). Although individual plantlets have these relatively short lives, well adapted clones probably survive, through successive vegetative generations, for very long periods. Indeed some *P. aquilinum* clones are thought to have existed since the Iron Age (Harper, 1977).

Abscission. This separation of "plantlets" from the parent plant by active cellular changes is controlled hormonally and seems to be a relatively uncommon adaptation among weed species. It is, however, found in:

(a) *Myriophyllum verticillatum* and other aquatic plants, the turions of which are shed, and dispersed in flowing water. Turions are sometimes formed in response to environmental stress (Sculthorpe, 1967), but more usually are specifically initiated environmentally to fulfil the summer or winter dormant phases of a life-cycle. For example, in *M. verticilliatum* they are initiated in autumn by short days and low temperatures (Weber and Noodén, 1976), while in *Potamogeton crispus* they are formed in summer (Sastroutomo *et al.*, 1979) by long days and warm temperatures (Sastroutomo, 1980).

(b) In apomictic "seed" producing species like *Rubus fruticosus* and *Taraxacum officinale.*

(c) Bulbil-forming apomicts like *Allium vineale*. Unlike the results of natural separation by senescence, where plantlets grow near to their parents, the plantlets released by abscission are usually dispersed appreciable distances.

2. *Fragmentation*

Regeneration following the fragmentation of a plant by soil disturbance, grazing animals etc., can only occur, if fragments have the means for surviving and making new growth. This poses few problems for aquatic and above-ground creeping stems (stolons, runners and decumbent tillers), which often produce roots at their nodes before fragmentation, but when underground organs are fragmented there is an essential need to re-establish aerial shoots from inactive buds or undifferentiated primordia, to maximize survival. This requires special adaptations.

Resistance to rotting. Although there are few detailed observations, it is clear that rhizome fragments of *Agropyron repens* strongly resist decay, remaining viable in warm, dark, moist environments for a year or more (Leakey *et al.*, 1978a). Fragments of other species, for example *Tussilago farfara*, seem less resistant. Little, if anything, is known about the mechanisms of this resistance.

Use of stored reserves. As already noted, most regenerative organs have appreciable accumulations of carbohydrates and other substances which can be mobilized when forming new shoots following fragmentation. The ability to regenerate seems to be related to seasonal factors. In *Agropyron repens*, for example, amounts and availability of endogenous reserves of carbohydrate and nitrogen have been investigated (Johnson and Buchholtz, 1962) and regenerative capacity *in vitro* was closely related to concentrations of nitrogen (Leakey *et al.*, 1977a, b). Other species showing similar variations in regeneration and rhizome reserves are *Sorghum halepense* (Boyd, 1967; Hull, 1970; Horowitz, 1972b), *Cynodon dactylon* (Horowitz, 1972a), *Solidago canadensis* (Bradbury and Hofstra, 1977) and *Tussilago farfara* (Bakker, 1960; Otzen and Koridon, 1970). *Achillea millefolium* does not, however, appear to have such marked seasonal variations in its regenerative capacity (Bourdôt *et al.*, 1979). It seems that the depletion of nitrogen reserves in *Agropyron repens* and the consequent "restricted regenerative capacity" in late spring can be attributed to the mobilization of nitrogen from the rhizome network to actively-growing apices (Leakey *et al.*, 1977a) and a subsequent inability to utilize the relatively plentiful carbohydrate reserves (Leakey, 1974). *In vivo*, small quantities of soil nitrogen are probably sufficient to offset these deficiencies of endogenous nitrogen reserves (Leakey *et al.*, 1977b).

The ability of rhizome fragments to regenerate is also influenced by their original position on the rhizome, apical fragments of *A. repens* rhizomes having greater regenerative capacity than basal fragments, this being directly related to their larger concentrations of nitrogen (McIntyre, 1972; Nigam and McIntyre, 1977; Leakey et al., 1977b). The same was true for *Cynodon dactylon* in spring, but in autumn regenerative capacity of basal buds was maximal (Moreira and Rosa, 1976).

In tuber-forming rhizomatous species like *Cyperus rotundus*, stored reserves are concentrated into the tuber, the rhizomes being purely connecting organs following the sloughing of cortical tissues (Holm et al., 1977).

Regeneration from root fragments is often poor (Henson, 1969) and in *Convolvulus arvensis* although most (83%) fragments produced shoots only 10% developed new roots (Swan and Chancellor, 1976). It is possible however that the ability to initiate new roots may be strongly season-dependent.

As in creeping stems, amounts of shoot growth and stored reserves in creeping roots vary seasonally, e.g. *Cirsium avense* (Bakker, 1960), *Sonchus arvensis* (Otzen and Koridon, 1970; Håkansson and Wallgren, 1972), *Convolvulus arvensis* (Swan and Chancellor, 1976) and *Euphorbia esula* (Raju et al., 1964). In *S. arvensis*, considerable mobilization of assimilates to the root system has been observed after flowering (Fykse, 1974), this presumably replenishing the amounts of stored carbohydrates. In tap root species like *Taraxacum officinale*, root fragments will in ideal laboratory conditions, regenerate at any time of the year (Hudson, 1955), although regeneration in the field is poor in the spring (Mann and Cavers, 1979). In this instance, reserves may have been utilized in the spring flush of growth and the early burst of flowering. Mann and Cavers (1979), found that positional effects in this species were related to fragment size, whereas Khan (1972) attributed them to differences in protein and nucleic acid synthesis during regeneration. In contrast to that in *T. officinale*, regeneration from tap roots of *Rumex crispus* and *Rumex obtusifolius* appear to be restricted to the root crown (Healey, 1953; Cavers and Maun, 1970; Monaco and Cumbo, 1972) where the latter has a brittle zone facilitating fragmentation.

Other factors which affect the amounts of reserves available for regeneration from fragments are:

(a) Fragment size, which affects the depth from which shoots can emerge.

(b) Frequency of parent plant defoliation prior to fragmentation, which depletes amounts of stored reserves (Turner, 1966, 1968a, 1969; Håkansson, 1967, 1968, 1969a, b).

(c) Development of dominance between shoots on the large fragments of rhizome and root formed by ploughing and other cultivations.

The latter, in multi-node *A. repens* rhizome fragments, is achieved in about three weeks, following a regular sequence of shoot growth, in which nearly all buds grow initially, but subsequently one of the apical group (usually that at nodes 2 or 3) suppresses all the others (Chancellor, 1974). Similar events seem to occur in rhizomes of *Sorghum halepense* (McWhorter, 1972), between sprouts of potato tubers (Goodwin, 1967) and the shoots from root fragments. In this latter instance roots grow from the apical (distal) ends and shoots from the other (proximal) end (Raju *et al.*, 1964; Hamdoun, 1972; Richardson, 1975).

This assertion of "dominance" among buds from tubers, root and multi-node rhizome fragments has two major adaptive features:

(a) By restricting growth to one shoot, emergence from greater depths becomes possible.

(b) By preventing the proliferation of shoots, reserves are conserved against further disturbance and refragmentation (Schwanitz, 1936; Chancellor, 1975).

In rhizome fragments the polarity of shoot growth at the apical end is thought to be due to an interaction between the competition for carbohydrate and nitrogen, and the inhibitors produced by rapidly-growing shoots (Leakey *et al.*, 1975; 1978a). The reverse polarity found in root fragments presumably reflects fundamental differences between root and shoot tissues. However, somewhat similar hormone balances appear to regulate root and shoot growth for when auxins were applied to the proximal end (especially if cytokinin was simultaneously applied to the distal end) it induced the atypical development of roots at the end of the root fragment, while, as expected, the distal application of auxin did not (Bonnett and Torrey, 1965).

Interestingly, if rhizome fragments of *A. repens* are shallowly buried, several shoots can emerge before the establishment of dominance. These aerial shoots are then no longer subject to mutual inhibition (Leakey, 1974), although those remaining below ground will be inhibited. Exposure of multi-node rhizome fragments to wavelengths of light between 690 and 720 nm prevented dominance development without reactivating inhibited buds (Leakey *et al.*, 1978b). Like these rhizome fragments the patterns of regeneration from isolated *Cyperus rotundus* tubers is affected by light (Musik and Cruzado, 1953; Aleixo and Valio, 1976). This advantageous adaptation obviously allows all emerged shoots to become independent plants.

In tuber-forming species like *Cyperus rotundus* rhizomes appear to have lost their ability to regenerate, being without stored reserves and axillary buds (Holm *et al.*, 1977). In some other species (e.g. *Equisetum arvense* and *Solanum tuberosum*) rhizomes remain unchanged and probably retain some regenerative capacity, although it is poor in *E. arvense* rhizome fragments.

C. REGENERATION FROM DAMAGED ROOT AND RHIZOME SYSTEMS

While cultivations often fragment roots and rhizomes in upper soil horizons, those at greater depths remain unharmed. Roots of *Convolvulus arvensis* have been observed at depths of *c.* 1·2 m within 13 weeks of establishment (Frazier, 1943), and like *Euphorbia esula*, spp. can emerge from great depth. Clearly regeneration of this sort demands considerable quantities of stored reserves, but as yet relatively little is known about this method of regeneration, which in some species is more important than the regeneration from detached fragments.

D. PERENNATION AND DORMANCY

As already mentioned many perennial weeds regenerate vegetatively, the organ of regeneration also being that for perennation. Commonly these organs are subject to bud dormancy, which delays regeneration either until the following growing season or until several years have elapsed.

The former is of considerable ecological importance as it ensures that attempts at establishment will be made in a variety of conditions. It is important to distinguish true bud dormancy, which occurs even when buds are isolated, from those that are inactive because of the influence of other buds, as in apical dominance or the effects of adverse autumn and winter environments. True dormancy is controlled by growth regulatory substances, the balance of which must be altered, often by chilling, before growth is resumed, as in the turions of *Myriophyllum verticillatum* and *Potamogeton nodosus* (Sculthorpe, 1967). Similarly, the dormancy of bulbs of *Oxalis latifolia* is broken by three weeks at 5°C, although in this instance a period of high temperature is also necessary before sprouting occurs at 23°C (Chawdhry and Sagar, 1974). Primary, unlike secondary offsets of *O. latifolia*, may sprout immediately after formation and while still attached to their parent plants. After separation, however, they become dormant.

Bulbs and major offsets of *Allium vineale* are usually dormant during summer and early autumn. The mechanism is not known but may be controlled by a balanced inhibitor/auxin/gibberellin complex, as in the cultivated onion (Thomas, 1969). In contrast, minor offsets of *A. vineale* may be dormant for six years, whereas its bulbils lose their dormancy like seeds during their first winter. This diversity of adaptations made *A. vineale* well suited to a variety of conditions (Lazenby, 1960).

Stem tubers of *Cyperus rotundus* are of two types: true tubers, which have the capacity for dormancy, and "basal bulbs", which do not. Characteristically, basal bulbs, the organ of immediate regeneration, are formed prior to tubers, the organs of perennation and dormancy. In *C. rotundus* tubers, bud dormancy appears to be imposed by abscisic acid (Teo *et al.*, 1974), a mechanism similar to that in potato tubers, where dormancy is regulated by the balance between inhibitors and promoters, possibly abscisic acid and gibberellin (Cutter, 1978), except that in *C. rotundus* cytokinins have found to break dormancy (Teo and Nishimoto, 1973). In the aquatic *Hydrilla verticillata*, tuber dormancy again seems to be of two types but here "Thiourea" breaks dormancy, encouraging sprouting in summer, but not in winter (Basiouny *et al.*, 1978).

In contrast to all the above examples, the growth of creeping stems and roots are rarely controlled by dormancy mechanisms. Indeed, only in rhizome fragments of *Agropyron repens*, which as mentioned earlier have some similarities with tubers of *Solanum tuberosum* and *Cyperus rotundus*, has dormancy been reported. In this instance, immediate exposure of inactive buds to light, prevented the outgrowth of axillary buds following fragmentation (Leakey *et al.*, 1978b), a mechanism possibly minimizing risks of desiccation.

III. GENETIC VARIATION WITHIN SPECIES

Very little is known of the varying ability within species to regenerate vegetatively, a possibly important facet of adaptive biology. For example, the ability to regenerate vegetatively from corms differs within *Arrhenatherum elatius*, perhaps indicating that the processes of developing the ability to regenerate vegetatively are in progress. In one arable population, only 50% of plants produced sufficiently large corms to regenerate successfully, 6% being completely non-bulbous (Ayres, 1977), while in permanent pastures, populations of corm-forming individuals can be scarce.

The growth form of *Ranunculus repens*, which is distributed from northernmost Norway to North Africa, is remarkably constant throughout its range (Coles, 1977), except for a tendency towards a greater proportion of flowering stems to stolons in plants of Finnish origin. In contrast, however, its leaf shape showed a systematic clinal trend from broadly-lobed leaves with a single petiolule in south-west Europe to finely dissected leaflets on leaves with two petiolules in the north-east, although this characteristic was also highly variable within any population (Coles, 1977). More importantly, however, when two populations of *R. repens*, one from grassland and the other from woodland, were grown collaterally, it was found that the former invested more dry matter in parent plant leaf, stem and root tissues than the woodland population, which in turn diverted more dry matter to daughter plantlets, while sustaining a greater turnover of leaves (Clegg, 1978).

In population studies of *Agropyron repens*, clones ranged from large, well-tillered, erect forms to those which were small and had a prostrate habit. In this instance, no relationship was found between growth habit and origin from grassland or arable communities (Neuteboom, 1975). This contrasts with other grasses and herbs in which severe grazing or mowing pressures appear to select individuals with a prostrate form (Warwick and Briggs, 1978, 1979). The production of daughter plantlets also differed appreciably among clones, presumably reflecting their differing abilities to invade new ground. As has also been found by Haddad and Sagar (1968) and Pooswang *et al.* (1972), that clones also differed in the distribution of dry matter between parent shoots, roots and rhizomes, and this may be of considerable importance in differing habitats, as indeed may clonal variations in rhizome carbohydrate content (Turner, 1968b). Interestingly, however, in the population study, numbers of daughter plantlets produced following defoliation were not related to rhizome size (dry weight/unit length), nor to the characteristics of the original habitat (Neuteboom, 1975). Similarly, when *Bellis perennis* from habitats with different mowing and grazing pressures were transplanted into a regularly mown lawn no simple relationship was found between mean numbers of daughter rosettes produced per plant and their habitat of origin, although some populations differ in their ability to regenerate vegetatively by rhizomes (Warwick and Briggs, 1980a). By contrast, in *Achillea millefolium*, which also spreads laterally by rhizomes, no evidence was found of populations differing in their ability to regenerate vegetatively in mown lawns (Warwick and Briggs, 1980b).

The most striking evidence of population differentiation is that of Solbrig and Simpson (1974, 1977) who investigated the competitiveness of two electrophoretically identified biotypes of the apomict *Taraxacum officinale*, grown in disturbed and undisturbed conditions. They found that one biotype, by virtue of its greater vegetative growth, was more competitive in undisturbed conditions, whereas the other was more successful because of its earlier flowering and greater "seed" production on disturbed sites.

The processes of evolution are clearly continuing actively among vegetatively regenerating weeds; "weak" clones being suppressed or eliminated by "aggressive" clones wherever population pressures develop, as for example in stands of *Festuca rubra* (Harberd and Owen, 1969) and *Trifolium repens* (Turkington *et al.*, 1979). Presumably, however, the potential for rapid adaptation is greater in plants combining sexual reproduction with vegetative regeneration, than in those relying on one or the other alone, for well-adapted parents will exist as extensive clones with high reproductive potential, and the regular establishment of new genotypes from seed (as observed in *Ranunculus repens* populations—Soane and Watkinson, 1979) will provide the essential diversity. It is likely, therefore, that the various adaptations found among vegetatively regenerating plants will become more specialized fairly rapidly, increasing the weed problems of the future.

IV. CONCLUSIONS

Many plant species have evolved methods of vegetative regeneration, some of which are highly specialized, requiring substantial morphological adaptations and changes in the physiological processes controlling growth and perennation. In many circumstances, these adaptations have enabled them to become agricultural weeds, often by lateral spread in closed communities or by survival of fragments on disturbed sites. Agricultural practices have undoubtedly intensified selective pressures and probably hastened the evolution of appropriately adapted within-species variants, with some of the latter, perhaps to their future detriment, becoming almost wholly dependent upon the production of vegetative propagules.

Vegetative regeneration has been achieved by adapting shoots, roots and inflorescences, and five fundamentally different methods of regeneration have been identified: (1) detached shoots; (2) creeping stems; (3) modified shoot bases; (4) root suckers; and (5) asexually-formed seeds and fruit.

The success of different weed species can often be attributed to a combination of several different methods of vegetative regeneration. For example, *Allium vineale* produces bulbs, offsets and bulbils, *Taraxacum officinale* combines the ability to regenerate from tap root fragments with the apomictic production of seed, whereas *Cynodon dacylon* forms both rhizomes and stolons. Undoubtedly, however, the combination of one or more of these different methods of vegetative regeneration with sexual reproduction is likely to confer the greatest potential for success as a weed.

Do these different methods of vegetative regeneration, the success of which vary within and between populations of a number of weed species, enable different weeds to colonize different habitats? One study, albeit rather limited, has attempted to quantify and compare the vegetative spread of 17 species of grasses and herbaceous plants (not necessarily weeds) in two different Prairie soils: loam and sand in Nebraska, USA (Mueller, 1941). The quantification of plants in such populations is difficult, necessitating the identification of different plant "units" of vegetatively regenerating plant complexes (Harper, 1977). In the American study, the ability to spread was determined by measuring the various ramifications of the plants into surrounding bare ground. Eleven of the species used were purely rhizomatous, four combined rhizomes with either creeping roots or stolons, one was purely stoloniferous and one wholly dependent on creeping roots. The latter was more successful than any of the rhizomatous species, while the stoloniferous grass was ranked fifth. Of the rhizomatous species those which produced long, branched rhizomes were the most invasive.

In Britain, there is no evidence among the dicotyledonous plants with the ability to regenerate vegetatively, to suggest that they occupy distinctive habitats because of their abilities to form stolons, runners, rhizomes or creeping roots (Clegg, 1978). However, vegetatively regenerating species occupy, on average, more habitats per species than non-clonal species. Thus it seems that vegetative regeneration, possibly by enabling the multiplication and selection of proven clones appears to give a better chance of successful establishment in diverse habitats than sexual reproduction. For example, vegetative propagules of *Ranunculus repens* are six times more likely to survive from one season to the next than seedlings (Sarukhán, 1976).

Acknowledgements

I would like to thank colleagues from the Weed Research Organization, the School of Plant Biology of the University College of North Wales and the Institute of Terrestrial Ecology for sharing their ideas and for their comments on the manuscript.

References

Alexio, M. de F. D. and Valio, I. F. M. (1976). Effect of light, temperature and endogenous growth regulators on the growth of buds of *Cyperus rotundus* L. tubers. *Z. Pflanzenphysiol* **80**, 336–347.

Amor, R. L. (1973). Ecology and control of blackberry (*Rubus fruticosus* L. agg.) 1. *Rubus* spp. as weeds in Victoria. *Weed Res.* **13**, 218–223.

Anon, (1958). Creeping buttercup (*Ranunculus repens*), *Turf Sport* **3**, 8–9.

Ayres, P. (1977). The growth of *Arrhenatherum elatius* var. *bulbosum* (Wild.) Spenn, in spring barley, as influenced by cultivation. *Weed Res.* **17**, 423–428.

Baker, H. G. (1965). Characteristics and modes or origin of weeds. *In* "The Genetics of Colonizing Species" (Baker, H. G. and Stebbins, G. L., eds), pp. 147–172. Academic Press, New York and London.

Baker, H. G. (1975). The evolution of weeds. *Ann. Rev. Ecol. & Syst.* **5**, 1–24.

Bakker, D. (1960). A comparative life-history study of *Cirsium arvense* (L.) Scop. and *Tussilago farfara* L., the most troublesome weeds in the newly reclaimed polders of the former Zuiderzee, *In* "The Biology of Weeds" (Harper, J. L., ed.), pp. 205–222. Blackwells Scientific Publications, Oxford.

Barnola, P. (1971). Recherches sur le déterminisme du marcottage de l'extrémité apical des tiges de la Ronce (*Rubus fruticosus* L.). *Rev. Gen. Bot.* **78**, 185–199.

Basiouny, F. M., Haller, W. and Garrard, L. A. (1978). The influence of growth regulators on sprouting of *Hydrilla* tubers and turions. *J. exp. Bot.* **29**, 663–669.

Bennett-Clark, T. A. and Ball, N. G. (1951). The diageotropic behaviour of rhizomes. *J. exp. Bot.* **2**, 169–203.

Betria, A. I. and Montaldi, E. R. (1974). Light effects on bulb differentiation and leaf growth in *Cyperus rotundus* L., *Phyton* **32**, 1–8.

Betria, A. I. and Montaldi, E. R. (1975). Tuber production by purple nutgrass (*Cyperus rotundus* L.) in darkness. *Weed Res.* **15**, 73–76.

Bonnett, H. T. (1972). Phytochrome regulation of endogenous bud development in root cultures of *Convolvulus arvensis*, *Planta* **106**, 325–330.

Bonnett, H. T. and Torrey, J. G. (1965). Chemical control of organ formation in root segments of *Convolvulus* cultured *in vitro*. *Plant Physiol.* **40**, 1228–1236.

Bonnett, H. T. and Torrey, J. G. (1966). Comparative anatomy of endogenous bud and lateral root formation in *Convolvulus arvensis* roots cultured *in vitro*. *Am. J. Bot.* **53**, 496–507.

Bourdôt, G. W., White, J. G. H. and Field R. J. (1979). Seasonality of growth and development in Yarrow, *Proc. 32nd N.Z. Weed Pest Control Conf.*, 49–54.

Boyd, F. M. (1967). A life history of the growth potential and carbohydrate food reserves of Johnsongrass rhizomes (*Sorghum halepense* (L.) Pers.) Ph.D. Thesis, University of California.

Bradbury, I. K. and Hofstra, G. (1976). The partitioning of net energy resources in two populations of *Solidago canadensis* during a single development cycle in southern Ontario. *Can. J. Bot.* **54**, 2449–2456.

Bradbury, I. K. and Hofstra, G. (1977). Assimilate distribution patterns and carbohydrate concentration changes in organs of *Solidago canadensis* during an annual development cycle. *Can. J. Bot.* **55**, 1121–1127.

Cavers, P. B. and Maun, M. A. (1970). Effects of fragmentation of the rootstock on winter survival and subsequent seed production in *Rumex crispus*. *Abst. Meet. of Weed sci. Soc. Am.* **15**.

Chadwick, M. J. and Obeid, M. (1963). The response to variation in nitrogen level of some weed species in Sudan. *Weed Res.* **3**, 230–241.

Chancellor, R. J. (1970). Biological background to the control, of three perennial broad-leaved weeds. *Proc. 10th Br. Weed Control Conf.*, 1114–1120.

Chancellor, R. J. (1974). The development of dominance amongst shoots arising from fragments of *Agropyron repens* rhizomes. *Weed. Res.* **14**, 29–38.

Chancellor, R. J. (1975). Further shoot regrowth of rhizome fragments of *Agropyron repens* after loss of the dominant shoot. *Proc. E.W.R.S. Symp. Status, Biology and Control of Grassweeds in Europe*, **1**, 69–76.

Chawdhry, M. A. (1974). Growth study of *Oxalis latifolia*, H. B. K. *E. Afr. Agri. For.* **7**. **39**, 402–406.

Chawdhry, M. A. and Sagar, G. R. (1974). Dormancy and sprouting of bulbs in *Oxalis latifolia* H.B.K. and *O. pes-capiae* L., *Weed Res.* **14**, 349–354.

Clapham, A. R. (1945). Studies in the depth adjustment of subterranean plant organs. 1. Raunkiaer's experiment on depth perception in *Polygonatum*. *New Phytol.* **44**, 105–109.

Clapham, A. R., Tutin, T. G. and Warburg E. F. (1962). "Flora of the British Isles" (second edition). Cambridge University Press, Cambridge.

Clegg, L. M. (1978). The morphology of clonal growth and its relevance to the population dynamics of perennial plants. Ph.D. Thesis University of Wales.

Coles, S. M. (1977). *Ranunculus repens* L. in Europe. *Watsonia* **11**, 353–366.

Cutter, E. G. (1978). Structure and development of the potato crop. *In* "The Potato Crop: The Scientific Basis for Improvement" (Harris, P. M., ed.), pp. 70–152. Chapman and Hall, London.

Davis, P. H. and Heywood, V. H. (1963). "Principles of Angiosperm Taxonomy". Oliver and Boyd, Edinburgh and London.

Davison, J. G. (1970). The establishment of *Convolvulus arvensis* in a non-competitive situation. *Proc. 10th Br. Weed Control Conf.*, 352–357.

Duke, S. O. and Williams, P. D. (1977). Phytochrome distribution in Johnsongrass rhizomes. *Weed Sci.* **25**, 229–232.

Fail, H. (1956). Effect of rotary cultivation on the rhizomatous weeds. *J. agric. Eng. Res.* **1**, 68–80.

Fisher, J. E. (1964). Evidence of circumnutational growth movements of rhizomes of *Poa pratensis* L. that aid in soil penetration. *Can. J. Bot.* **42**, 293–299.

Fiveland, T. J., Erickson, L. C. and Seeley, C. I. (1972). Translocation of ^{14}C-assimilates and 3-amino-1,2,4-triazole and its metabolites in *Agropyron repens*. *Weed Res.* **12**, 155–163.

Fletcher, W. W. and Kirkwood, R. C. (1979). The bracken fern (*Pteridium aquilinum* L. Kuhn); its biology and control. *In* "The Experimental Biology of Ferns" (Dyer, A. F., ed.), pp. 591–626. Academic Press, London and New York.

Forde, B. J. (1966a). Translocation in grasses. 1. Bermuda grass. *N.Z. J. Bot.* **4**, 479–495.

Forde, B. J. (1966b). Translocation in grasses. 2. Perennial ryegrass and couch-grass. *N.Z. J. Bot.* **4**, 496–514.

Frazier, J. C. (1943). Nature and rate of development of root system of *Convolvulus arvensis*. *Bot. Gaz.* **104**, 417–425.

Fykse, H. (1974). [Research on *Sonchus arvensis* L., 1. Translocation of ^{14}C-labelled assimilates]. *Weed Res.* **14**, 305–312.

Giddens, J., Perkins, H. F. and Walker, L. C. (1962). Movement of nutrients in Coastal Bermudagrass. *Agron. J.*, **54**, 379–382.

Ginzo, H. D. and Lovell, P. H. (1973a). Aspects of the comparative physiology of *Ranunculus bulbosus* L. and *Ranunculus repens* L. I. Response to nitrogen. *Ann. Bot.* **37**, 753–764.

Ginzo, H. D. and Lovell, P. H. (1973b). Aspects of the comparative physiology of *Ranunculus bulbosus* L. and *Ranunculus repens*. II. Carbon dioxide assimilation and distribution of photosynthates. *Ann. Bot.* **37**, 765–776.

Goodwin, P. B. (1967). The control of branch growth on potato tubers. II. The pattern of sprout growth. *J. exp. Bot.* **54**, 87–99.

Grime, J. P. (1979). "Plant Strategies and Vegetation Processes". J. Wiley & Sons, Chicester.

Haddad, S. Y. (1968). Studies on clones of *Agropyron repens* (L.) Beauv. with particular reference to herbicide response. M.Sc. Thesis, University of Wales.

Haddad, S. Y. and Sagar, G. R. (1968). A study of the response of four clones of *Agropyron repens* (L.) Beauv. to root and shoot application of aminotriazole and dalapon. *Proc. 9th Br. Weed Control Conf.*, 142–148.

Håkansson, S. (1963). *Allium vineale* L. as a weed with special reference to the conditions in south-eastern Sweden. Agricultural College of Sweden, Department of Plant Husbandry Publication.

Håkansson, S. (1967). Experiments with *Agropyron repens* (L.) Beauv. 1. Development and growth, and the response to burial at different developmental stages. *Lantbr-Hogsk. Annlr* **33**, 823–873.

Håkansson, S. (1968). Experiments with *Agropyron repens* (L.) Beauv. III. Production of aerial and underground shoots after planting rhizome pieces of different lengths at varying depths. *Lantbr-Hogsk. Annlr.* **34**, 31–51.

Håkansson, S. (1969a). Experiments with *A. repens* (L.) Beauv. IV. Response to burial and defoliation repeated with different intervals. *Lantbr-Hogsk. Annlr.* **35**, 61–78.

Håkansson, S. (1969b). Experiments with *A. repens* (L.) Beauv. VI. Rhizome orientation and life length of broken rhizomes in the soil, and reproductive capacity of different underground shoot parts. *Lantbr-Hogsk. Annlr.* **35**, 869–894.

Håkansson, S. and Wallgren, B. (1972). Experiments with *Sonchus arvensis* L. II. Reproduction, plant development and response to mechanical disturbance. *Swed. J. agric. Res.* **2**, 3–14.

Hamdoun, A. M. (1972). Regenerative capacity of root fragment of *Cirsium arvense*. *Weed Res.* **12**, 128–136.

Harberd, D. J. and Owen, M. (1969). Some experimental observations on the clone structure of a natural population of *Festuca rubra* L., *New Phytol* **68**, 93–104.

Harper, J. L. (1957). Biological flora of the British Isles. *Ranunculus acris, Ranunculus repens* and *Ranunculus bulbosus*, *J. Ecol.* **45**, 289–342.

Harper, J. L. (1977). "Population Biology of Plants". Academic Press London and New York.

Harper, J. L., Lovell, P. H. and Moore, K. G. (1970). The shapes and sizes of seeds. *Annu. Rev. Ecol. Syst.* **1**, 327–356.

Hartmann, H. T. and Kester, D. E. (1968). "Plant Propagation—Principles and Practices" (second edition). Prentice-Hall, New Jersey.

Harvey, H. J. (1970). Patterns of assimilate transport in *Trifolium repens*. White Clover Research Occasional Symposium No. 6, British Grassland Society, 181–186.

Healey, A. (1953). Control of docks. *N.Z. Jl Sci. Technol.* (section A), **34**, 473–475.

Henson, I. E. (1969). Studies on the regeneration of perennial weeds in the glasshouse. 1. Temperate species. *Tech. Rep. A.R.C. Weed Res. Org.* **12**.

Heslop-Harrison, Y. (1959). Natural & induced rooting of the stem apex in *Rubus*. *Ann. Bot.* **23**, 307–318.

Holm, L. G., Plucknett, D. L., Pancho, J. V. and Herberger, J. P. (1977). "The World's Worst Weeds: Distribution and Biology". University Press of Hawaii, Hawaii.

Horowitz, M. (1972a). Development of *Cynodon dactylon* (L.) Pers. *Weed Res.* **12**, 207–220.

Horowitz, M. (1972b). Seasonal development of established Johnson grass. *Weed Sci.* **20**, 392–396.

Hudson, J. (1955). Propagation of plants by root cuttings. 2. Seasonal fluctuation of capacity to regenerate from roots. *J. Hort. Sci.* **30**, 242–251.

Hull, R. J. (1970). Germination control of Johnsongrass rhizome buds. *Weed Sci.* **18**, 118–121.

Jackson, E. I. (1960). A growth study of *Oxalis latifolia* HBK. *N.Z. J. Sci.* **3**, 600–609.

Johnson, B. G. and Buchholtz, K. P. (1962). The natural dormancy of vegetative buds on the rhizomes of *Agropyron repens*. *Weeds* **10**, 53–57.

Khan, M. I. (1972). Metabolic studies of the regeneration of the roots of *Taraxacum officinale, Pak. J. Biochem.* **5**, 61–65.

Kirby, K. J. (1980). Experiments on vegetative reproduction in bramble (*Rubus vestitus*). *J. Ecol.* **68**, 513–520.

Lazenby, A. (1960). Adaptability in *Allium vineale. In* "The Biology of Weeds" (Harper, J. L., ed.), pp. 223–235. Blackwells Scientific Publications, Oxford.

Leakey, R. R. B. (1974). Factors affecting the growth of shoots from fragmented rhizomes of *Agropyron repens* (L.) Beauv. Ph.D. Thesis, University of Reading.

Leakey, R. R. B., Chancellor, R. J. and Vince-Prue, D. (1975). Parental factors in dominance of lateral buds on rhizomes of *Agropyron repens* (L.) Beauv. *Planta* **123**, 267–274.

Leakey, R. R. B., Chancellor, R. J. and Vince-Prue, D. (1977a). Regeneration from rhizome fragments of *Agropyron repens* (L.) Beauv. 1. The seasonality of shoot growth and rhizome reserves in single-node fragments. *Ann. appl. Biol.* **87**, 423–431.

Leakey, R. R. B., Chancellor, R. J. and Vince-Prue, D. (1977b). Regeneration from rhizome fragments of *Agropyron repens* (l.) Beauv. II. The breaking of "Late Spring Dormancy" and the influence of chilling and node position on growth from single-node fragments. *Ann. appl. Biol.* **87**, 433–441.

Leakey, R. R. B., Chancellor, R. J. and Vince-Prue, D. (1978a). Regeneration from rhizome fragments of *Agropyron repens* (L.) Beauv. III. Effects of nitrogen and temperature on the development of dominance amongst shoots on multi-node fragments. *Ann. Bot.* **42**, 197–204.

Leakey, R. R. B., Chancellor, R. J. and Vince-Prue, D. (1978b). Regeneration from rhizome fragments of *Agropyron repens* (L.) Beauv. IV. Effects of light on bud dormancy and the development of dominance amongst shoots on multi-node fragments. *Ann. Bot.* **42**, 205–212.

Lutman, P. J. W. (1977). Investigations into some aspects of potatoes as weeds. *Weed Res.* **17**, 123–132.

Mann, H. and Cavers, P. B. (1979). The regenerative capacity of root cuttings of *Taraxacum officinale* under natural conditions. *Can. J. Bot.* **57**, 1783–1791.

Marsden-Jones, E. M. (1935). *Ranunculus ficaria*, life history and pollination. *J. Linn. Soc. (Bot.)* **50**, 39–55.

McIntyre, G. I. (1965). Some effects of the nitrogen supply on the growth and development of *Agropyron repens* L. Beauv. *Weed Res.* **5**, 1–12.

McIntyre, G. I. (1967). Environmental control of bud and rhizome development in the seedling of *Agropyron repens* L. Beauv. *Can. J. Bot.* **45**, 1315–1326.

McIntyre, G. I. (1969). Apical dominance in the rhizome of *Agropyron repens*. Evidence of competition for carbohydrate as a factor in the mechanism of inhibition. *Can. J. Bot.* **47**, 1189–1197.

McIntyre, G. I. (1970). Studies on bud development in the rhizome of *Agropyron repens*. 1. The influence of temperature, light intensity and bud position on the pattern of development. *Can. J. Bot.* **48**, 1903–1909.

McIntyre, G. I. (1971). Apical dominance in the rhizome of *Agropyron repens*. Some factors affecting the degree of dominance in isolated rhizomes. *Can J. Bot.* **49**, 99–109.

McIntyre, G. I. (1972). Studies on bud development in the rhizome of *Agropyron repens*. II. The effect of the nitrogen supply. *Can. J. Bot.* **50**, 393–401.

McIntyre, G. I. (1976). Apical dominance in the rhizome of *Agropyron repens*: the influence of water stress on bud activity. *Can. J. Bot.* **54**, 2747–2754.

McIntyre, G. I. (1977). The role of nutrition in apical dominance. *In* "Integration of Activity in the Higher Plant" (Jennings, D. H., ed.), pp. 251–273. Cambridge University Press, Cambridge.

McIntyre, G. I. and Hunter, J. H. (1975). Some effects of the nitrogen supply on growth and development of *Cirsium arvense, Can. J. Bot.* **53**, 3012–3021.

McWhorter, C. G. (1972). Factors affecting Johnsongrass rhizome production and germination. *Weed Sci.* **20**, 41–45.

Monaco, T. J. and Cumbo, E. L. (1972). Growth and development of Curly Dock and Broadleaved Dock. *Weed Sci.* **20**, 64–67.

Montaldi, E. R. (1970). *Cynodon dactylon*: possible causa de su diageotropismo. *Rev. Investi. Agropec. Ser. 2, Buenos Aires* **7**, 67–88.

Moore, R. J. (1975). The biology of Canadian weeds, 13, *Cirsium arvense* (L.) Scot. *Can. J. Pl. Sci.* **55**, 1033–1048.

Moreira, I. (1975). Some aspects of the biology of torpedograss. (*Panicum repens* L.). *Proc. Symp. on Status, Biology and Control of Grassweeds in Europe* **1**, 59–68.

Moreira, I. and Rosa, M. L. (1976). Influencia da posicao do nó nó abrolhamento de gemas de rizomas de *Cynodon dactylon*, II. *Simposio Nacional de Herbologia*, 37–43.

Mueller, I. M. (1941). An experimental study of rhizomes of certain prairie plants. *Ecol. Monog.* **11**, 167–188.

Musik, T. J. and Cruzado, H. J. (1953). The effect of 2,4-D on sprout formation in *Cyperus rotundus*. *Am. J. bot.* **40**, 507–512.

Myerscough, P. J. and Whitehead, F. H. (1967). Comparative biology of *Tussilago farfara* L., *Chamaenerion augustifolium* (L.) Scot., *Epilobium montanum* L. and *Epilobium adenocaulon* Hausskn. II. Growth and ecology. *New Phytol.* **66**, 785–823.

Neuteboom, J. H. (1975). *Variability of Elytrigia repens* (L.) Dsv. (Syn. *Agropyron repens* (L.) P.B.) on Dutch agricultural soils. *Meded. Landb-Hoogesch. Wageningen*, **75–7**, 1–29.

Nigam, S. N. and McIntyre, G. I. (1977). Apical dominance in the rhizome of *Agropyron repens*. The relation of amino acid composition to bud activity. *Can. J. bot.* **55**, 2001–2010.

Nyahoza, F., Marshall, C. and Sagar, G. R. (1973). The inter-relationship between tillers and rhizomes of *Poa pratensis* L.—an autoradiographic study. *Weed Res.* **13**, 304–309.

Nyahoza, F., Marshall, C. and Sagar, G. R. (1974). Some aspects of the physiology of the rhizomes of *Poa pratensis* L., *Weed Res.* **14**, 329–336.

Ogden, J. (1974). The reproductive strategy of higher plants 2. The reproductive strategy of *Tussilago farfara*. *J. Ecol.* **62**, 291–320.

Otzen, D. and Koridon, A. H. (1970). Seasonal fluctuations of organic food reserves in underground parts of *Cirsium arvense* and *Tussilago farfara*. *Acta. bot. neerl.* **19**, 495–502.

Palmer, J. H. (1956). The nature of the growth response to sunlight shown by certain stoloniferous and prostrate tropical plants. *New Phytol.* **55**, 346–355.

Palmer, J. H. (1958). Studies in the behaviour of the rhizome of *Agropyron repens* (l.) Beauv. I. The seasonal development and growth of the parent plant and rhizome. *New Phytol.* **57**, 145–159.

Palmer, J. H. (1962). Studies in the behaviour of the rhizome of *Agropyron repens*

(L.) Beauv. II. Effect of soil factors on the orientation of the rhizome. *Physiol. Plant.* **15**, 445–451.

Palmer, J. H. and Sagar, G. R. (1963). Biological flora of the British Isles, *Agropyron repens* (L.) Beauv. (*Triticum repens* L.: *Elytrigia repens* (L.) Nevski.) *J. Ecol.* **51**, 783–794.

Penfound, W. T. and Earle, T. T. (1948). The biology of the water hyacinth. *Ecol. Monogr.* **18**, 447–472.

Phillips, I. D. J. (1969). Apical dominance. *In* "Physiology of Plant Growth and Development" (Wilkins, M. B., ed.), pp. 165–202. McGraw-Hill, London.

Phillips, I. D. J. (1975). Apical dominance. *A. Rev. Pl. Physiol.* **26**, 341–367.

Pooswang, W., Huxley, P. A. and Buckley W. R. (1972). Difference in growth of four clones of *Agropyron repens* (L.) Beauv. *Proc. 11th Br. Weed Control*, 38–45.

Qureshi, F. A. and Spanner, D. C. (1971). Unidirectional movement of tracers along the stolon of *Saxifraga sarmentosa*, *Planta* **101**, 133–146.

Qureshi, F. A. and Spanner, D. C. (1973a). The effect of nitrogen on the movement of tracers down the stolon of *Saxifraga sarmentosa*, with some observations on the influence of light. *Planta* **110**, 131–144.

Quershi, F. A. and Spanner, D. C. (1973b). Movement of (^{14}C) sucrose along the stolon of *Saxifraga sarmentosa*, *Planta* **110**, 145–152.

Raju, M. V. S., Steeves, T. A. and Coupland, R. T. (1964). On the regeneration of root fragments of leafy spurge (*Euphorbia esula* L.). *Weed Res.* **4**, 2–11.

Richardson, R. G. (1975). Regeneration of Blackberry (*Rubus procerus*, P. J. Muell) from root segments. *Weed Res.* **15**, 335–337.

Richens, R. H. (1947). Biological flora of the British Isles. *Allium vineale* L., *J. Ecol.* **34**, 209–226.

Rogan, P. G. and Smith, D. L. (1974). Patterns of translocation of ^{14}C-labelled assimilates during vegetative growth of *Agropyron repens* (L.) Beauv. *Z. Pflanzenphysiol.* **73**, 405–414.

Rogan, P. G. and Smith, D. L. (1975). The effect of temperature and nitrogen level on the morphology of *Agropyron repens* (L.) Beauv. *Weed Res.* **15**, 93–99.

Rogan, P. G. and Smith, D. L. (1976). Experimental control of bud inhibition in rhizomes of *Agropyron repens* (l.) Beauv. *Z. Pflanzenphysiol.* **78**, 113–121.

Rogers, C. F. (1929). Canada thistle and Russian Knapweed and their control. *Bull. Colo. St. University agric. Exp. Stn*, 384.

Sagar, G. R. (1960). *Agropyron repens* – an introduction. *Proc. 5th Br. Weed Control Conf.*, 259–263.

Sagar, G. R. (1961). Some thoughts on the control of *Agropyron repens* by herbicides. *N.A.A.S. q. Rev.* **53**, 10–16.

Sagar, G. R. and Rawson, H. M. (1964). The biology of *Cirsium arvense* (L.). *Proc. 7th Br. Weed Control Conf.*, 553–562.

Sarukhán, J. (1971). Studies on plant demography. Ph.D. Thesis, University of Wales.

Sarukhán, J. (1974). Studies on plant demography: *Ranunculus repens* L., *R. bulbosus* L. and *R. acris* L. II. Reproductive strategies and seed population dynamics, *J. Ecol.* **62**, 151–177.

Sarukhán, J. (1976). On selective pressures and energy allocation in populations of *Ranunculus repens* L.; *R. bulbosus* L.: & *R. acris* L. *Ann. Mo. bot. Gdn.* **63**, 290–308.

Sarukhán, J. and Harper, J. L. (1973). Studies on plant demography: *Ranunculus repens*, L., *R. bulbosus* L. & *R. acris* L., *J. Ecol.* **61**, 675–716.

Sastroutomo, S. S., Ikusima, I., Numata M. and Iizumi, S. (1979). The importance of turions in the propagation of pondweed (*Potamogeton crispus* L.). *Ecol. Rev. (Sendai)* **19**, 75–88.

Sastroutomo, S. S. (1980). Environmental control of turion formation in curly pondweed (*Potamogeton crispus*). *Physiol. Plant.* **49**, 261–264.

Schwanitz, F. (1936). Beiträge zur analyse de pflanzenlichen polariatät. *Beih bot. Zbl.* **54A**, 520–530.

Sculthorpe, C. D. (1967). "Biology of Aquatic Vascular Plants". Edward Arnold, London.

Soane, I. D. and Watkinson, A. R. (1979). Clonal variation in populations of *Ranunculus repens*. *New Phytol.* **82**, 557–573.

Solbrig, O. T., and Simpson, B. B., (1974). Components of regulation of a population of dandelions in Michigan. *J. Ecol.* **62**, 473–486.

Solbrig, O. T. and Simpson, B. B. (1977). A garden experiment on competition between biotypes of the common dandelion (*Taraxacum officianale*). *J. Ecol.* **65**, 427–430.

Swan, D. G. and Chancellor, R. J. (1976). Regenerative capacity of field bindweed roots. *Weed Sci.* **24**, 306–308.

Teo, C. K. H. and Nishimoto, R. K. (1973). Cytokinin—enhanced sprouting of purple nutsedge as a basis for control. *Weed Res.* **13**, 138–141.

Teo, C. K. H., Nishimoto, R. K. and Tang, C. S. (1974). Bud inhibition of *Cyperus rotundus* L. tubers by inhibitor β or abscisic acid and the reversal by N-6-benzyladenine. *Weed Res.* **14**, 173–179.

Tepfer, D. A. and Bonnett, H. T. (1972). The role of phytochrome in the geotropic behaviour of roots of *Convolvulus arvensis*. *Planta* **106**, 311–324.

Tildesley, W. T. (1931). A study in the changes of the growth and food reserves in the underground parts of *Sonchus arvensis* L. and *Agropyron repens* (L.) Beauv. during the growing season 1931–1932. Ph.D. Thesis, University of Manitoba.

Thomas, R. G. (1970). The role of basal and apical factors in the co-ordination of growth in the stems of white clover (*Trifolium repens*). *In* "Plant Growth Substances 1970". (Carr D. J., ed.), pp. 624–632. Proc. 7th Conf. Plant Growth Subs. Springer Verlag, Berlin.

Thomas, T. H. (1969). The role of growth substances in the regulation of onion bulb dormancy. *J. exp. Bot.* **20**, 124–137.

Tomlinson P. B. and Gill A. M. (1973). Growth habits of tropical trees: some guiding principles. *In* "Tropical Forest Ecosystems in Africa and South America: A Comparative Review" (B. J. Meggers, E. S. Ayensu and W. D. Duckworth, eds), pp. 129–143. Smithsonian Institution Press, Washington DC.

Torrey, J. G. (1958). Endogenous bud and root formation by isolated roots of *Convolvulus* grown *in vitro*. *Plant Physiol.* **33**, 258–263.

Turkington, R. and Harper J. L. (1979a). The growth, distribution and neighbour relationships of *Trifolium repens* in a permanent pasture. I. Ordination, pattern and contact. *J. Ecol.* **67**, 201–218.

Turkington R. and Harper J. L. (1979b). The growth, distribution and neighbour relationships of *Trifolium repens* in a permanent pasture. II. Inter and intra-specific contact. *J. Ecol.* **67**, 219–230.

Turkington, R. and Harper J. L. (1979c). The growth, distribution and neighbour relationships of *Trifolium repens* in a permanent pasture. IV. Fine-scale biotic differentiation. *J. Ecol.*, **67**, 245–254.

Turkington, R. Cahn, M. A., Vardy, A. and Harper J. L. (1979). The growth, distribution and neighbour relationships of *Trifolium repens* in a permanent pasture. III.

The establishment and growth of *Trifolium repens* in natural and perturbed sites. *J. Ecol.* **67**, 231–243.

Turner, D. J. (1966). A study of the effects of rhizome length, soil nitrogen and shoot removal on the growth of *Agropyron repens* (L.) Beauv. *Proc. 8th Br. Weed Control Conf.*, 538–545.

Turner, D. J. (1968a). *Agropyron repens* (L.) Beauv.—some effects of rhizome fragmentation, rhizome burial and defoliation. *Weed Res.* **8**, 298–308.

Turner, D. J. (1968b). The response to defoliation of different strains of *Agropyron repens* and *Agrostis gigantea*. *Proc. 9th Br. Weed control Conf.*, 149–155.

Turner, D. J. (1969). The effects of shoot removal on the rhizome carbohydrate reserves of couch grass (*Agropyron repens* (L.) Beauv.). *Weed Res.* **9**, 27–36.

Valentine, D. and Richards, A. J. (1967). Sexuality and apomixis in *Taraxacum* and *Nature* **214**, 114.

Wareing, P. F. and Nasr. T. A. A. (1958). Gravimorphism in trees: effects of gravity on growth, apical dominance and flowering in fruit trees. *Nature* **182**, 379–381.

Wareing, P. F. and Nasr. T. A. A. (1961). Gravimorphism in trees. I. Effects of gravity on growth and apical dominance in fruit trees. *Ann. Bot.* **25**, 321–340.

Warwick, S. I. and Briggs, D. (1978). The genecology of lawn weeds. 1. Population differentiation in *Poa annua* L. in a mosaic environment of bowling green lawns and flower beds. *New Phytol.* **81**, 711–723.

Warwick, S. I. and Briggs, D. (1979). The genecology of lawn weeds. III. Cultivation experiments with *Achillea millefolium* L., *Bellis perennis* L., *Plantago lanceolata* L.; *Plantago major* L. and *Prunella vulgaris* L. collected from lawns and contrasting grassland habitats. *New Phytol.* **83**, 509–536.

Warwick, S. I. and Briggs, D. (1980a). The genecology of lawn weeds. IV Adaptive significance of variation in *Bellis perennis* L. as revealed in a transplant experiment. *New Phytol.* **85**, 275–288.

Warwick, S. I. and Briggs, D. (1980b). The genecology of lawn weeds. VI The adaptive significance of variation in *Achillea millefolium* L. as investigated by transplant experiments. *New Phytol.* **85**, 451–460.

Watt, A. S. (1947). Contributions to the ecology of bracken (*Pteridium aquitinum*). IV. The structure of the community. *New Phytol* **46**, 97–121.

Watt, T. A. and Haggar, R. J. (1980). The effect of defoliation upon yield, flowering, and vegetative spread of *Holcus lanatus* growing with and without *Lolium perenne*. *Grass & Forage Science* **135**, 227–234.

Weber, J. A. and Noodén, L. D. (1976). Environmental and hormonal control of turion formation in *Myriophyllum verticillatum*. *Pl. Cell Physiol.* **17**, 721–731.

Wells, G. J. (1974). The biology of *Poa annua* and its significance in grassland. *Herb. Abstr.* **44**, 385–389.

Williams, E. D. (1979). Studies on the depth distribution and the germination and growth of *Equisetum arvense* L. (field horsetail) from tubers. *Weed Res.* **19**, 25–32.

Williams, G. H. and Foley, A. (1976). Seasonal variations in the carbohydrate content of bracken. *Linn Soc. (Bot)* **73**, 87–93.

Woolley D. J. and Wareing, P. F. (1972). The role of roots, cytokinins and apical dominance in the control of lateral shoot form in *Solanum andigena*; *Planta* **105**, 33–42.

Young, D. P. (1958). *Oxalis* in the British Isles. *Watsonia* **4**, 51–69.

The Impact of Impoundments on the Downstream Fisheries and General Ecology of Rivers

M. P. BROOKER[1]

UWIST Field Centre,
Newbridge-on-Wye, Powys,
Wales

I. Introduction

The construction of dams and barrages by man to impound freshwaters dates back almost 5000 years and the oldest dam which remains today is at Sadd el-Kafara in Egypt, built about 2800 B.C. to supply water for drinking and stone extraction (Smith, 1971; Henderson-Sellers, 1979). There is also evidence that about 1000 years later a complex system of diversion dams was built on the Rivers Tigris and Euphrates and used for irrigation purposes, but it was not until the time of the

[1] Present address: Welsh Water Authority, Nash Area Laboratory, West Nash, Newport, Gwent, Wales.

Roman civilization that the full concept of the reservoir dam and water supply system was pioneered.

In Europe, of course, there were few irrigation needs to stimulate dam construction and there was little major water resource development until the nineteenth century and the emergence of industrialized societies which were dependent upon efficient transport. At this time networks of canals, constructed both in Europe and the USA as transport systems, were supplied by reservoirs, the dams generally being constructed of earth up to a height of about 20 m with low level outlets (Smith, 1971).

Subsequently the use of water power necessitated a constant supply of water to drive mills and more reservoirs were built specifically for this purpose. Indeed, mill operators would demand water only during working hours and discharges from reservoirs were stopped entirely at other times irrespective of the damaging effects on other downstream resources. However, the long retention times of many large rivers caused this widespread practice to be stopped since all mills could not be accommodated. Later, rules were developed for reservoirs in an endeavour to protect the downstream environment: 66% of the total annual yield was allocated for direct uses and 33% discharged uniformly downstream (Walters, 1966). Such practices still remain today, chiefly for direct supply reservoirs, but little effort has been devoted to assessing any impact of such procedures on the river downstream or their effectiveness in protecting aquatic resources. Water power was, of course, replaced by steam power but recent developments in the balance between energy supply and demand may see a substantial review of the potential of water for generating electricity (e.g. Freeman, 1979). The first hydroelectric scheme was on the Fox River, Wisconsin (1882) and currently there are many nations—Norway, Switzerland, Canada, Sweden, New Zealand and Italy—which derive the majority of their electrical power from hydroelectric schemes (Beckinsale, 1969; Lillehammer and Saltveit, 1979).

In Britain the rate of dam building was greatest in the nineteenth century during the period of rapid industrial development but the world-wide peak of major structures (greater than 15 m high) was from 1945 to 1971 (Fig. 1) when a total of over 8000 major dams were completed, 548 in 1968 alone. Most of these have been built in North America, especially the USA, where over 200 major dams were commissioned each year during this period. The rate of dam construction has slowed in the last decade (Fig. 1), but it has been estimated that about 10% of the world's stream flow is now regulated by impoundments (Croome et al., 1976), with many dams of massive proportions

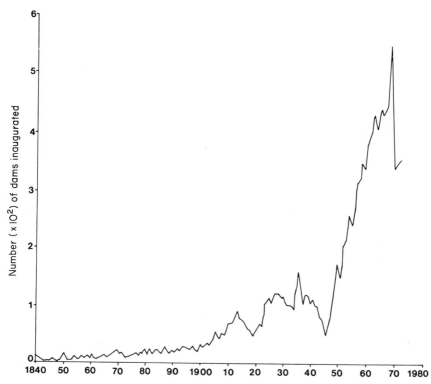

Fig. 1. Distribution of world dam construction since 1840. Redrawn from Beaumont (1978).

(Table I). Such regulation is especially marked in America where about 20% of stable runoff (that derived principally from ground water and therefore fluctuating little from year to year) is impounded (Table II). In Europe and Asia the figures are 15% and 14% respectively and only in South America (4%) and Australasia (6%) does man appear to have had a relatively small impact on river regimes (Beaumont, 1978), although in Australasia certain rivers, particularly the Murray-Darling, are severely regulated for irrigation purposes (see p. 101).

The potential gross effects of dams on downstream resources was not lost on earlier civilizations. The Assyrians, during their campaign against the Babylonians in 689 B.C., dammed the Euphrates to build up a reservoir of water and breached the impoundment to destroy the city of Babylon. In more recent years the disastrous downstream effects

TABLE I. World's ten highest dams arranged in rank order. From Mermel (1979)

Rank	Name	Year[a] completed	River or Basin	Country	Height above lowest foundation (m)	Reservoir Capacity (m^3 × 10^6)
1	Rogun	(1985)	Vakhsh	USSR	330	11700
2	Nurek	(1985)	Vakhsh	USSR	317	10400
3	Grand Dixence	1962	Dixence	Switzerland	285	400
4	Inguri	(1985)	Inguri	USSR	272	1100
5	Chiocasen	1957	Grijalva	Mexico	264	1660
6	Vaiont	1961	Vaiont	Italy	262	169
7	Mica	1972	Columbia	Canada	242	24670
8	Sayan-Shushen	(1980)	Yenisei	USSR	242	31300
9	Mauvoisin	1975	Drangede Bagnes	Switzerland	237	180
10	Chivor	1977	Bata	Colombia	237	815

[a] projected completion in parentheses

TABLE II. River runoff (km^3) of the continents of the world regulated by reservoirs. From Beaumont (1978).

Continent	Stable runoff[a]	Total runoff	Regulated by reservoirs	Regulated reservoirs as % of total stable runoff[a]
Africa	1905	4233	400	21·0
North America	2380	5950	490	20·6
Europe	1325	3081	200	15·1
Asia	4005	13350	560	14·0
Australasia	495	1980	30	6·1
South America	3900	10263	160	4·1

[a] Stable runoff is defined as river flow resulting largely from groundwater discharge

of dam failure, such as the loss of the Morri Dam No. 2 in Gujarat, India in 1979 when up to 10000 people died, have always, rightly, received considerable attention. However, the less dramatic implications of impoundment construction on the ecology of downstream areas has been seriously neglected, the planning and execution of such major water resource developments concentrating generally on technical and social problems. For example, on the African continent there are 23 large man-made lakes on 12 major river systems in 13 countries regulating the flow of about 92000 km of river but the downstream effects on fisheries and general ecology have not been measured or have received limited consideration (Ewer, 1966; Davies, 1979). Similar neglect has been highlighted in Europe (Armitage, 1979, Lillehammer and Saltveit, 1979), North America (Stanford and Ward, 1979) and Australia (Cadwallader, 1978; Dexter, 1978). Those environmental and ecological changes occurring at the site of impoundment have generally attracted greater attention than downstream effects (Neel, 1963; Lagler, 1969; Baxter, 1977), the centre of such investigations being the development of commercial fisheries (Lagler, 1969). In addition, there are few situations where reasonably comprehensive river water quality data have been collected before impoundment has occurred (R. Zambezi— Hall *et al.*, 1976; R. Nile—Ramadan, 1972; R. Wye—Oborne *et al.*, 1980) so that subsequent effects might be assessed.

It is only in recent years that the full implications of impoundment on the downstream environment have received full cognizance. However, challenges, under the Endangered Species Act, 1973 (USA), to the completion of the Tellico Dam on the Little Tennessee River, by an environmental lobby on the basis of the discovery of a relatively rare fish species, when the dam was half completed (Anon., 1978) perhaps reduce the credibility of those striving for a proper assessment of the environmental impact of such schemes. The ecologist should provide the relevant information for balanced assessments of the use of impoundments, and whilst recognition of the commercial and social need for dams is not his professional concern, identification of the ecological effects of their use should involve gains as well as losses.

This review considers the impact impoundments have on the downstream aquatic environment and the effects of such changes on downstream fisheries and general ecology.

II. The Effects of Impoundments on the Downstream Aquatic Environment

A. GENERAL CONSIDERATIONS

The construction of a dam to impound a river or stream may affect the downstream biota, both directly and indirectly, in two principal ways:

(1) The impoundment may dramatically alter the downstream flow regime.

(2) Storage of water in reservoirs may lead to considerable changes in physical and chemical water quality which may substantially influence the downstream environment.

Such effects may be *local* or *extensive* (Edwards and Crisp, 1980) and the type and use of the impoundment will influence considerably the magnitude of environmental changes downstream of the dam. Hull (1967) recognized at least eight functions served by impoundment and associated river regulation:

(1) Navigation.
(2) Flood control.
(3) Hydroelectric power.
(4) Water supply.
(5) Recreation.
(6) Fish and wildlife improvement.
(7) Salinity control.
(8) Waste dilution and disposal.

Clearly each reservoir may serve one or more functions and currently in North America (Stanford and Ward, 1979) and elsewhere, reservoirs chiefly serve the purposes of flood control, hydroelectric power generation and water supply. For example, the Tennessee Valley river system, which has a watershed of about 100 000 km², was first impounded in the 1930s in order to alleviate severe flooding and improve navigation. A system of nine large multipurpose dams on the main river controlled flooding and guaranteed the required depth for navigation. Subsequently, further impoundment of tributaries restored large areas to agricultural production and substantial generation of hydroelectric power was possible although not without major environmental changes (Krenkel et al., 1979).

In addition to reservoir use downstream environmental effects may depend on the location of dams, on the main river channel or on tributaries, and upon the refill sources, from the same river catchment or by diversion from remote gathering grounds. The import/ export of water between separate catchments may be particularly important in the transfer of biologically active materials, e.g. parasites, plant nutrients, and is an extensive subject worthy of separate review.

Several authors (Neel, 1963; Ackermann et al., 1973; Ridley and Steel, 1975; Baxter, 1977) have reviewed the environmental changes resulting from impoundment but place the greatest emphasis on changes in the reservoir basin, only briefly considering the downstream effects. The major physio-chemical changes downstream from impoundments are related to the sedimentation and stratification processes occurring in the reservoir and the modification of flow regime imposed by the function of the reservoir and these are considered below.

B. SEDIMENTATION

Damming of any river system results in the immediate settlement of sediment carried in suspension and impoundments permanently store practically all the sediment load of the river, discharging clear water downstream of the dam (Leopold *et al.*, 1964; Nilsson, 1976; Petts, 1978): indeed some sedimentation effects may be recorded in channels upstream of the reservoir. Serruya (1969) estimated that the Rhone carried about $2·7 \times 10^6$ t per annum of suspended matter into Lake Geneva where it was deposited. The River Nile carried about 132×10^6 t per annum of silt into the impoundment formed by the Aswan High Dam and the direct loss of these sediments had severe consequences downstream where its considerable value to agriculture on the food plain had to be recompensed by application of artificial fertilizer (Turner, 1971). A direct comparison of the Rivers Ystwyth and Rheidol in West Wales, which are of similar catchment area and topography, indicated that the sediment yield of the naturally flowing Ystwyth was 7–16 times greater than that of the Rheidol, which has about 54% of its catchment impounded (J. T. Lewin, pers. comm.).

Such sedimentation also leads to loss of storage capacity of reservoirs. Glymph (1973) concluded that average storage loss in USA reservoirs ranged from 0·16–3·50% per annum depending on size and he estimated that 50% of the smaller reservoirs (capacity, less than $1·235 \times 10^4$ m^3) would be half filled with sediment in 33 years or less. Brown (1944) suggested that sluicing and venting may reduce trap efficiency by as much as 10% but no consideration was made of possible downstream ecological effects and clearly such operations should be discouraged if ecological damage is to be avoided: Nisbet (1961) described how periodic discharges of sediment from the dam at Verbois turned the upper Rhone into a stream of mud.

C. RIVER FLOW

River flows downstream of dams, and their subsequent effects on the river channel, reflect the purposes for which impoundments are used. Reservoirs built to provide a direct supply of water, usually by pipeline or tunnel, generally discharge a constant "compensation" flow downstream whilst impoundments designed for hydroelectric generation or for irrigation purposes impose a very variable downstream flow regime, fluctuating from high to low flows depending on the demand for electricity or irrigation supply.

The magnitude of the discharge will be of considerable importance in controlling downstream river bed erosion because river channels tend towards a steady state morphology in which the channel is adjusted to the input of water and the sediment supplied by the drainage basin (Petts, 1978). Since sedimentation in the reservoir results in a discharge of clear water below dams then, where the reservoir outflow has sufficient tractive force to initiate movement of sediment, considerable river bed degradation may occur (Mostafa, 1955; Leopold *et al.*, 1964; Komura and Simmons, 1967; Hales *et al.*, 1970; Buma and Day, 1977). Such degradation will be recorded when sediment load in the outfall of the reservoir is lower than the carrying capacity of the stream, but the magnitude of entrainment may not be easily predicted since it is unlikely to be a simple function of particle weight (Laronne and Carson, 1976; Reynolds, 1976).

Where river bed degradation occurs—and this can be to a depth of 1 m (Leopold *et al.*, 1964; Factorovitch, 1967; Makkaveyev, 1970)—the finer fractions of bed material are transported downstream and the degraded reaches nearest to the dam become armoured with coarse bed material (Komura and Simmons, 1967), changes which are clearly of importance to river biota. For example, downstream of the Hoover

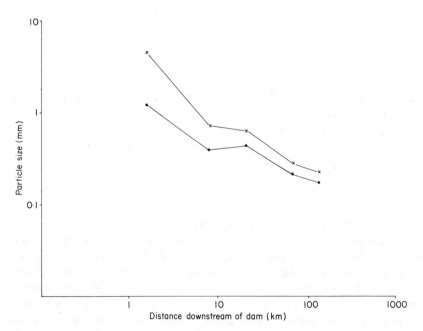

Fig. 2. Changes in river bed particle size downstream of the Hoover Dam, 1935–38. Data from Mostafa (1955). x——x 65% finer; ●——● 35% finer.

Dam on the Lower Colorado median particle size (d_{50}) after closure was approximately equal to the range d_{80}–d_{95} before closure and bed material was relatively finer at sites downstream (Fig. 2).

Hey (1975) considered the likely effects of flow regulating releases to the River Severn in mid-Wales and concluded that the magnitude of such discharges was unlikely to cause major changes in channel morphology and was therefore acceptable. However, such analyses are based upon bed load movement (i.e. the movement of coarse particles along the river bed) and it is probable that changes resulting from the redistribution of finer particles at lower flows are of ecological significance (Webster *et al.*, 1979).

Where discharges from dams are reduced below the threshold for sediment transport, and this is generally the case where reservoirs are built for direct supply purposes, the channel perimeter forms a natural protective armoured layer preventing erosion and channel morphology adjusts primarily to hydrologic regime (Petts, 1978). Gregory and Park (1974) reported a reduction of channel capacity downstream of Clatworthy Reservoir in the UK and this persisted for 11 km—a relatively local effect—until the catchment area contributing to the river was at least four times that draining the reservoir. Such changes result principally from reductions in peak flows and associated scour and channel capacity reductions ranging from 34–73% have been associated with a reduction in the magnitude of mean annual flood of 25–60% (Northrup, 1965; Gregory and Park, 1974; Gregory, 1976; Gregory and Park, 1976). Reductions in flow may also cause the deposition of tributary sediment loads within the main channel and this may be of local importance to the biota. *208635*

The wide variety of effects that reservoirs have upon streamflow has been summarized (Rutter and Engstrom, 1964) and related principally to the structural design and the use made of the reservoir. Moore (1969) estimated that specifically designed flood water retarding structures resulted in a 35–67% reduction in peak discharge, but all reservoirs perform some flood control function and Anderson (1975) reported that the construction of a dam on the R. Blythe in England caused a halving of the annual amplitude of discharge. Such changes are generally measured immediately downstream of the impoundment however and Bussell (1979), in a preliminary analysis of the flow regime of flow regulated rivers in the UK, concluded that, except for reaches of river immediately below upland reservoirs, the principal detectable effect in downstream parts of the catchments was that of increasing minimum flow. There was little evidence of flow regulation having changed the incidence or pattern of spates at lowland sites in either

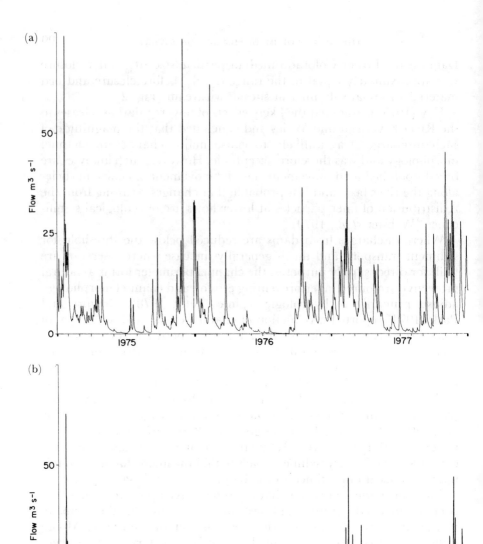

FIG. 3. Seasonal changes in flow in (a) the upland reaches of the naturally flowing River Wye, Wales and (b) the nearby impounded (for direct supply) tributary, the River Elan.

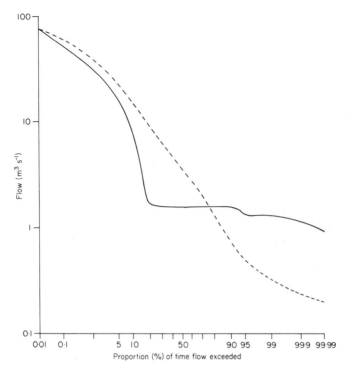

Fig. 4. Flow duration curve for the upland reaches of the naturally flowing River Wye, catchment area 174 km², (– – – –) and for the nearby impounded tributary, River Elan, catchment area 184 km², (———).

winter or summer. A comparison of flows in the naturally flowing upper R. Wye and immediately downstream of an impoundment in its tributary, the R. Elan, illustrates the substantial local changes in high and particularly low flows in the impounded system (Figs 3 and 4) but such effects were not easily detectable in lowland reaches.

Crisp (1977) compared the flow in the natural flowing Trout Beck with that of the nearby impounded R. Tees, which was regulated by the Cow Green Reservoir, constructed for augmentation of river abstractions, and concluded that, locally, there was a virtual elimination of daily minimum discharges of less than 0·1 times the annual mean compared with an occurrence of about 15–25% of such values in the unregulated tributary. In addition, daily maxima more than eight times the annual mean were not recorded and there was a considerable reduction in the frequency of values more than five times the annual mean in the regulated river.

In the Murray–Darling river system in Australia, which is regulated over a distance of 2200 km, the use of Hume Dam to provide water for irrigation reduced winter and spring flows and increased summer and autumn flows. Like the upper Wye catchment (Fig. 4) the flow duration curve was modified with a reduction in high flows and an increase in lower flows and overall the long term average flow was reduced to 78% of that before regulation (Walker, 1979).

Hydroelectric schemes may have the most dramatic short-term effects on flow (Krenkel *et al.*, 1979) and these fluctuations may be of considerable ecological significance. However, where reservoirs are used to divert water to hydroelectric sets they may lead to relatively constant or reduced downstream flows (Lillehammer and Saltveit, 1979). Comparison of flows in the adjacent Rivers Ystwyth (naturally flowing) and Rheidol (impounded for hydroelectric generation) indicates that natural flows in the Ystwyth show a smooth recession following floods while flow patterns in the impounded Rheidol are distorted by stored water being released through the turbines (Fig. 5). The overall effect of impoundment in this case is to lower the high flows and increase the low flows but with the occurrence of rapid increases in flows, peaks being achieved in 3–12 h.

D. WATER QUALITY

Changes in water quality downstream of reservoirs may result from two principal causes—the location of the basin in the catchment and

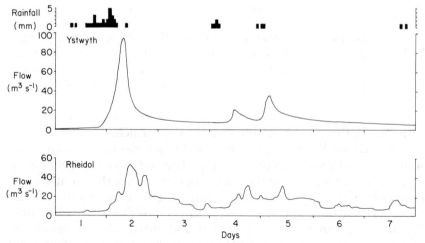

FIG 5. Conparison of flood events in the adjacent catchments of the River Ystwyth (natural flow) and the River Rheidol (impounded for hydroelectric generation). Data from J. T. Lewin, pers. comm.

the direct effects of storage on water quality. In the first case water quality changes imposed on downstream rivers by supplementation of flow from upland, generally nutrient-poor reservoirs, during periods of low natural flow and high solute concentrations are likely to be detectable even in lowland reaches. Using simulation techniques Edwards and Crisp (1980) concluded that modest flow regulation of the R. Wye in the UK during the summer would result in mean concentrations in lowland reaches being reduced by up to 44% (PO_4–P). In addition, simulated 90 percentile values were reduced as a result of flow regulation; by 53% (alkalinity), 42% (Cl), 89% (PO_4–P) and 51% (total hardness).

Changes in water quality resulting from storage result principally from the integrative effect of the collection of long-period stream flow, which will dampen oscillations in water quality, and *in situ* physiochemical and biological modifications of quality. For example, Crisp (1977) reported that fluctuations in ionic content of the outflow from Cow Green Reservoir in the north of England were generally smaller than fluctuations in the reservoir inflow, illustrating the buffering effect of storage on water quality.

However, Beeton and Sikes (1978) found that the chemical quality of Skadar Lake was similar to that of its major source river indicating that the source supply was relatively stable in quality or that the buffering effect of reservoir storage was small.

The basin morphology of reservoirs is likely to influence considerably those processes leading to *in situ* changes in water quality, and the chemical and thermal changes resulting from impoundment are generally similar to those reported for natural lakes (Hutchinson, 1975; Wetzel, 1975). Such changes have been considered in some detail elsewhere (Neel, 1963; Ackermann *et al.*, 1973; Ridley and Steel, 1975; Baxter, 1977) and only those most pertinent to downstream river ecology are highlighted here.

Even short-term storage of water, pumped from rivers, can result in substantial changes in chemistry (Slack, 1978), and Toms *et al.* (1975) concluded that in Grafham Water, a shallow reservoir in the UK receiving water pumped from a nearby river, there were reductions in the concentrations of the major plant nutrients NO_3–N, PO_4–P and dissolved silica and in calcium concentration. Water passing through the Bighorn Lake on Bighorn River underwent considerable reduction in concentrations of most dissolved constituents (Soltero *et al.*, 1973) and the nutrient concentrations and ionic composition of the Zambezi appear to have been altered by both the Kariba and Kafue Gorge Dams (Davies, 1979).

In hot climates certain changes in water quality sometimes result from evaporation processes which concentrate salts in reservoirs and the resulting high salinity is likely to influence both the downstream biota and crops supported by irrigation. However, the major impact of water quality on downstream reaches is likely to result from the development of thermal and chemical stratification in relatively deep reservoirs, the hypolimnion (Hutchinson, 1975) displaying reductions in dissolved oxygen and temperature which are often associated with increases in concentrations of iron, manganese and hydrogen sulphide (Pfitzer, 1954; Wunderlich and Elder, 1967; Krenkel and Parker, 1969; Burns, 1971; Tyler and Buckley, 1974; Walker, 1979). Where stratification occurs the depth from which water is withdrawn is of critical importance in determining the temperature and water quality of water discharged downstream of the impoundment. Even in relatively shallow (max. depth 40 m) and exposed reservoirs, like Caban Coch Reservoir in mid-Wales, stratification may occur and differences of up to 11°C have been recorded between surface reservoir water and water drawn from a depth of about 25 m (Thompson, 1954; Hopper, 1978) (Fig. 6a). In Fontana Reservoir, Tennessee Valley the temperature at 90 m was about 6°C compared with about 20°C at the surface.

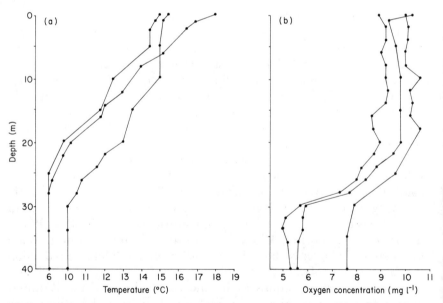

Fig. 6. Profiles of (a) temperature and (b) oxygen in Caban Coch Reservoir on three occasions during July–August 1977. Data from Hopper (1978).

Ward and Stanford (1979a) recognized six categories of thermal modification downstream of impoundments:
(1) Increased diurnal constancy.
(2) Increased seasonal constancy.
(3) Summer depression.
(4) Summer elevation.
(5) Winter elevation.
(6) Thermal pattern change.

Often mean annual stream temperatures are not greatly modified by impoundment (Jaske and Goebel, 1967; Lavis and Smith, 1972; Ward, 1976a): nevertheless changes in thermal patterns, perhaps seasonal amplitude or time of maximum temperature, may be of considerable importance to the biota. The annual range of temperature 5 km downstream from a Japanese reservoir was reduced from 21°C to 12°C (Nishizawa and Yamabe, 1970) reflecting a reduction in the amplitude of temperature changes. Lavis and Smith (1972) reported that maximum stream temperatures below an upland reservoir in Britain were depressed by as much as 12°C in summer.

Generally where impoundments are used to supplement low natural flows in temperate climates water temperatures are consistently colder in summer and warmer in winter than unregulated streams (see Fig. 7) (Neel, 1963; Wright, 1967; Pearson et al., 1968; Hilsenhoff, 1971; Lehmkuhl, 1972). In Australia Hume Dam reduced the annual temperature range in the Murray River by about 6°C, changing from 7–24°C to 10–21°C, and the impoundment delayed seasonal changes in temperature, trends in spring, summer and autumn being retarded by about one month (Walker, 1979).

Severe thermal fluctuations over short periods, resulting from power generating releases, have been recorded by Pfitzer (1967), summer temperatures rapidly changing by 6–8°C below deep release dams as power releases were required. Changes in outlet valves may also produce rapid thermal changes below reservoirs.

Releases of water from the hypolimnion as well as influencing thermal regime may be substantially deoxygenated and rich in manganese and iron and minerals (Fig. 6b). Nevertheless, thermal and chemical effects are generally restricted to those reaches of river immediately downstream of the impoundment. In hot climates water temperatures equilibrate rapidly with air temperatures (Coche, 1968; Balon and Coche, 1974) and this is generally the case in temperate climates where the water passes through rapidly-flowing, turbulent reaches. In the upper R. Wye catchment, mid-Wales, the release of "compensation" water from a direct supply reservoir on the R. Elan, a tributary of the

M. P. BROOKER

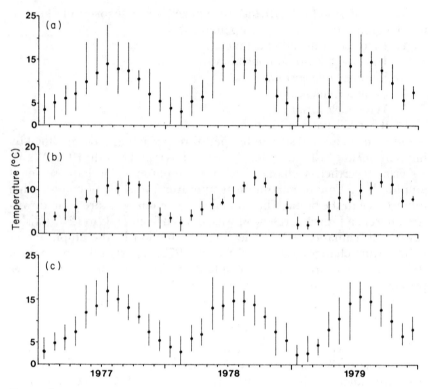

Fig. 7. Seasonal changes in monthly mean (with ranges) water temperatures (a) in the naturally flowing River Wye in upland Wales and (b) immediately downstream of the dam in a nearby impounded tributary, River Elan and (c) 11 km downstream of the confluence of the rivers and about 20 km from the impoundment.

Wye, had little effect on the temperature regime of the main river, about 20 km downstream of the impoundment (Fig. 7). However, Ward (1974) noted the effects of a deep release reservoir 8·5 km below the dam, and Walker (1979) concluded that the effects of an impoundment on river thermal regime were still apparent 200 km downstream.

With respect to oxygen, Edwards and Crisp (1980) concluded that, in typical shallow and fast upland rivers, the magnitude of mass-transfer coefficients was sufficiently large that rapid equilibration of de-oxygenated water, derived from the hypolimnion of reservoirs, would occur and any effects would be local. However, reaeration of water discharged from Cherokee Dam in the Tennessee Valley was extremely slow and oxygen concentrations increased from less than 1 mg l^{-1} (10%

saturation with respect to air), at the dam outlet, to only about 3 mg 1^{-1} (35%) 16 km downstream and about 6 mg 1^{-1} (70%) 70 km downstream (Fig. 8). This contrasted sharply with water in the Little Tennessee River downstream of Calderwood Dam where oxygen concentrations increased from 65% air saturation at the dam to 93% air saturation within 6 km.

Such physico-chemical changes depend, therefore, upon limnological phenomena occurring in the reservoir, the location of the dam outlets and the nature of the river channel downstream of the impoundment. The use of multilevel draw-offs is now widely practised to minimize the release of poor quality water and, in addition, consider-

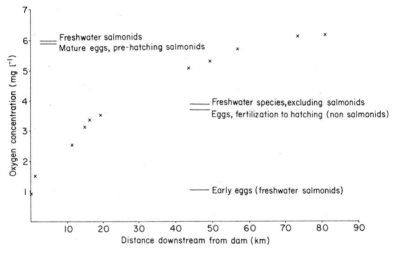

Fig. 8. Changes in dissolved oxygen concentration downstream of Cherokee Dam, Tennessee Valley. Data from Pfitzer (1954). Recommended thresholds (x) for fish survival from Davis (1975).

able effort has been directed at developing engineering procedures to cause destratification in reservoirs (Ridley *et al.*, 1966; Symons *et al.*, 1967; Steichen *et al.*, 1979): Middleton (1967) concluded that the control of the temperature of water discharged from multipurpose reservoirs is now an accepted goal. However, other chemical changes may occur where water passes over high dams into a deep pool, the entrainment of air sometimes resulting in supersaturation of oxygen and nitrogen (Ebel, 1969; Beiningen and Ebel, 1970), and this may have local effects on biota.

III. Effects of Impoundments on Downstream Fisheries

A. BARRIERS TO MOVEMENT

As early as 1215 the Magna Charta demanded that obstructions to fish movements in rivers and streams should not be constructed: such advice has been comprehensively disregarded and dams now either form complete barriers to the upstream or downstream movement of fish, delay their passage or damage them as they pass through the structure (Brett, 1957; McGrath, 1958; Royal and Cooper, 1958; Somme, 1958; Zalumi, 1970; Mills and Shackley, 1971). Clearly where free passage is required for anadromous fish to pass upstream from the sea to spawn, or local movements are a prerequisite of spawning behaviour, the effects on fisheries will be apparent both upstream and downstream of impoundments. Holden (1979) lists a large number of fish species, e.g. Pacific salmon (*Oncorhyncus* sp,), Atlantic salmon (*Salmo salar*), shad (*Alosa* sp.) and bass (*Morone* sp.) which are likely to be affected in this way and, in general, it is the anadromous salmonid fish which have received the most attention because of their commercial importance.

In addition to limiting fish movement, the flooding of considerable areas suitable for spawning by fish, such as salmon, upstream of impoundments leads to a substantial loss of spawning potential (Hourston, 1958). It has been estimated recently that the Columbia River system once supported runs of anadromous salmon and trout into 80% of the drainage basin but currently, because of a series of mainstream dams, less than 30% of the original habitat remains accessible to spawners (Turner, 1971; Stanford and Ward, 1979).

There are several ways in which the upstream and downstream movement of anadromous fish may be aided and these have been reviewed extensively elsewhere (Dill and Kesteven, 1960; Andrew and Geen, 1960; Bell, in press). However, where runs of salmon are not large enough to be regarded as commercially important they may not warrant the provision of adequate passage facilities and this may lead to a substantial decline of populations (Elder, 1964).

Such facilities range from a simple notch cut in the crest of a weir to several types of sophisticated fishways, fishlocks and fishlifts: Rogers and Cane (1979) reported that salmon successfully negotiated a diversion tunnel (2·2 km long) designed for such purposes as part of a pumped storage scheme in Wales. Generally the establishment of facilities to ensure the passage of adult salmon over low dams in small rivers is quite feasible but high dams cause substantial problems (Larkin, 1971) and certain structures, like the Grand Coulee and John Day Dams on the Columbia River, may contain no fish pass facilities (Stanford and Ward, 1979). In such cases alternative methods must be sought and these often involve the physical transportation of adults or artificially propagated fry around dams in order to sustain the fishery.

The largest river flow-regulating scheme in the UK is that on the R. Dee where 15% of the total catchment is regulated by Llyn Tegid, Llyn Celyn and Llyn Brenig (Iremonger, 1970; Edwards and Crisp, 1980). When Llyn Celyn was constructed about 6 km of the spawning area of Atlantic Salmon was flooded and access to a further 10 km of river was prevented by the dam. Such losses were to be made good by trapping and stripping fish immediately below the dam, supplying $0.5–1.0 \times 10^6$ salmon eggs each year for hatching and artificial propagation, but the success of such procedures was never established (Iremonger, 1970). In this case there has been a significant reduction in the number of fish caught in the trap below the dam after closure and it has been suggested that adults may have been moving from the main river to tributaries.

Alternative methods for ameliorating the effects of dams on fish movement involve "trapping and trucking" adult fish and such methods have been used in the UK (Howells and Jones, 1972) and in North America (Andrew and Geen, 1960; Geen, 1975). Norris et al. (1960) have reviewed the history of fish transportation and describe the practices available for maintaining the condition of fish in transit. Such methods as these are, of course, impracticable when large numbers of fish need to be moved. It has been estimated (Andrew and Geen, 1960; Geen, 1975) that the minimum numbers of sockeye salmon (Oncorhynchus nerka) requiring transportation on the Fraser River if a proposed (Moran) dam were built would be 0.75×10^6 per day with a daily maximum of 1.5×10^6. Fishways on the Bonneville Dam on the Columbia River have only passed 0.58×10^6 salmon in a whole year with a maximum of only about 0.05×10^6 per day (Andrew and Geen, 1960) and it is clear that fish passes and the various methods of "trapping and trucking" seriously restrict the full run of fish past dams to their spawning locations.

In addition, delays in migration caused by barriers may be critical to the spawning success of certain species, e.g. sockeye salmon, where energy resources are limited (Clemens, 1958). Schoning and Johnson (1956) recorded that the migration of chinook salmon (O. tshawytscha) may be delayed at dams and reservoirs by up to three days but Raleigh and Ebel (1967), who released sonically tagged adult chinook salmon in Brownlee Reservoir and in upstream tributaries, found no significant differences in the numbers caught on the spawning beds and concluded that the passage through the reservoir caused little delay.

Although the effect of flows on fish are considered later (see p. 111), it is important to highlight the necessity for maximizing the movement of fish through fish passes to avoid delay. In many instances, because the amount of flow available for attracting, for example, salmon to fish pass facilities may only be a small proportion of the total streamflow the fish may attempt to pass upstream through a spillway or turbine discharge. Telemetry has been used to assess the optimum conditions for the entry of salmon to a fish ladder on the R. Tuloma (Poddubnyi et al., 1971).

Dill and Kesteven (1960) concluded that there was no device for conveying downstream migrants, principally juveniles of anadromous species, over high dams and considerable losses of fish may result from such movements which are essential for the completion of the life cycle. Detrimental effects on migratory fish result from two causes, delays during passage through the impoundment and associated predation and direct mortality or injury inflicted during passage over the barrier. Geen (1975) concluded that young salmon were likely to be delayed by 11–25 days in the proposed Moran Reservoir during downstream migration and Raymond (1969) reported that, in the Columbia River, chinook salmon (O. tshawytscha) moved only about a third as fast through impoundments as through free flowing stretches of river. Such delays are considered to lead to substantial losses from predation by birds and other fish (Pyefinch, 1966).

In Scotland Mills and Shackley (1971) captured over 10 000 juvenile Atlantic salmon (S. salar) (smolts) in traps in the R. Bran upstream of hydroelectric reservoirs, where heavy mortalities were known to occur, and transported them below the impoundments with a total mortality of only 0·2%. Howells and Jones (1972) have also reported the use of smolt traps in Wales to capture downstream migrants for "trucking" downstream of a reservoir, but such procedures are not regarded as being very successful on a large scale (Geen, 1975).

Any damage inflicted on juvenile salmon moving downstream through turbines in hydroelectric sets must also influence their survival downstream of the reservoir. Calderwood (1945) subjected young trout (Salmo trutta) to pressures in excess of those generated by turbines in a hydroelectric set, and concluded that salmon smolts (S. salar) passing through the turbines were unlikely to be damaged. However, Pyefinch

(1966) reported that, in Scotland, 10–20% of salmon smolts were damaged as they passed through turbines and Cramer and Oligher (1964) reported that turbine mortality may vary from 9–55%. Evidence for large mortalities has been found by other workers (e.g. Ellis, 1956).

Fish can be prevented from entering turbine intakes by the provision of screens but these require regular cleaning, any blockage causing loss of generating head. Bates and Visonhaler (1957) reported success in guiding downstream migrants past intakes using louvre screens.

B. RIVER FLOW

Depending on the use of an impoundment and the nature of the river channel, changes in downstream river flow will alter a variety of, often interrelated, physico-chemical factors. Changes in flow result in changes in velocity, depth and wetted area of river and these together with some of their consequences, such as transport of food and dissolved oxygen, are important in governing the survival and distribution of fish species. Salmonid species, for example, have morphological adaptations for survival in fast-flowing, high gradient rivers whilst coarse fish species are better adapted to slow-flowing river systems.

Many workers (Hartman, 1963; Gibson, 1966; Everest and Chapman, 1972; Bustard and Narver, 1975a, b) have identified specific differences in flows utilized by salmonids. Symons and Heland (1978) concluded that, on the basis of 20 years of data from the Northwest Miramichi River, the highest densities of underyearling and yearling Atlantic salmon were recorded at water velocities of 50–60 cm s^{-1} and that different age classes had distinct depth and substratum preferenda.

Similarly, Chambers (1956) reported preferenda for spawning salmonids, in this case Pacific salmon (Table III). Such surface selection of spawning sites may be related to the sub-surface velocity of

TABLE III. Spawning limits of Pacific salmon (*Oncorhynchus* spp) (from Chambers, 1956)

	Depth (cm)	Velocity (cm s^{-1})
Chinook (*O. tshawytscha*)		
Spring run	38– 69	46– 76
Autumn run	38–198	84–114
Coho (*O. kisutch*)	30– 46	8– 61
Sockeye (*O. nerka*)	23– 69	38– 76

water through gravels which is known to influence the egg survival
of salmonids (Wickett, 1954; Peters, 1962; Cooper, 1974) and any
changes in flow resulting from impoundment may therefore be of considerable importance in determining the development of fish eggs.
However, Brannon (1965) concluded that embryos of sockeye salmon
were unaffected by velocity but that survival was determined by the
degree of exposure to diffuse daylight and the magnitude of dissolved
oxygen concentrations—which, of course, may be related to velocity.

Any reduction in wetted channel area resulting from the restriction
of flow by impoundments could also severely limit those habitats most
intensively used for egg and juvenile development. The space requirements of Atlantic salmon have received considerable attention (Elson,
1975; Symons and Heland, 1978; Symons, 1979) and Symons (1979),
who distinguished nursery areas (habitat suitable for underyearlings)
from rearing areas (suitable for all ages up to smolting), concluded
that for the maximum production of smolts, 2 + years-old, nursery
areas need to comprise 18–35% of the total stream area utilized by
juveniles.

Natural changes in flow are known to be important in determining
the survival of some fish species and Elwood and Waters (1969) reported that brook trout (*Salvelinus fontinalis*) were almost eliminated
from a stream by four severe floods over a short period (1965–1966).
Other workers (Gangmark and Broad, 1956; Gangmark and Bakkala,
1960) concluded that flooding was the most important factor governing the mortality of salmon (*O. tshawytscha*) populations in freshwater:
50% of deposited eggs were washed out by floods and there were indirect losses resulting from deposition of sand and silt. Elliott (1976a)
found that water velocity also influenced the rate at which eggs of
salmonids, such as trout (*S. trutta*), were washed out of spawning sites.
Any stabilization of flow resulting from impoundment which particularly diminished the effects of flood events is, therefore, likely to be
beneficial to the survival of such populations.

Flow is inextricably linked, via water velocity, with substrate structure which may influence the viability of developing salmonid embryos
and alevins within gravels. In turn, substrate structure and surface
velocities determine inter-gravel flow which influences oxygen supply
rates and loss rates of products of metabolism such as ammonia (Hamor
and Garside, 1976).

For migratory, anadromous fish which spawn in freshwaters the
magnitude and pattern of flow changes in rivers may be important in
providing a stimulus for movement from estuaries to spawning areas.
The use of artificial freshets released from impoundments has been

successful in stimulating the ascent of salmon in some river systems (Calderwood, 1921; Hayes, 1953; Banks, 1969), but natural spates are generally more successful than artificial releases and fish are known to move without freshets or even during reduced flows. Iremonger (1970) concluded that in the Dee Atlantic salmon were reluctant to migrate upstream when confronted with artificial flows and Pyefinch (1966) reported that releases from a hydroelectric scheme in Scotland had little effect on the numbers of salmon caught on rod and line, a natural flood in a small tributary being much more successful. However, the response of fish to an artificial lure may not directly reflect their patterns of movement or activity and much further work needs to be carried out on the behaviour of fish in response to river flow changes. The use of ultrasonic tags to track fish migrations will enable better interpretation of the relationships between natural and artificial flow changes and movement.

Hellawell (1976) considered the possibility that the overall pattern of migration was constant in Atlantic salmon and only secondarily modified by flow. If this be the case identification of those periods when flow might be beneficial to movement would be invaluable to any management scheme designed to manipulate the migration of fish in freshwater.

Little attempt has been made to utilize actively impoundments in the management of downstream fisheries. Indeed, early impoundments were developed on a single purpose basis with little regard for other uses or for the damage inflicted on downstream resources. The impact on downstream fisheries varies, of course, depending on the type of flow pattern resulting from impoundment but Holden (1979) concluded that 60% of fish species on the endangered or threatened list in the USA were riverine and that dams were one of the major causes of decline.

Kroger (1973) noted that abrupt reductions in flow below Jackson Lake (Wyoming) stranded fish in the Snake river and Allan (1976) also identified this as a potential hazard where reservoirs are used for flow regulation. Bishop and Bell (1978) reported that a temporary stoppage of water flow below Tallowa Dam in New South Wales resulted in the stranding and mortality of a large number of the rare species Australian grayling (*Prototroctes maraena*). In Australia changes in flow regime in the Murray-Darling river system affected fish survival indirectly by substantially reducing flood events which, pervious to impoundment, had acted as a triggering mechanism for spawning and were essential to the subsequent survival of eggs and young (Walker, 1979).

However, such effects can be ameliorated: Mellquist (1979) reported that the ecological impact of reductions in flow downstream of high altitude dams was minimized by building weirs in the river channel and trout populations trebled in this stabilized environment. Lister and Walker (1966) conducted experiments on the production of Pacific salmon in natural and regulated flows: survival during four years of natural flow conditions varied from 5–17% for chum salmon (*O. keta*), being inversely related to peak daily flow during the incubation period, and from 24·5–25·2% during the two years of controlled flow. This supports Somme's (1958) conclusion that stocks of salmon and sea trout often increased, and fluctuations in stock decreased, in well regulated rivers. However, Havey (1974) found no significant difference in the mean standing crop of Atlantic salmon (*S. salar*) before and after regulation by a low-head water control dam which stabilized flow, with minimum flows 16 times greater than the lowest flows before construction. Similar results were reported by Lillehammer and Saltveit (1979) who found that, where the regulation of rivers for hydroelectric purposes resulted in relatively constant flows, there were no major changes in population size.

Dominy (1974), reviewing the status of the salmon stocks of the St John river in Canada, suggested that a reduction of about 40% resulted, in part, from hydroelectric developments. There is no doubt that such schemes are likely to have the greatest impact of all river regulation operations on fisheries, releases of water to support power peaks resulting in daily fluctuations in water level of up to 2 m (Holden, 1979). Plans to extend the use of the recently commissioned river regulating Kielder Dam in the north-east of England (Freeman, 1979) to include power generation in order to supply electricity to the national grid, deserve a substantially different ecological evaluation than that already undertaken to assess the effects of the original proposals for flow-regulation to supplement low natural flows for abstraction.

Gorman and Karr (1978) reported that stream habitat complexity was correlated with fish species diversity and there is considerable other evidence that the diversity and biomass of fish in river channels is influenced by the physical configuration of the channel. Any engineering works undertaken downstream of reservoirs to allow the release of substantial discharges of water are therefore likely to affect the fish community. Moyle (1976) found that the channelization of a stream in California reduced the overall biomass of rainbow trout (*Salmo gairdneri*) by more than 65%. Indeed changes in populations of brook trout (*Salvelinus fontinalis*) resulting from channel modifications were not manifested until three to six years after treatment (Hunt, 1976), emphasizing the need for long-term evaluation of ecological changes. In addition, the substantial potential for change in channel capacity and substratum below reservoirs (see p. 98) is likely to influence fisheries in a similar way.

TABLE IV.   Flows proposed by Baxter (1961) for Atlantic salmon (*S. salar*) in rivers and streams of England and Scotland

Month	Flow (% of ADF)	
	Smaller rivers	Larger rivers
October	15–12·5	15–12·5
November	25	15
December	25–12·5	15–10
January	12·5	10
February	12·5	10
March	20	15
April	25	20
May	25	20
June	25–20	20–15
July	20–15	15–12.5
August	15	15–12·5
September	15–12·5	15–12·5

Although in the past considerable attention has been paid to the development of fishery resources in new reservoir schemes (e.g. Lake Kariba, Lake Volta) little effort has been directed into assessing the probable downstream effects of changes in stream flow or to determining the magnitude or pattern of flows which should be maintained below dams (Fraser, 1972a, b, 1975).

The determination of stream flow downstream of impoundments for navigation, abstraction and industrial use is much simpler than setting flows for fish and fisheries, the requirements of which are complex. Initially, flows of water were unrelated to biological requirements and based entirely on water yield (see p. 92) but Baxter (1961) proposed variable compensation flows below dams based on the seasonal needs of fish in 15 rivers in the UK. He recommended flows in different reaches and sizes of rivers which would support a variety of requirements, such as spawning and egg development of Atlantic salmon and these were expressed in proportions of the "average daily flow" (ADF) (Table IV). In addition, be suggested that, for upstream movement in spring, salmon required flows of 30–50% and 70% of the ADF in the lower/middle reaches and upper reaches respectively and concluded that such releases of water from impoundments need not be longer than 18 h of which 12 h should be at full rate, being reduced gradually over the last 6 h. As a minimum flow during hot weather Baxter (1961) recommended 0·125 ADF.

Such pioneering recommendations were based on knowledge of the life history of the fish, an essential requirement in any operating proposals, and the experience of the researcher but have not been generally applicable to other river systems (Fraser, 1975). Fraser (1975) reviewed in detail the requirements of discharges for fluvial resources and considered the Montana method, described in Elser (1972), as the best of the simple methods for deriving acceptable stream flows for fish: 10% of the annual mean flow was recommended as a short-term "survival" flow and 30% and 60% of the annual mean flow was recommended for resident salmonids during October–March and April–September respectively.

Application of the concept of minimum acceptable flows for fishery resources has generally proved impractical, quantitative assessments of the flow requirements of fish generally having been confounded by other factors (Giger, 1973). However, over the last two decades, considerable effort has been devoted to a better understanding of the relationships between flow and fish populations. Stalnaker (1979) used multiple regression techniques to establish significant relationships between observed fish biomass and arbitrarily defined "usable area units" of stream habitat: such relationships are likely to be useful as management tools when changes in habitat occur.

Changes in the rate of operational discharges from impoundments need close scrutiny and are generally based upon subjective assessment of flow requirements of fish. In the R. Dee the operational strategy of releases from reservoirs was designed to ensure that the rate of change of the discharge to the stream channel was "ecologically acceptable" (Iremonger, 1970). The maximum allowable rate of reduction, about $22 \times 10^3 \, \mathrm{m}^3 \, \mathrm{h}^{-1}$, is achieved in uniformly spaced steps not greater than about $1·9 \times 10^3 \, \mathrm{m}^3 \, \mathrm{h}^{-1}$ (Dec.–Feb.) and during the remainder of the year, when there is a greater risk of stranding fish, the maximum rate of reduction is $3·8 \times 10^3 \, \mathrm{m}^3 \, \mathrm{h}^{-1}$ for lesser flows, in uniform steps not greater than $2·3 \times 10^3 \, \mathrm{m}^3$. To provide this protection for salmon required the design of the generating plant and outlet structures to accommodate a wide range of flows $(0·4 - - 2·9 \, \mathrm{m}^3 \, \mathrm{s}^{-1})$ under varying heads (Crann, 1968). It is suggested that these rules are approximate simulations of the rate of recession of a natural river after rain (Blezard et al., 1970) and, therefore, unlikely to influence adversely the distribution and survival of fish.

In addition to flow pattern changes in rivers other factors may be important in governing patterns of migratory behaviour. In particular, the quality of river water is of considerable significance in acting as an olfactory cue for the homing of fish (Hasler and Wisby, 1951; Hasler, 1954; Leggett, 1977) and such behaviour appears to be under genetic influence (Bams, 1976). The effect of water storage and the pattern of flow release from reservoirs on the downstream efficiency of homing and migratory stimuli is not known but any effects are likely to be greater when water is transferred between river catchments (MAFF, 1976).

C. WATER TEMPERATURE

The physiological and behavioural functions of fish, which are poikilo-thermic organisms, are related directly to temperature and any change in temperature immediately downstream of a reservoir may influence their distribution and survival.

Behavioural responses to temperature may be very specific and Bardach and Bjorklund (1957) have reported that fish are able to sense temperature changes as small as $0.05°C$, changes which are difficult to measure in the field and extremely small in relation to those changes likely to occur immediately downstream from a reservoir releasing water from the hypolimnion (see p. 104). Gibson (1973) found that juveniles (parr) of Atlantic salmon and brook trout in stream tanks sheltered at $10°C$ and disappeared into the substratum at $9°C$. Other workers have reported that temperature may be an important trigger to behaviour, either locally or in relation to extensive migrations (Allen, 1940; Saunders and Henderson, 1969; McCleave, 1978; Solomon, 1978), although Lantz (1970) suggested that fish are able to acclimatize readily to seasonal and minor changes, particularly increases, in temperature.

The development of salmonid eggs is temperature dependent and in temperate climates the higher winter temperatures downstream of impoundments compared with natural river waters may lead to enhanced development (Embody, 1934; Hayes et al., 1953; Peterson et al., 1977) although there is little field evidence. Generally salmonid eggs can tolerate a range $0-16°C$ but Peterson et al. (1977) concluded that for the initial phase of development optimum temperature was about $6°C$, deviations leading to abnormal development and both immediate and delayed mortality.

In contrast, the rate of development of coarse fish eggs may be lowered by the reduction of summer temperatures downstream of impoundments and in the UK recruitment success in such species has been shown to be related to temperature, lower summer temperatures reducing year class strength (Kipling and Frost, 1970; Mann, 1976).

The relationship between temperature, feeding rates, conversion efficiencies and growth rates are complex (Brocksen and Bugge, 1974; Brett and Shelbourne, 1975; Elliot, 1975a, b, 1976b; Wurtsbaugh and Davis, 1977) but are of considerable importance in understanding the effects of thermal change downstream of impoundments. It may be expected from studies of Elliott (1975a, b) that salmonids benefit from seasonal stability of temperature regime with optima, varying with species, between 13 and 16·5°C. The critical importance of temperature in accounting for population differences in growth rate has been suggested by recent work of Edwards et al. (1979). Edwards (1980) used temperature data from the upper Wye catchment (see Fig. 7) to model trout growth at various sites with different temperature regimes (Fig. 9). Such an approach is likely to be useful in predicting the effects of impoundment on fish growth and has already been used (Crisp, 1977)

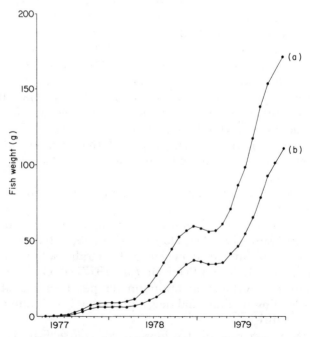

FIG. 9. Computer simulation, based on water temperature, of the growth of trout (*Salmo trutta*) in (a) the natural upper River Wye and (b) the River Elan immediately below an impoundment. Redrawn from Edwards (1980).

to assess the possible effects of changes in water temperature on trout growth in the R. Tees below Cow Green Reservoir.

Changes in temperature may also influence the type of diseases affecting fish and hence their survival (Snieszko, 1974) and some cognisance of this should be made when considering the likely effects of thermal changes downstream of reservoirs. For example, in salmonid fish of the Pacific coast of North America *Ctophaga psychrophilia* causes cold water disease at 5–10°C, temperatures often recorded in summer immediately downstream of reservoirs, but the disease is effectively controlled at temperatures of 10–15°C, closer to natural summer water temperatures.

It is clear that thermal changes in downstream rivers resulting from impoundments are likely to have considerable, although probably local (see p.105), effects on fisheries. Geen (1975) considered the overall effects of the proposed Moran Dam on the Fraser river and concluded that higher autumn temperatures would result in decreased spawning success of pink (*O. gorbusha*) and chum (*O. keta*) salmon because relatively high temperatures were associated with poor returning runs of fish—probably because increased temperature increased the rate of development of eggs and premature emergence of fry would result in seaward migration before adult marine food resources became available.

Parsons (1955) reported a decrease in the number of warm-water fish, generally common in the river system, below Dale Hollow Dam and the cold tail-waters, with an annual temperature range of about 8–15°C (monthly variation, about 2°C), were stocked successfully with rainbow trout (*Salmo gairdneri*) which provided an alternative sport fishery. Edwards (1978) found only 18 species of fish downstream of Canyon Reservoir (average temperature, 19·7°C; range 17–22°C) compared with 22 species in reaches upstream of the reservoir (average, 29·6°C; range 27–36°C) and attributed the differences to temperature conditions. Similarly the release of cold hypolimnion water from a flood control dam, which caused a lag of about four weeks in the spring rise of water temperature and resulted in a maximum summer temperature 7°C lower than upstream of the reservoir, was considered to be responsible for the absence of 4 warm-water species at the limit of their most northerly distribution (Spence and Hynes, 1971a).

There are many other reports of thermal changes governing fish distribution immediately downstream of reservoirs (Wiebe, 1960; Penaz *et al.*, 1968, Krenkel *et al.*, 1979; Zalumi, 1970; Cadwallader, 1978; Lloyd and Beaver, 1978; Walker, 1979). Generally, such temperature changes are likely to be local (Edwards and Crisp, 1980) but there are examples of more extensive effects: for example, Holden and Stalnaker (1975) concluded that the Colorado River below Glen Canyon Dam in the Grand Canyon remained too cold for most native fishes for over 400 km.

D. WATER CHEMISTRY

Except in a few exceptional cases the most dramatic effects of downstream changes in water chemistry resulting from impoundment have a relatively local impact. However, the construction of the Aswan High Dam on the Nile prevented the transport of substantial quantities of nutrients to the Eastern Mediterranean and the sardine catch in coastal waters is reported to have been reduced by 95% (Turner, 1971). Generally, effects are more restricted, principally resulting from the entrainment of gases spilling from high dams (see p. 107) or from the release of water from the hypolimnion (see p. 104).

Supersaturation of water with gases has been recorded immediately downstream of many high dams (Ebel, 1969; Beiningen and Ebel, 1970; Dominy, 1974; Holden, 1979) and may lead to gas bubble disease in fish, a subject reviewed by Rucker (1972) and, more recently, by Weitkamp and Katz (1977). The symptoms of the disease, similar to those of decompression-induced diver's bends (D'Aoust and Smith, 1974), include bubbles under the skin, in the fins, tail, mouth, behind the eyeballs and in the vascular system.

Gas bubbles disease has been reported to cause major mortality in young salmon and trout in the Pacific Northwest (Holden, 1979). For oxygen to cause gas bubble disease concentrations must be 350% saturated with respect to air but much lower concentrations of nitrogen (115–118%) will cause mortality (Rucker, 1972; D'Aoust and Smith, 1974). Beiningen and Ebel (1970) reported mortality of juvenile and adult salmon (*S. salar*) and steelhead trout (*S. gairdneri*) below the John Day Dam on the Columbia River, with heavy spillway discharge resulting in 123–143% supersaturation of nitrogen: such effects were probably enhanced by delays caused by the dam (see p. 109). Snieszko (1974) suggested that more than 50% of fish were lost to gas embolisms in the Columbia River and that where mortality was not caused directly, lesions resulting from gas blisters led to microbial infection.

Nebeker et al. (1978) reported that different life stages of steelhead trout had different susceptibilities to nitrogen supersaturation: eggs and embryos and newly hatched fry were unaffected at about 127% nitrogen, but at about 16 days post-hatch there was 99% mortality at 127% nitrogen and about 45% mortality at 115% of nitrogen. Chinook salmon, showing symptoms of distress at nitrogen concentrations of about 140% in the Columbia River below Grand Coulee Dam, were subjected to reduced temperatures or increased air pressure and the symptoms were relieved (Ebel, 1969).

There is no evidence that the passage of water through turbines increases concentrations of dissolved nitrogen and Beiningen and Ebel (1970) suggested that increasing flow through turbines, where available, rather than allowing substantial discharges of water over spillways, would help reduce supersaturation below dams: in addition, every possible effort should be made to reduce any delay in the passage of salmon over dams. Spillway deflectors, which prevent water from being carried deep into plunge basins, have also been suggested as a means of reducing supersaturated concentrations of gases (Ward and Stanford, 1979b).

In the UK problems of gas bubble disease resulting from supersaturation of gases below impoundments have not been reported. However, the height of dams in the UK is generally low (maximum Brenig Dam: 80 m) compared with the highest dams in the world (see Table I). Geen (1975) forecast that gas bubble disease would develop below the proposed high head hydroelectric Moran Dam on the Fraser River, developing a head of 215 m, if it was ever built and it is clear that such effects result from spillage over much higher dams than those in the UK.

Releases of water from the hypolimnion of dams may be low in concentrations of dissolved oxygen and rich in hydrogen sulphide and this may lead to direct mortalities of fish immediately downstream of impoundments—indeed the effects may be widespread where equilibration of gases in water and air is slow. Concentrations of dissolved oxygen below the Cherokee Lake in the Tennessee Valley resulted in oxygen concentrations below the recommended criteria for the survival of salmonids (Davis, 1975) for up to 40 km (see Fig. 8) and Pfitzer (1954) attributed the reduction of fish populations, in part, to reduced oxygen concentrations. Such effects are likely to be very local in upland rivers where turbulence enhances the rate of transfer of oxygen from the atmosphere.

Where the spawning beds of salmonids are downstream of hypolimnial releases the oxygen supply to the eggs and alevins may also be critical to fish survival. The full oxygen requirements of fish have been reviewed by Davis (1975) (see Fig. 8). For Atlantic salmon critical dissolved oxygen concentrations have been established for different life stages—early eggs, 0·76 mg litre^{-1} (Lindroth, 1942); eyed ova, 3·1 mg litre^{-1} (Hayes et al., 1951); pre-hatching, 5·8 mg litre^{-1} (Lindroth,

1942) and hatching, 7·1–10·0 mg litre^{-1} (Lindroth, 1942; Hayes *et al.*, 1951). The sublethal effects of hypoxia on early stages are important since they may influence survival: these effects include, reduced growth rate and efficiency of yolk utilization (Johansen and Krough, 1914; Alderdice *et al.*, 1958; Shumway *et al.*, 1964; Garside, 1966; Hamor and Garside, 1976), delayed hatching, premature hatching, reduced size at hatching (Shumway *et al.*, 1964; Garside, 1966) and morphological changes (Alderdice *et al.*, 1958; Garside, 1959).

In extreme cases hydrogen sulphide may be generated in the hypolimnion. Wright (1967) reported that concentrations of hydrogen sulphide were sufficient to cause fish kills below Clark Canyon Reservoir on Beaverhead River in Montana and Begg (1973), cited in Davies (1979), reported that discharges from an impoundment on the Zambezi contained large quantities of hydrogen sulphide which had drastic effects on the biota.

Stratification of reservoirs may also lead to releases of high concentrations of dissolved iron and manganese from the hypolimnion and downstream precipitation of the oxidized salts may lead to silting of spawning beds or may affect fish directly. Lloyd and Beaver (1978) reported that releases of water from the hypolimnion of a Malaysian impoundment, rich in dissolved iron and manganese, led to staining of the river bed for 2–3 km downstream and may, in part, have seriously affected the fishery.

In some water resource developments impoundments may store water generally more acid than the remaining catchment (Truesdale and Taylor, 1978) or water from the hypolimnion may have a lower pH than the natural receiving water and this may influence fish survival (Daye and Garside, 1975; Menendez, 1976; Robinson *et al.*, 1976; Trojnar, 1977; Lloyd and Jordan, 1964). In some studies pH less than 3·5 was found to be lethal to salmonid eggs and adults and Menendez (1976) found that continual exposure to pH less than 6·5 considerably reduced the hatching success of eggs of brook trout. Trojnar (1977) reported that the hatching success of eggs of the same species, incubated at a range of pH between 4·6 and 8·0, varied from 76 to 91%.

Where rivers receive polluting discharges downstream of impoundments the retention of clean diluting water in the reservoir may considerably alter the survival of fish in downstream reaches. In Wales the development of a hydroelectric scheme on the Afon Rheidol endangered trout (*S. trutta*) and salmon (*S. salar*) populations by reducing the amount of diluting water from effluents, containing zinc and lead, from old mine workings (Treharne, 1963). Treatment facilities were required to reduce the amount of metal discharged to the river below the hydroelectric impoundments. Where rivers receive water from the hypolimnion of reservoirs which are low in oxygen the biological oxygen demand resulting from· industrial and domestic effluents may not be fully satisfied because of the limited oxygen sources available (Kittrell, 1959).

IV. Effects of Impoundments on Downstream Macroinvertebrates

A. GENERAL CONSIDERATIONS

Like fish, macroinvertebrates inhabiting flowing water are adapted morphologically, behaviourally and physiologically to their environment (Ambuhl, 1959, 1961, 1962; Edington, 1965, 1968; Philipson, 1969; Hynes, 1970) and therefore perturbations, as a result of impoundments, of those factors such as temperature, flow, substrate, vegetation, dissolved minerals, food and biotic reactions, will alter the composition and abundance of stream benthos.

With so many environmental attributes changing the identification of specific causal factors has generally proved difficult. Ward and Stanford (1979a) regarded temperature, flow and substrate, and their indirect and interacting effects, as the major factors controlling macroinvertebrate distribution and survival and the same authors have reviewed the general effects of stream regulation on macroinvertebrates

TABLE V.   A summary of the effects of river regulation by impoundment on the relative abundance of macroinvertebrates. From 23 studies reviewed by Ward and Stanford (1979a)

Taxon	Number of studies where effects on relative abundance recorded		
	Increased	Decreased	No change
Amphipoda	8	–	–
Isopoda	3	–	2
Mollusca	8	4	1
Turbellaria	3	2	1
Oligochaeta	6	–	1
Plecoptera	1	13	5
Ephemeroptera	3	12	7
Trichoptera	9	10	3
Coleoptera	4	8	–
Diptera	14	2	1

utilizing data from over 20 studies (Table V). Most (18) studies had low level (hypolimnetic) draw-offs and in nine of these there was an increase in standing crop downstream of the reservoir, in eight there was a decrease in standing crop and in one no change. In all three studies with releases of water from upper levels standing crop increased. In general, except for Plecoptera, Ephemeroptera and Coleoptera, most major taxa were relatively more abundant downstream of reservoirs than at control sites on nearby rivers or upstream of the impoundments (Table V). The response of the Trichoptera was the most variable.

Since the major factors likely to govern, directly or indirectly, the survival and distribution of macroinvertebrates downstream of reservoirs are likely to be flow, temperature and water chemistry it is convenient to consider each separately, accepting that any changes in community structure are likely to result from the interaction of several environmental factors.

B. RIVER FLOW

The effects of changes in flow regime may reflect changes in water velocity, depth and wetted area (Chutter, 1969) and these will depend upon channel configuration. However, many stream invertebrates are not exposed to appreciable water currents even though they reside in rapid streams (Jaag and Ambuhl, 1964) and, in many instances, the direct effects of flow are inseparable from the indirect effects resulting from changes in the substratum (see p. 98).

Badcock (1953) concluded that wooded streams with more stable substrate and lower water velocities supported higher densities of certain Ephemeroptera and other invertebrates such as *Polycentropus* and *Gammarus*. A similar fauna might be expected to develop in rivers where river flow is stabilized by impoundment. Rivers which are subject to variable, natural flows may have reduced benthic populations as a result of bed load movement (Petran and Kothe, 1978) and any stabilization of flow resulting from impoundment should prevent such effects.

Kraft (1972) and Curtis (1959) concluded that current velocity was generally the factor most affected below dams but time of travel studies at low to medium steady flows in the R. Wye using a chemical tracer have shown that whilst flow is related principally to average water velocity in the lowland catchment of the river, in the upper reaches changes in the cross-sectional wetted area are also substantial (Edwards and Crisp, 1980). Where cross-sectional area is affected, supplementation of low natural flows to support abstraction is likely to increase the area of shallow riffles, which generally support greater (Rabeni and Minshall, 1977) or similar (Logan, 1979) densities of macroinvertebrates than deeper pools and lead to greater overall productivity in the system. However, Trotsky and Gregory (1974) observed that riffle-pool sequences in the Kennebec River, recorded downstream of an impoundment at low flows (7 m³ s⁻¹), were engulfed at the highest discharges (170 m³ s⁻¹), and concluded that this might lead to a reduction in overall population numbers. The technique of canalizing rivers below reservoirs in order to provide capacity for maximum releases may also substantially reduce total macroinvertebrate biomass: Morris *et al.* (1968) reported that, although average standing crop per unit area was similar in canalized and uncanalized sections in the impounded Missouri River, there was a reduction of 67% in wetted area in the canalized section.

The most adverse effects of flow regulation are likely to result from substantial, intermittent flow variations periodically exposing large areas of channel bed (Neel, 1963). In the Snake river, Wyoming, irrigation demands produced violent fluctuations in the water level of the river, with flows being reduced from 2·8–0·3 m³ s⁻¹—equivalent to a water level fall of 0·3 m—in less than 5 min (Kroger, 1973). It was estimated, from samples of surface substratum, that such changes resulted in the destruction of about 3×10^9 macroinvertebrates in a 3 km length of river, although no cognizance was taken of the probable behavioural movement of organisms into the substratum (Coleman and Hynes, 1970; Poole and Stewart, 1976; Morris and Brooker, 1979). In particular, those organisms, such as the stonefly (*Pteronarcella badia*), which migrated to the shallow edges of the river prior to emergence were most frequently observed stranded.

In a detailed study by Fisher and Lavoy (1972), the distribution of macroinvertebrates on a bar 425 m long and 70 m wide and 10 km below Turner's Falls hydroelectric dam in Massachusetts were described in detail. The bar, which was subjected to substantial fluctuations in water level, was completely submerged under high flow and exposed during low flow, representing a "tidal" amplitude of about 1 m. Benthic communities were less diverse and had lower biomass, total numbers and numbers of taxa in zones subjected to drying out compared with those areas of river bed continuously exposed, although the zone exposed for only 13% of the time was generally similar to the flooded zone. Few insects, except Chironomidae (Diptera), were recorded in the "tidal" zone (see Table V).

Ward (1976b) reviewed the effects of flow on macroinvertebrates downstream of impoundments and concluded that, in general, flow constancy below impoundments results in enhanced numbers or bio-mass of benthic macroinvertebrates, even when short-term fluctuations are imposed on that type of flow regime. However, the composition of the benthos in receiving streams is often greatly altered. Diptera, particularly chironomids and simuliids, Oligochaeta, Amphipoda and Gastropoda are usually favoured below dams whilst Trichoptera and Ephemeroptera may be enhanced or reduced: Plecoptera are usually severely reduced by such conditions.

In Norway weirs are built to ameliorate the effects of rapid flow changes resulting from hydroelectric schemes (Mellquist, 1979; Raasted, 1979) and Simuliidae (Diptera) larvae are numerically dominant at the outlets of such structures, comprising 80–90% of peak densities of 160000 m^{-2}, whilst areas downstream of the weirs princi-pally support Chironomidae (Diptera) larvae, which comprise up to 85% of total density. Changes in the composition of invertebrate com-munities below reservoir outfalls, which frequently involve filter-feeders, and particularly hydropsychid trichopterans, have generally been at-tributed to the contribution of lake plankton as a food source (Briggs, 1948; Fremling, 1960; Müller, 1962; Oswood, 1979). Boon (1979), who studied *Amphisyche meridiana* downstream of a tropical reservoir, con-cluded that this hydropsychid was well adapted to both low and high flows resulting from the impoundment.

Boon (1978) undertook a pre-impoundment survey in the River Tyne, UK which supported earlier conclusions from laboratory ex-periments (Philipson and Moorhouse, 1974) that two species of the trichopteran genus, *Hydropsyche*, coexisted partly by exhibiting different flow preferences: therefore, their distribution may be influenced by the use of impoundments to support low natural river flows, higher velocity, cooler water being more favourable to *H. siltalai* than *H. pellucidula*.

Despite the general reduction of plecopteran populations below im-poundments, *Alloperla* spp., a small slender organism inhabiting the hyporheic zone, is able to survive rapid flow increases downstream of irrigation dams (Ward and Short, 1978) and hydroelectric power supply dams (Radford and Hartland-Rowe, 1971; Trotsky and Gregory, 1974). Certainly there is evidence, collected during natural floods, that various taxa may move downwards in response to high flows (Williams and Hynes, 1974) and, following high flows in the impounded Brazos River, Texas, Poole and Stewart (1976) recorded increased numbers of net-spinning trichopterans, *Cheumatopsyche* sp. and *Hydropsyche simulans*, and the mayfly, *Neochoroterpes mexicanus*, at depths of 10–30 cm.

Sprules (1947) reported that a seven-day flood, resulting from the

bursting of a beaver dam, had similar effects on benthic macroinvertebrates as natural high flood events in an unregulated river:
numbers of Ephemeroptera, Plecoptera and Chironomidae (Diptera)
were considerably reduced whilst Trichoptera and Simuliidae
(Diptera) were less seriously affected. The effects of natural floods or
artificial releases from reservoirs on the downstream displacement of
drifting macroinvertebrates (Waters, 1972) have not been widely
studied, most investigations describing the daily and seasonal fluctuations in drift density. Brooker and Hemsworth (1978) recorded increases in both the total number and also the density of drifting
benthic macroinvertebrates as a result of an artificial freshet from an
impoundment but it is not certain whether such increases would be
sustained over long periods (Fig. 10). Two types of effect were recorded; an immediate increase in the total numbers and density of the
tubicolous chironomid *Rheotanytarsus* spp., and, several hours after the
increase in flow and during darkness, a delayed increase in the numbers

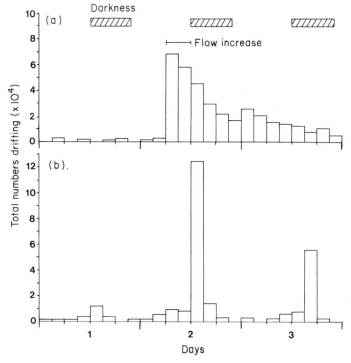

FIG. 10. The effect of a release of water ($3m^3s^{-1}$) from an impoundment about 20 km
upstream on the numbers of (a) *Rheotanytarsus* (Chironomidae) and (b) *Ephemerella
ignita* drifting in the River Wye. Natural river flow, about $0.7m^3s^{-1}$.

and density of drifting macroinvertebrates, principally the mayfly, *Ephemerella ignita*.

Increased drift of invertebrates in response to flow reduction has been observed in several rivers receiving releases of water for hydro-electric purposes (Pearson *et al.*, 1968; Radford and Hartland-Rowe, 1971) and as a result of dam closure Gore (1977) recorded massive drift at about 3.7 m^3 s^{-1} when flow fell from 7.0 m^3 s^{-1}. Since drifting macroinvertebrates are important in the diet of several species of fish (Kalleberg, 1958; Elliot, 1970; Mundie, 1974) any increase in avail-ability might be regarded as beneficial to fisheries, provided increased loss rates from the benthos did not lead to inadequate population re-plenishment. From the proportion of a population of invertebrates drifting—generally less than 0.5%—and assuming that such drifting organisms travel at the average velocity of water in the river channel it is possible to calculate net downstream population displacement (Hemsworth and Brooker, 1979). It would appear that most cohorts are displaced less than 10 km during a generation time in sub-stantially unregulated rivers and it seems unlikely that, with normal dispersive mechanisms, particularly of species with aerial flight stages, such displacement will lead to general depopulation.

Armitage (1977, 1978) recorded differences in the drift and benthos of the regulated R. Tees and the nearby unregulated Maize Beck (UK). He concluded that, as a result of increased stability of sub-stratum and vegetation in the regulated channel, there was an increase in Naididae (Oligochaeta) and the chironomids, Orthocladiinae (Diptera). However, general seasonal fluctuations in the drift were similar in the two rivers. In addition, the reservoir on the R. Tees was the source of large numbers of microcrustaceans and *Hydra* (Armitage and Capper, 1976) which were also collected in drift samples: in one year microcrustaceans represented up to 99% of both numbers and weight of the total drift fauna and this was reflected in the diet of fish below the impoundment (Crisp *et al.*, 1978).

The generation of organic particles, either phytoplankton or zoo-plankton, in reservoirs may lead to the development of high densities of macroinvertebrate filter-feeders below impoundments but the rate of removal of particles downstream may be rapid and their effect local. Several workers (Chandler, 1937; Cushing, 1963; Cowell, 1967; Elliott and Corlett, 1972; Ward, 1975) have studied the fate of zooplankton below reservoirs: Ward (1975) found that the downstream persistence of zooplankton was closely related to the magnitude of river flow and the size of the zooplankter, distance travelled being negatively cor-related with the species size. Chandler (1937) reported that plankton density 8 km from a lake outfall to the Huron River varied considerably from 12–16898 litre^{-1}, and in Sweden, Müller (1956) found that the influence of lake plankton was still significant 30 km from a lake outfall with a flow of 200 m^3 s^{-1}. Hopper (1978) studied zooplankton losses from the water column in an upland river downstream of a reservoir in Wales and found rapid rates of removal (Fig. 11). Only 12–32%, depending on species, of the total estimated biomass remained at the confluence with another river 6·5 km downstream from the dam. It is clear that rates of removal will depend on such factors as river flow, turbulence and the presence of vegetation in the channel (Chandler, 1937).

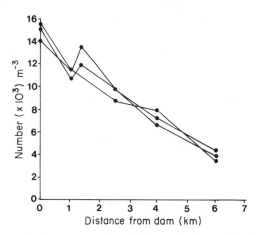

Fig. 11. Changes in the concentration of the zooplankter, *Ceriodaphnia*, downstream of an impoundment, Caban Coch Reservoir, in mid-Wales on three occasions during July–August. River flow, about 1·3 m^{3}s^{-1}. From Hopper (1978).

C. WATER TEMPERATURE

The development of many stages in the life cycles of macroinvertebrates is known to be temperature dependent (Macan, 1960; Elliott, 1972, 1978; Rupprecht, 1975) and the downstream effects of impoundments on temperature regime, which can be of six types (Ward and Stanford, 1979a, see p. 105) are likely to be of considerable importance in determining benthic communities even where the mean annual stream temperature may not be greatly modified (Jaske and Goebel, 1967; Lavis and Smith, 1972; Ward, 1976a).

Experimentation to date, generally restricted to the consideration of the effects of constant temperatures on development (e.g. Elliott, 1972), suggests that changes in temperature regime below impoundments may influence the survival and development of invertebrates (see Table VI). Ward and Stanford (1979a) concluded that the postulated link between diurnal constancy and low growth efficiency of lotic macroinvertebrates may have some validity and much more experimental study, reproducing natural or imposed fluctuations in temperature, is required before the effects of temperature pattern changes can be predicted satisfactorily.

Species requiring seasonal cycles of temperature to cue developmental stages may be eliminated below dams and invertebrates requiring cold winter diapause may be adversely affected by warmer winter temperatures below dams (Lehmkuhl, 1972). Some invertebrates, such as *Gammarus lacustris*, which are able to complete their life cycle under constant thermal conditions may be favoured (Ward and Stanford, 1979a).

TABLE VI. The time taken (days) for 50% of the eggs of two generations (April–May and July) of *Baetis rhodani* to hatch immediately below Caban Coch dam (A) in mid-Wales and a site about 30 km downstream (B) 1977. Based on developmental data from Elliott (1972)

Time of year	Average water temperature		Time for 50% hatching (days)	
	A	B	A	B
April–May	5·6	7·2	56	35
July	9·3	14·8	28	15

Hilsenhoff (1971) reported that *Simulium*, *Gammarus* and Chironominae replaced Baetidae, *Hydropsyche* and Elmidae downstream of a release from a hypolimnion and attributed this to changes in thermal regime. Similar changes have been reported from other rivers receiving cold water discharges with Amphipoda, Chironomidae and Oligochaeta generally replacing Coleoptera and Ephemeroptera (Blanz *et al.*, 1969; Brown *et al.*, 1967; Lehmkuhl, 1972).

Temperature range changes from 0·5–21·0°C to 2·0–12·5°C, following impoundment of Green River by Flaming Gorge Dam, resulted in the complete failure of a summer generation of *Baetis* (Ephemeroptera) and this was attributed to the low water temperatures and the high flows (Pearson *et al.*, 1968). However, evidence linking changes in natural temperature regimes with life-cycle patterns is equivocal. Nebeker

(1971) reported that natural stream temperatures, influenced by altitude, could cause delays in the emergence of adult stoneflies by up to six months and Macan and Maudsley (1966) showed that the onset of emergence in some Odonata was related to the first date on which certain temperatures were reached. In contrast, Gledhill (1960) concluded that the onset of insect emergence was not related to temperature in the Lake District.

Water temperature does influence the hatching of some insects (e.g. *Baetis rhodani*) from the egg stage (Elliott, 1972) and comparison of water temperatures immediately below a reservoir (Caban Coch) in mid-Wales and about 20 km downstream suggest that the time for 50% of the eggs of the two generations (April–May and June) of *B. rhodani* to hatch would have been considerably different at the two sites (Table VI). Unfortunately other factors influenced the distribution of this species in the impounded river below the dam and an empirical assessment of any differences resulting from water temperature was not possible (Brooker and Morris, 1980).

D. WATER CHEMISTRY

Changes in water quality downstream of reservoirs will reflect the original quality of the stored surface water and those changes occurring in the reservoir itself (see p. 102). Even modest regulation of river flow from reservoirs may have local or sometimes far-reaching effects on water quality which could influence the survival and distribution of macroinvertebrates.

Gross deoxygenation and the generation of hydrogen sulphide in the hypolimnion of an impoundment will severely affect invertebrates immediately downstream of deep release reservoirs (Davies, 1979) but generally equilibration will be rapid and the effects local. Iron and manganese, both soluble and associated with particles, are commonly discharged at relatively high concentrations from impoundments where there is a substantial oxygen deficit. Caban Coch Reservoir in mid-Wales had hypolimnetic oxygen concentrations above 5 mg litre^{-1} (see Fig. 6) but total iron concentrations sometimes exceed 0·7 mg litre^{-1} downstream of the dam considerably higher than concentrations of 0·07 mg litre^{-1} in natural streams of the catchment. Local settlement of the particulate phase, exacerbated by similar precipitates discharged from a water treatment works, forms deposits, associated with organic material, on the river bed and probably accounts for differences in the macroinvertebrate fauna of this river (Elan), downstream of the impoundment, compared with the nearby unregulated R.

Wye (Brooker and Morris, 1980). Thus, the Elan supports few Trichoptera, Ephemeroptera Coleoptera or they are absent completely, features of other rivers polluted by ferric hydroxide (Letterman and Mitsch, 1978). Such effects result, in part, from the relatively constant summer flows downstream of the impoundment (see Figs 3 and 4) and the use of artificial freshets of water to scour deposits from the bed of the river (Eustis and Hillen, 1954) is likely to improve considerably the quality of habitat for macroinvertebrates.

Where soft, nutrient-poor upland water is stored in order to support natural low flows, when ionic concentrations in river water are likely to be at their highest, considerable dilution and softening of the natural drainage may influence the distribution of macroinvertebrates, e.g. Amphipoda, Mollusca, which have a minimum requirement for chemicals such as calcium (Edwards et al., 1978; Edwards and Crisp, 1980). However, much more detailed study of the limiting requirements of such organisms is necessary.

V. Effects of Impoundments on Downstream Aquatic Plants

There are few studies (see review by Lowe, 1979) of the effects of flow regulation on aquatic plants although changes in flow regime and physical and chemical water quality resulting from impoundment are likely to be of considerable influence on such organisms.

Reservoirs may act as refuges for aquatic plants characteristically found in still waters but able to survive in flowing conditions and such refuges may be important in "seeding" downstream rivers, particularly if the impoundment stabilizes previously torrential reaches. In addition, the transfer of stored water between catchments, containing different species of plants, may also lead to changes in species composition of rivers (Holmes and Whitton, 1977).

Changes in current velocity and other physical and chemical characteristics are known to be of considerable importance in determining the growth and species composition of algae and macrophytes (McIntire, 1964; Peltier and Welch, 1969; Wong et al., 1978) and changes in such factors are likely to influence plant communities downstream of reservoirs. In addition, reductions or increases in the amplitude of river height will influence riparian vegetation for such plants have a characteristic vertical zonation (Merry, 1979).

The specific nature of changes in vegetation will depend upon the flow regime imposed upon the river by the impoundment. Hall and Pople (1968) recorded considerable increases in the growth of Potamogeton and Vallisneria below the Volta Dam and attributed this to a reduction of scouring resulting from lower flows. Increases in macro-phyte growths below impoundments have also been recorded by Ward (1974) and Hilsenhoff (1971) and other workers reported that stabilized flows resulted in heavy growths of algae (Pearson *et al.*, 1968; Spence and Hynes, 1971b; Ward, 1974; Armitage, 1976) and mosses (Jonsson and Sandelund, 1979). Such growths may considerably influence the distribution and abundance of macroinvertebrates by supporting sub-stantial numbers of invertebrates (Percival and Whitehead, 1929; Hynes, 1961; Minckley, 1963) or may support qualitatively different faunas from those on nearby gravel habitats (Whitcomb, 1963; Harrod, 1964). However, the destruction of algae, and their associated fauna, downstream of dams by periodic dessication has been observed (Powell, 1958; Radford and Hartland-Rowe, 1971; Kroger, 1973) although dense mats of vegetation may afford protection to other biota during periods of temporary dessication (Brusven *et al.*, 1974).

The scouring effects of irregular high flows below hydroelectric and irrigation dams may reduce the abundance of epilithic algae and higher plants (Ward and Short, 1978) although Nilsson (1978) was unable to identify major vegetational changes along the R. Umëalven, regulated for hydroelectric purposes. In addition, the hydrology of rivers developed for hydroelectric purposes in Sweden, resulting from the "stair-cases" of hydroelectric power plants and associated reservoirs, has resulted in a succession of riverine to terrestrial vegetation in those reaches now dry as a result of regulation.

Lowe (1979) has recently reviewed the major effects of stream regulation by reservoirs, which are principally related to flow, water temperature, turbidity and nutrient changes and these are summarized in Table VII.

TABLE VII.   A summary of the effects of stream regulation from deep-release reservoirs on lotic aquatic plants. From Lowe (1979)

Downstream environmental change	Effects on plants
Temperature—warmer winter cooler summer	Increased standing crop. Favours cool-water stenotherms like *Ulothrix* and *Hydrurus*.
Flow—stabilization	Increased standing crop of green filamentous algae, *Callitriche, Myriophyllum, Ranunculus, Sparganium, Potomogeton* and *Zannichellia*
Flow widely fluctuating (power and irrigation dams)	Decreased standing crop
Turbidity—decreased by reservoirs	Increased standing crop
Nutrients—increased	Increased standing crop. May favour species such as *Cladophora*

The displacement downstream of both attached and planktonic algae in rivers will be closely related to water velocity which may be influenced by releases from upstream impoundments. In addition, residence time, which is inversely proportional to average water velocity, will limit the potential growth of planktonic algae in the river. However, predictions of the effects of flow regulation on the kinetics of growth, based on residence time changes, are likely to be considerably influenced by other factors, such as the light-climate, temperature and nutrient concentrations, which are also rate limiting. Other aquatic plants which may provide surfaces for algal colonization and which create frictional resistance in river channels (Brooker *et al.*, 1978; Dawson, 1978)—hence determining retention time—will also be important. Baker and Baker (1979) considered the effects of temperature and current velocity on the concentration and photosynthetic activity of phytoplankton in the upper Mississippi and found that peaks of concentration of chlorophyll *a* were highest in a dry year with low flows (high retention times) compared with a wet year with high flows. In general, when other chemical and physical factors are not limiting, similar effects are likely to be seen in regulated rivers, supplementation of low natural flows reducing the residence time and the concentrations of algae.

VI. GENERAL CONCLUSIONS

(1) The major expansion in the construction of impoundments was initiated by the emergence of industrialized societies during the nineteenth century but the peak of dam building, for single and multi-purpose uses such as direct water supply, hydroelectricity and river flow regulation, occurred during the period 1945–1971. It has been estimated that about 10% of the world's streamflow is now regulated by reservoirs, greatest developments being in Africa and North America.

(2) The enormous potential for damage, particularly to human populations, of the stored water in impoundments has always been acknowledged, but the less dramatic implications of impoundments on the ecology of downstream areas were seriously neglected during the period of intensive dam building, the planning and execution of water resource developments concentrating generally on technical and social problems. In many instances, although major changes in ecology have obviously occurred downstream of impoundments, a full assessment of such effects will never be made since few or no base-line data describing water quality and biota before impoundment are available. A similar

lack of foresight may limit our interpretation of major inter-regional water transfers, many emanating from impoundments, if proper study programmes are not implemented at an early stage (Golubev and Biswas, 1979).

(3) The major downstream ecological effects of impoundments relate to changes in the aquatic environment resulting from three principal modifications (Fig. 12):

(i) The barrier to migrations, upstream and downstream, imposed by the dam and reservoir: this relates chiefly to fish.

(ii) Changes in the downstream water quality caused by storage of water in the reservoir.

(iii) Changes in the river flow regime downstream of the reservoir.

Clearly such changes, particularly (ii) and (iii), do not act upon the biota independently, the interaction of the different factors and the wide variety of reservoirs and their uses making it difficult to provide specific forecasts of the effects of impoundments on downstream ecology. However, the pathways of such effects can be highlighted from published studies and the general pattern of changes identified (Fig. 12).

(4) Where dams have formed barriers to the migration of fish, particularly the anadromous salmon, the effects have sometimes been dramatic, resulting in a substantial decline in populations. In some instances the provision of structures for the passage of fish, upstream and downstream, has ameliorated the problem but generally such facilities are not totally effective where runs of fish are substantial or dams are high.

(5) Reservoir storage both reduces major oscillations in water quality of stream sources and changes physical and chemical quality. Most changes result from stratification in the reservoir and when water is released from the hypolimnion substantial differences in water temperature and concentrations of, for example, dissolved oxygen have been recorded between the reservoir and natural receiving waters. Such effects, which are generally of a *local* nature because of the often rapid equilibration with atmospheric gases and temperature, are known to influence considerably the growth and survival of fishes and macroinvertebrates. Other changes in water quality—for example, softening of receiving waters—may be more extensive but are unlikely to have major ecological consequences. Clearly any appraisal of such downstream phenomena depends on a knowledge of the basin morphology and water dynamics of the reservoir, the location of the dam outlets and the nature of the downstream channel.

(6) The wide range of types and uses of impoundments ensure that the effects of flow modifications on downstream biota are diverse. In

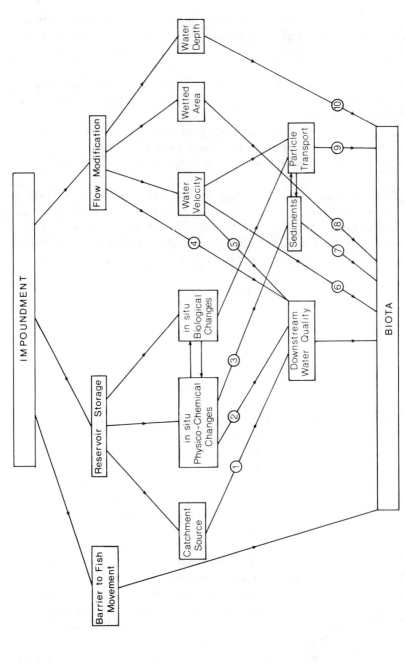

FIG. 12. A summary of the potential downstream ecological effects of impoundments. The numbers represent the types or targets of effects and generally can be illustrated by examples. Key— 1, e.g. dilution; 2, e.g. stratification; 3, e.g. sedimentation; 4, e.g. volume of discharge; 5, e.g. residence time; 6, e.g. fish behaviour; 7, e.g. fish spawning habitat; 8, e.g. changes in productive area; 9, e.g. food of invertebrates; 10, e.g. changes in riffle-pool ratio.

general, impoundments reduce the oscillations in flow recorded in natural rivers and provide a more stable environment. However, flow changes downstream of dams resulting from the generation of hydro-electricity may vary widely and rapidly and, particularly when they are associated with short-term changes in quality associated with fluctuations in drawn-off location, generally lead to the harshest conditions downstream of impoundments. Flow modifications may be manifested in changes in a wide variety of environmental characteristics, such as velocity, depth and width, all of which are of considerable importance in determining the survival and distribution of fish, macroinvertebrates and plants. The response of the biota to such modifications cannot be specifically forecast without a full knowledge of its present status and location and the proposed operating scheme in an impoundment.

(7) There has been little emphasis on the possible positive effects of impoundments on river systems, the recent preoccupation being with the possible detrimental effects of reservoir management—a natural reaction after a long period of neglect of such considerations. Stabilization of flow and water-quality, providing a more homogenous habitat, has proved beneficial to the growth of some plants, invertebrates and fish: salmonid fish spawning may also be more successful when major flood events are prevented. The possibility of using reservoirs as water banks for the stimulation of the upstream movement of fish, such as salmon, has received some assessment but the results are equivocal. The practical application of such management techniques is therefore currently limited by our knowledge and further experimentation is required before firm conclusions can be drawn.

(8) The current extent of the regulation of river flows by impoundments has now focused considerable attention on such systems and, although proper assessment of the full effects of impoundment may never be possible in some instances, detailed consideration of the pathways highlighted in Fig. 12 will allow forecasts of the effects of reservoir schemes for specific locations and lead to a better fundamental understanding of aquatic systems.

Acknowledgements

Much of the information utilized in this review was collected whilst I was contracted by the Department of Applied Biology, UWIST, Cardiff to undertake an ecological study of the R. Wye, Wales, under the sponsorship of the Welsh Water Authority with funds from the Craig Goch Joint Committee and Central Water Planning Unit, Department of the Environment. I an indebted to many people for help in completing this review and I apologize to those inadvertently omitted from the following. Drs R. Abel, J. Scullion and N. J. Milner allowed access to unpublished manuscripts and J. Beaver, Dr A. S. Gee, D. L. Morris and Dr M. D. Newson provided data and valuable discussion about the subject. Library facilities were provided by UWIST and the Welsh Water Authority and I thank Mrs S Majury particularly for help in obtaining information. Professor R. W. Edwards provided a wealth of ideas and information and his critical review of the draft text improved it considerably. The manuscript was typed with great patience by Mrs M. Thomas and Mrs V. Hughes and Mr P. Logan drew the figures.

References

Ackermann, W. C., White, G. F. and Worthington, E. B. (1973). Man-made lakes: their problems and environmental effects. *In* "Geophysical Monographs", Vol. 17.

Alderdice, D. F., Wickett, W. P. and Brett, J. R. (1958). Some effects of temporary exposure to low dissolved oxygen levels on Pacific salmon eggs. *J. Fish. Res. Bd. Can.* **15**, 229–249.

Allan, I. R. H. (1976). The implications of water transfers, with special reference to fish. *Chemy. Ind.* **4**, 136–138.

Allen, K. R. (1940). Studies on the biology of the early stages of the Salmon (*Salmo salar*) I. Growth in the River Eden. *J. Anim. Ecol.* **9**, 1–23.

Ambühl, H. von (1959). Die Bedentung der stromung als oekologischer Faktor. *Schweiz. z. Hydrol.* **21**, 133–164.

Ambühl, H. von (1961). Die Strömung als physiologischer und oekologischer Faktor. Experimentelle Untersuchungen an Bachtieren. *Verh. Internat. Verein Limnol.* **13**, 40–48.

Ambühl, H. von (1962). Die besonderheiten der Wasserstroemung in physikalischer, chemischer und biologischer Hinsicht. *Rev. Suisse Hydrol.* **24**, 367–382.

Anderson, M. G. (1975). Demodulation of stream-flow series. *J. Hydrol.* **26**, 115–121.

Andrew, F. J. and Geen, G. H. (1960). Sockeye and pink salmon production in relation to proposed dams in the Fraser River system. *International Pacific Salmon Fisheries Commission.* Bulletin XI.

Anon. (1978). Dam-busting fish endangers US environment act. *New Scientist* **78**, 646.

Armitage, P. D. (1976). A quantitative study of the invertebrate fauna of the River Tees below Cow Green Reservoir. *Freshwat. Biol.* **6**, 229–240.

Armitage, P. D. (1977). Invertebrate drift in the regulated River Tees and an unregulated tributary Maize Beck, below Cow Green Dam. *Freshwat. Biol.* **7**, 167–183.

Armitage, P. D. (1978). Downstream changes in the composition, numbers and biomass of bottom fauna in the Tees below Cow Green Reservoir and an unregulated tributary Maize Beck in the first five years after impoundment. *Hydrobiologia* **58**, 145–156.

Armitage, P. D. (1979). Stream regulation in Great Britain. *In* "The Ecology of Regulated Streams", (Ward, J. V. and Stanford, J. A. eds), pp. 165–181. Plenum Publishing Corporation, New York.

Armitage, P. D. and Capper, M. H. (1976). The numbers, biomass and transport downstream of micro-crustaceans and *Hydra* from Cow Green Reservoir (upper Teesdale). *Freshwat. Biol.* **6**, 425–432.

Badcock, R. M. (1953). Comparative studies in the populations of streams. *Rep. Inst. Freshwat. Res. Drottningholm* **35**, 38–50.

Baker, A. L. and Baker, K. K. (1979). Effects of temperature and current discharge on the concentration and photosynthetic activity of phytoplankton in the upper Mississippi River. *Freshwat. Biol.* **9**, 191–198.

Balon, E. K. and Coche, A. G. (1974). Lake Kariba, a man-made tropical eco-system in Central Africa. Monographiae Biologicae, 24 June, The Hague p. 767. Cited in Davies (1979).

Bams, R. A. (1976). Survival and propensity for homing as affected by presence or absence of locally adapted paternal genes in two transplanted populations of pink salmon (*Oncorhynchus gorbuscha*). *J. Fish. Res. Bd. Can.* **33**, 2716–2725.

Banks, J. W. (1969). A review of the literature on the upstream migration of adult salmonids. *J. Fish. Biol.* **1**, 85–136.

Bardach, J. E. and Bjorklund, R. G. (1957). The temperature sensitivity of some American freshwater fishes. *Amer. Nat.* **91**, 233–251.

Bates, J. M. and Visonhaler, R. (1957). Use of louvers for guiding fish. *Trans. Am. Fish. Soc.* **86**, 38–57.

Baxter, G. (1961). River utilization and the preservation of migratory fish life. *Proc. Inst. Civ. Eng.* **18**, 225–244.

Baxter, R. M. (1977). Environmental effects of dams and impoundments. *Ann. Rev. Ecol. Syst.* **8**, 255–283.

Beaumont, P. (1978). Man's impact on river systems: a world-wide view. *Area* **10**, 38–41.

Beckinsale, R. P. (1969). Human responses to river regimes. *In* "Water, Earth and Man" (Chorley, R. J. ed.), pp. 487–509. Methuen, London.

Beeton, A. M. and Sikes, S. (1978). Influence of aquatic macrophytes on the chemistry of Skadar Lake, Yugoslavia. *Verh. Internat. Verein Limnol.* **20**, 1055–1061.

Begg, G. W. (1973). The biological consequences of discharge above and below Kariba Dam. Commission Internationale des Grands Barrages, Eleventh Congress, Madrid, 421–430. Cited in Davies (1979).

Beiningen, K. T. and Ebel, W. J. (1970). Effect of John Day Dam on dissolved nitrogen concentrations and salmon in the Columbia River, 1968. *Trans. Am Fish. Soc.* **99**, 664–671.

Bishop, K. A. and Bell, J. D. (1978). Observations on the fish fauna below Tallowa Dam (Shoathaven River, New South Wales) during river flow stoppages. *Aust. J. mar. Freshwat. Res.* **29**, 543–549.

Blanz, R. E., Hoffman, C. E., Kilambi, R. V. and Liston, C. R. (1969). Benthic macroinvertebrates in cold tailwaters and natural streams in the State of Arkansas. *Proc. Am. Conf. S.E. Fish. Game Comm.* **23**, 281–292.

Blezard, N., Crann, H. H. Iremonger, D. J. and Jackson, E. (1970). Conservation of the environment by river regulation. *Association of River Authorities Annual Conference* 1970, 70–111.

Boon, P. J. (1978). The pre-impoundment distribution of certain Trichoptera larvae in the North Tyne river system (Northern England), with particular reference to current speed. *Hydrobiologia* **57**, 167–174.

Boon, P. J. (1979). Adaptive strategies of *Amphipsyche* larvae (Trichoptera: Hydropsychidae) downstream of a tropical impoundment. *In* "The Ecology of Regulated Streams" (Ward, J. V. and Stanford, J. A., eds), pp. 237–255. Plenum Publishing Corporation, New York.

Brannon, E. L. (1965). The influence of physical factors on the development and weight of sockeye salmon embryos and alevins. International Pacific Salmon Fisheries Commission, 24 pp.

Brett, J. R. (1957). Implications and assessments of environmental stress. *In* "The Investigation of Fish-Power Problems", (Larkin, P. A., ed.). pp. 69–820. Symposium, University of British Columbia, April 1957.

Brett, J. R. and Shelbourn, J. E. (1975). The growth rate of young sockeye salmon, *Oncorhynchus nerka*, in relation to fish size and ration level. *J. Fish. Res. Bd. Can.* **32**, 2103–2110.

Briggs, J. C. (1948). The quantitative effects of a dam upon the bottom fauna of a small California stream. *Trans. Am. Fish. Soc.* **78**, 70–81.

Brocksen, R. W. and Bugge, J. P. (1974). Preliminary investigations on the influence of temperature on food assimilation by rainbow trout *Salmo gairdneri* Richardson. *J. Fish. Biol.* **6**, 93–97.

Brooker, M. P. and Hemsworth, R. J. (1978). The effect of the release of an artificial discharge of water on invertebrate drift in the R. Wye, Wales. *Hydrobiologia* **59**, 155–163.

Brooker, M. P. and Morris, D. L. (1980). A survey of the macroinvertebrate riffle fauna of the R. Wye. *Freshwat. Biol.* **10**, 437–458.

Brooker, M. P., Morris, D. L. and Wilson, C. J. (1978). Plant-flow relationships in the R. Wye catchment. *Proc EWRS 5th Symp. on Aquatic Weeds*, 63–70.

Brown, C. B. (1944). Sedimentation in reservoirs. *Trans. Amer. Soc. Civ. Engrs.* **109**, 1085.

Brown, J. D., Liston, C. R. and Dennie, R. W. (1967). Some physiochemical and biological aspects of three cold tailwaters in Northern Arkansas. *Proc. Ann. Conf. S.E. Assoc. Game Comm.* **21**, 369–381.

Brusven, M. A., MacPhee, C. and Biggam, R. (1974). Effects of water fluctuations on benthic insects. *In* "Anatomy of a River", pp. 67–79. Pacific Northwest River Basins Commission Report, Vancouver, B.C.

Buma, P. G. and Day, J. C. (1977). Channel morphology below reservoir storage projects. *Environmental Conservation* **4**, 279–284.

Burns, J. W. (1971). The carrying capacity of juvenile salmonids in some northern California streams. *Calif. Fish and Game* **57**, 44–57.

Bussell, R. B. (1979). Changes of river regime resulting from regulation which may affect ecology. A preliminary approach to the problem. Central Water Planning Unit, Reading, UK.

Bustard, D. R. and Narver, D. W. (1975a). Aspects of the winter ecology of juvenile coho salmon *Oncorhynchus kisutch* and steelhead trout (*Salmo gairdneri*). *J. Fish. Res. Bd. Can.* **32**, 667–680.

Bustard, D. R. and Narver, D. W. (1975b). Preferences of juvenile coho salmon (*Oncorhynchus kisutch*) and cutthroat trout (*Salmo clarki*) relative to simulated alteration of winter habitat. *J. Fish. Res. Bd. Can.* **32**, 681–687.

Cadwallader, P. L. (1978). Some causes of the decline in range and abundance of native fish in the Murray-Darling River system. *Proc. Roy. Soc. Vict.* **90**, 221–224.

Calderwood, W. L. (1921). "The Salmon Rivers and Lochs of Scotland." London. Cited in Huntsman (1945).

Calderwood, W. L. (1945). Passage of smolts through turbines. *Salm. Trout Mag.* **115**, 214–221.

Chambers, J. S. (1956). Research relating to study of spawning grounds in natural areas. *In* "Annual Report to U.S. Army Corps of Engineers, North Pacific Division (Fisheries Engineering Research Program)." Washington Department of Fisheries. Cited in Fraser (1975).

Chandler, D. C. (1937). Fate of typical lake plankton in streams. *Ecol. Monogr.* **7**, 447–479.

Chutter, F. M. (1969). The distribution of some stream invertebrates in relation to current speed. *Int. Rev. ges Hydrobiol.* **54**, 413–422.

Clemens, W. A. (1958). The Fraser River salmon in relation to potential power development. *In* "The Investigation of Fish-Power Problems", (Larkin, P. A., ed.), pp. 3–10. Symposium, University of British Columbia, April 1957.

Coche, A. G. (1968). Description of physico-chemical aspects of Lake Kariba, an impoundment, in Zambia-Rhodesia. *Fish. Res. Bull. Zambia* **5**, 200–267. Cited in Davies (1979).

Coleman, M. J. and Hynes, H. B. N. (1970). The vertical distribution of the invertebrate fauna in the bed of a stream. *Limnol. Oceanogr.* **15**, 31–40.

Cooper, A. C. (1974). Physical aspects of sediment in relation to survival of salmon eggs. *In* "Symposium on Stream Ecology," pp. 1–13. Cent. Cont. Educ. Univ. B. C. FP 2407.

Cowell, B. (1967). The Copepoda and Cladocera of a Missouri river reservoir: a comparison of sampling in the reservoir and discharge. *Limnol. Oceanogr.* **12**, 125–136.

Cramer, F. K. and Oligher, R. C. (1964). Passing fish through hydraulic turbines. *Trans. Am. Fish. Soc.* **93**, 243–259.

Crann, H. H. (1968). Llyn Brenig. Part I: the concept and its promotion. *J. Inst. Wat. Engrs. Sci.* **32**, 279–287.

Crisp, D. T. (1977). Some physical and chemical effects of Cow Green (upper Teesdale) impoundment. *Freshwat. Biol.* **1**, 109–120.

Crisp, D. T., Mann, R. H. K. and McCormack, J. C. (1978). The effects of impoundment and regulation upon the stomach of fish at Cow Green, Upper Teesdale. *J. Fish. Biol.* **12**, 287–301.

Croome, R. L., Tyler, P. A., Walker, K. F. and Williams, W. D. (1976). A limnological survey of the River Murray in the Albury-Wodonga area. *Search* **7**, 14–17.

Curtis, B. (1959). Changes in a river's physical characteristics under substantial reductions in flow due to hydro-electric diversion. *Calif. Fish Game* **45**, 181–188.

Cushing, C. F. (1963). Filter feeding insect distribution and planktonic food in the Montreal river. *Trans. Am. Fish. Soc.* **92**, 216–219.

D'Aoust, B. G. and Smith, L. S. (1974). Bends in fish. *Comp. Biochem. Physiol.* **49A**, 311–321.

Davies, B. R. (1979). Stream regulation in Africa: a review. *In* "The Ecology of Regulated Streams" (Ward, J. V. and Stanford, J. A., eds), pp. 113–142. Plenum Publishing Corporation, New York.

Davis, J. C. (1975). Minimal dissolved oxygen requirements of aquatic life with emphasis on Canadian species: a review. *J. Fish. Res. Bd. Can.* **32**, 2295–2332.

Dawson, F. H. (1978). The seasonal effects of aquatic plant growth on the flow of water in a stream. *Proc. EWRS 5th Symp. on Aquatic Weeds*, 1978, 71–78.

Daye, P. G. and Garside, E. T. (1975). Lethal levels of pH for brook trout, *Salvelinus fontinalis* (Mitchill). *Can. J. Zool.* **53**, 639–641.

Dexter, B. D. (1978). Silviculture of the River Red Gum Forests of the Central Murray food plain. *Proc. Roy. Soc. Vict.* **90**, 175–191.

Dill, W. A. and Kesteven, G. L. (1960). Methods of minimizing the deleterious effects of water- and land-use practices on aquatic resources. *Proc. Int. Conf. Prot. Nat.*, **4**, 271–306.

Dominy, C. L. (1974). Recent changes in Atlantic salmon (*Salmo salar*) runs in the light of environmental changes in the Saint John River New Brunswick, Canada. *Biological Conservation* **5**, 105–113.

Ebel, W. J. (1969). Supersaturation of nitrogen in the Columbia River and its effects on salmon and steelhead trout. *U.S. Dept. of Interior Fishery Bull.* **68**, 1–11.

Edington, J. M. (1965). The effect of water flow on populations of net-spinning Trichoptera. *Verh. Internat. Verein Limnol.* **13**, 40–48.

Edington, J. M. (1968). Habitat preferences in net-spinning caddis larvae with special reference to the influence of water velocity. *J. Anim. Ecol.* **37**, 675–692.

Edwards, R. J. (1978). The effect of hypolimnion reservoir releases on fish distribution and species diversity. *Trans. Am. Fish. Soc.* **107**, 71–77.

Edwards, R. W. (1980). Predicting the environmental impact of a major reservoir development. Presented to the British Ecological Society Symposium on Environmental Impact Assessment. April 1980, University of East Anglia.

Edwards, R. W. and Crisp, D. T. (1980). Ecological implications of river regulation in the U.K. Presented to an International Workshop on Engineering Problems in the Management of Gravel Beds in Rivers. 23–28 June 1980, Gregynog, Newtown, Wales.

Edwards, R. W., Oborne, A. C., Brooker, M. P. and Sambrook, H. T. (1978). The behaviour and budgets of selected ions in the Wye catchment. *Verh. Internat. Verein. Limnol.* **20**, 1418–1422.

Edwards, R. W., Densem, J. W. and Russell, P. A. (1979). An assessment of the importance of temperature as a factor controlling the growth rate of brown trout in streams. *J. Anim. Ecol.* **48**, 501–507.

Elder, H. Y. (1964). Biological effects of water utilization by hydroelectric schemes in relation to fisheries, with special reference to Scotland. *Proc. R. Soy. Edin.* B, LXIX, 246–271.

Elliott, J. M. (1970). Diel changes in invertebrate drift and the food of trout (*Salmo trutta* L.). *J. Fish Biol.* **2**, 161–165.

Elliott, J. M. (1972). Effect of temperature on the time of hatching of *Baetis rhodani* (Ephemeroptera: Baetidae). *Oecologia* (Berl.) **9**, 47–51.

Elliott, J. M. (1975a). The growth rate of brown trout (*Salmo trutta* L.) fed on maximum rations. *J. Anim. Ecol.* **44**, 805–821.

Elliott, J. M. (1975b). Number of meals in a day, maximum weight of food consumed in a day and maximum rate of feeding for brown trout, *Salmo trutta* L. *Freshwat. Biol.* **5**, 287–303.

Elliott, J. M. (1976a). The downstream drifting of eggs of brown trout, *Salmo trutta* L. *J. Fish Biol.* **9**, 45–50.

Elliott, J. M. (1976b). The energetics of feeding, metabolism and growth of brown trout (*Salmo trutta* L.) in relation to body weight, water temperature and ration size. *J. Anim Ecol.* **45**, 923–948.

Elliott, J. M. (1978). Effect of temperature on the hatching time of eggs of *Ephemerella ignita* (Poda) (Ephemeroptera: Ephemerellidae). *Freshwat. Biol.* **8**, 51–58.

Elliott, J. M. and Corlett, J. (1972). The ecology of Morecambe Bay. IV. Invertebrate drift into and from the R. Leven. *J. appl. Ecol.* **9**, 195–206.

Ellis, C. H. (1956). Tests on hauling as a means of reducing downstream migrant salmon mortalities on the Columbia River. *Fish. Res. Pap. St. Wash.* **1**, 46–48. Cited in Elder (1964).

Elser, A. A. (1972). A partial evaluation and application of the 'Montana Method' of determining stream flow requirements. *In* "Transactions and Proceedings of the Instream Flow Requirements Workshop," p. 3–8. Sponsored by Pacific Northwest Basins Commission, Vancouver, Washington.

Elson, P. F. (1975). Atlantic salmon river, smolt production and optimal spawning: an overview of natural production. *Int. Atl. Salmon Found. Spec. Publ. Ser.* **6**, 96–119.

Elwood, J. W. and Waters, T. F. (1969). The Foyle fisheries: new bases for rational management. Special Report of the Foyle Fisheries Commission, Londonderry, Northern Ireland.

Embody, G. C. (1934). Relation of temperature to the incubation periods of eggs of four species of trout. *Trans. Am. Fish. Soc.* **64**, 281–292.

Everest, F. H. and Chapman, D. W. (1972). Habitat selection and spatial interaction of juvenile chinook salmon and steelhead trout in two Idaho streams. *J. Fish. Res. Bd. Can.* **29**, 91–100.

Eustis, A. B. and Hillen, R. H. (1954). Stream sediment removal by controlled reservoir releases. *Progv. Fish Cult.* **16**, 30–35.

Ewer, D. W. (1966). Biological investigations on the Volta Lake, May 1964 to May 1965. *In* "Man-made Lakes" (Lowe-McConnell, R. H. ed.). Symposia of Institute of Biology No. 15, pp. 21–31. Academic Press, London and New York.

Faktorovitch M. E. (1967). Transformations de lits à l'aval des usines hydroelectriques de grande puissance de l'U.R.S.S. International Union of Geodesy and Geophysics. General Assembly of Bern, 23 September–7 October. Commission on Surface Waters Symposium on River Morphology. Publication No. 75, pp. 401–403.

Fisher, S. G. and Lavoy, A. (1972). Differences in littoral fauna due to fluctuating water levels below a hydroelectric dam. *J. Fish. Res. Bd. Can.* **29**, 1427–1476.

Fraser, J. C. (1972a). Regulated discharge and the stream environment. *In* "River Ecology and Man" (Oglesby, R. T., Carlson, C. A. and McCann, J. M. eds), pp. 262–285. Academic Press, New York and London.

Fraser, J. C. (1972b). Regulated stream discharge for fish and other aquatic resources–an annotated bibliography. FAO Fisheries Technical Paper No. 112 (FIRI/T112).

Fraser, J. C. (1975). Determining discharges for fluvial resources. FAO Fisheries Technical Paper No. 143 (FIRS/T143).

Freeman, L. (1979). Kielder electricity? *Water* **28**, 11–12.

Fremling, C. R. (1960). Biology and possible control of nuisance caddisflies of the upper Mississippi River. Agricultural and Home Economics Experimental Station, Iowa State University of Science and Technology *Res. Bull.* **483**, 856–879.

Gangmark, H. A. and Bakkala, R. G. (1960). A comparative study of unstable and stable (artificial channel) spawning streams for incubating king salmon at Mill Creek. *Calif. Fish Game* **46**, 151–164.

Gangmark, H. A. and Broad, R. D. (1956). Experimental hatching of king salmon in Mill Creek, a tributary of the Sacramento River. *Calif. Fish Game* **41**, 233–242.

Garside, E. T. (1959). Some effects of oxygen in relation to temperature on the development of lake trout embryos. *Can. J. Zool.* **37**, 689–698.

Garside, E. T. (1966). Effects of oxygen in relation to temperature on the development of embryos of brook trout and rainbow trout. *J. Fish. Res. Bd. Can.* **23**, 1037–1134.

Geen, G. H. (1975). Ecological consequences of the proposed Moran Dam on the Fraser River. *J. Fish. Res. Bd. Can.* **32**, 126–135.

Gibson, R. J. (1966). Some factors influencing the distributions of brook trout and young Atlantic salmon. *J. Fish. Res. Bd. Can.* **23**, 1977–1980.

Gibson, R. J. (1973). Interactions of juvenile Atlantic salmon (*Salmo salar* L.) and brook trout (*Salvelinus fontinalis* (Mitchill) *Int. Atl. Salmon Found. Spec. Publ. Ser.* **4**, 181–202.

Giger, R. D. (1973). Stream flow requirements of salmonids. Oregon Wildlife Comm., Federal Aid Progress Report, Project No. AFS-62-1.

Gledhill, T. (1960). The Ephemeroptera, Plecoptera and Trichoptera caught by emergence traps in two streams during 1958. *Hydrobiologia* **15**, 179–188.

Glymph, L. M. (1973). Sedimentation of reservoirs. *In* "Man-made lakes: their problems and environmental effects. Geophysical Monographs" Vol. 17, pp. 342–348. (Ackermann, W. C., White, G. F. and Worthington, E. B. eds). Am. Geophys. Union, Washington D. C.

Golubev. G. N. and Biswas, A. K. (eds) (1979). "Interregional Water Transfers" Vol. 6, Water Development, Supply and Management series. Pergamon Press, Oxford.

Gore, J. A. (1977). Reservoir manipulations and benthic invertebrates in a prairie river. *Hydrobiologia* **55**, 113–123.

Gorman, O. T. and Karr, J. R. (1978). Habitat structure and stream fish communities. *Ecology* **59**, 507–515.

Gregory, K. J. (1976). Drainage basin adjustments and man. *Geographica Polonica* **34**, 155–173.

Gregory, K. J. and Park, C. C. (1974). Adjustments of river channel capacity downstream from a reservoir. *Wat. Resour. Res.* **10**, 870–873.

Gregory, K. J. and Park, C. C. (1976). Stream channel morphology in north west Yorkshire. *Revue Geomorph. Dyn.* **25**, 63–72.

Hales, Z. L., Shindale, A. and Denson, K. H. (1970). River bed degradation prediction. *Water Resour. Res.* **6**, 549–556.

Hall, J. B. and Pople, W. (1968). Recent vegetational changes in the lower Volta River. *Ghana J. Sci.* **8**, 24–29.

Hall, A., Davies, B. R. and Valente, I. (1976). Cabora Bassa: some preliminary physico-chemical and zooplankton pre-impoundment survey results. *Hydrobiologia* **50**, 17–25.

Hamor, T. and Garside, E. T. (1976). Developmental rates of Atlantic salmon, *Salmo salar* L. in response to various levels of temperature, dissolved oxygen and water exchange. *Can. J. Zool.* **54**, 1912–1917.

Harrod, J. (1964). The distribution of invertebrates on submerged aquatic plants in a chalk stream. *J. Anim. Ecol.* **33**, 335–348.

Hartman, G. F. (1963). Observations on behaviour of juvenile brown trout in a stream aquarium during winter and spring. *J. Fish. Res. Bd. Can.* **20**, 769–787.

Hasler, A. D. (1954). Odour perception and orientation in fishes. *J. Fish. Bd. Can.* **11**, 107–129.

Hasler, A. D. and Wisby, W. J. (1951). Discrimination of stream odors by fishes and its relation to parent stream behaviour. *Am. Nat.* **65**, 223–238.

Havey, K. A. (1974). Effects of regulated flows on standing crops of juvenile salmon and other fishes at Barrows stream, Maine. *Trans. Amer. Fish. Soc.* **103**, 1–9.

Hayes, F. R. (1953). Artificial freshets and other factors controlling the ascent and population of Atlantic salmon in the Le Have River *N.S. Bull. Fish. Res. Bd. Can.* **99**, 47pp.

Hayes, F. A., Pelluet, D. and Gorham, E. (1953). Some effects of temperature on the embryonic development of the salmon (*Salmo salar*) *Can. J. Zool.* **31**, 42–51.

Hayes, F. A., Wilmot, I. R. and Livingstone, D. A. (1951). The oxygen consumption of the salmon egg in relation to development and activity. *J. Exp. Zool.* **116**, 377–395.

Hellawell, J. M. (1976). River management and the migratory behaviour of salmonids. *Fish. Mgmt.* **7**, 57–60.

Hemsworth, R. J. and Brooker, M. P. (1979). The rate of downstream displacement of macro-invertebrates in the upper Wye, Wales. *Holarctic Ecology* **2**, 130–136.

Henderson-Sellers, B. (1979). "Reservoirs." Macmillan, London.

Hey, R. D. (1975). Response of alluvial channels to river regulation. Proceedings 2nd World Congress, International Water Resources Association, New Delhi, **V**, 183–188.

Hilsenhoff, W. L. (1971). Changes in the downstream insect and amphipod fauna caused by an impoundment with a hypolimnion drain. *Ann. ent. Soc. Am.* **64**, 743–746.

Holden, P. B. (1979). Ecology of riverine fishes in regulated stream systems with emphasis on the Colorado River. *In* "The Ecology of Regulated Streams" pp. 57–74, (Ward, J. V. and Stanford, J. A. eds). Plenum Publishing Corporation, New York.

Holden, P. B. and Stalnaker, C. B. (1975). Distribution and abundance of main-stream fishes of the middle and upper Colorado River basins, 1967–1973. *Trans. Am. Fish. Soc.* **104**, 217–231.

Holmes, N. T. H. and Whitton, B. A. (1977). The macrophytic vegetation of the River Tees in 1975: observed and predicted changes. *Freshwat. Biol.* **7**, 43–60.

Hopper, F. N. (1978). Faunal studies on the River Elan. Unpublished M.Sc. Thesis. University of Wales.

Hourston, W. R. (1958). Power developments and anadromous fish in British Columbia. *In* "The Investigation of Fish Power Problems". Symposium University of British Columbia, April 1957, pp. 15–23, (Larkin, P. A., ed.).

Howells, W. R. and Jones, A. N. (1972). The River Towy regulating reservoir and fishery protection scheme. *J. Inst. Fish. Mgmt.* **3**, 5–19.

Hull, C. H. J. (1967). River regulation. *In* "River Management". (Isaac, P. C. G., ed.), pp. 86–123. Maclaren, London.

Hunt, R. L. (1976). A long-term evaluation of trout habitat developments and its relation to improving management-related research. *Trans. Am. Fish. Soc.* **105**, 361–364.

Hutchinson, G. E. (1975). "A Treatise on Limnology". Vol. 1. Geography and Physics of Lakes. John Wiley & Sons, London.

Hynes, H. B. N. (1961). The invertebrate fauna of a Welsh mountain stream. *Arch. Hydrobiol.* **57**, 344–388.

Hynes, H. B. N. (1970). "The Ecology of Running Waters". Liverpool University Press.

Iremonger, D. J. (1970). River regulation and fisheries. Salmon and Trout Association Conference. 25 November 1970, 64–79.

Jaag, O. and Ambuhl, H. (1964). The effect of the current on the composition of biocoenoses in flowing water streams. *Adv. Wat. Poll. Res.* **1**, 31–49.

Jaske, J. T. and Goebel, J. B. (1967). Effects of dam construction on temperature of Columbia River. *J. Amer. Wat. Wks. Assoc.* **59**, 935–942.

Johansen, A. C. and Krough, A. (1914). The influence of temperature and certain other factors upon the rate of development of the eggs of fishes. Conseil permanent international pour l'exploration de la mer. Pub. de Circonstance No. 68.

Jonsson, B. And Sandelund, O. T. (1979). Environmental factors and life histories of isolated river stocks of brown trout (*Salmo trutta* m. *fario*) in Sore Osa river system, Norway. *Env. Biol. Fish.* **4**, 43–54.

Kalleberg, H. (1958). Observations in a stream tank of territoriality and competition in juvenile salmon and trout. *Inst. Freshwater Res. Drott. Swed. Rept* **39**, 55–98.

Kipling, C. and Frost, W. E. (1970). A study of the mortality, population numbers, year-class strengths, production and food consumption of pike, *Esox lucius* L in Windermere from 1944 to 1962. *J. Anim. Ecol.* **39**, 115–157.

Kittrell, F. W. (1959). Effects of impoundments on dissolved oxygen resources. *Sew. Ind. Wast.* **31**, 1065–1078.

Komura, S. and Simmons, D. B. (1967). River-bed degradation below dams. *Proc. A.S.C.E.* HY4 **93**, 1–14.

Kraft, M. E. (1972). Effects of controlled flow reduction on a trout stream. *J. Fish. Res. Bd. Can.* **29**, 1405–1411.

Krenkel, P. A., Lee, G. F. and Jones, R. A. (1979). Effects of TVA impoundments on downstream water quality and biota. *In* "The Ecology of Regulated Streams" pp. 289–306. (Ward, J. V. and Stanford, J. A. eds). Plenum Publishing Corporation, New York.

Krenkel, P. A. and Parker, F. L. (1969). Engineering aspects, sources and magnitude of thermal pollution. *In* "Biological Aspects of Thermal Pollution", pp. 10–52. (Krenkel, P. A. and Parker, F. L. eds). Vanderbilt University Press, New York.

Kroger, R. L. (1973). Biological effects of fluctuating water levels in the Snake River, Grand Teton National Park, Wyoming. *Am. Midl. Nat.* **89**, 478–481.

Lagler, K. F. (1969). "Man-made Lakes—Planning and Development." F.A.O., Rome.

Lantz, R. L. (1970). Influence of water temperature on fish survival, growth and behaviour. Proceedings of a Symposium on Forest Land Uses and Stream Environment, Oregon State University Press, Corvallis, October 19–20, 182–193.

Larkin, P. A. (1971). The environmental impact of hydropower. *In* "Energy and the Environment", pp. 162–175. H. R. MacMillan Lectures for 1971 (Efford, I. E. and Smith, B. M. eds).

Laronne, J. B. and Carson, M. A. (1976). Interrelationships between bed morphology and bed-material transport for a small, gravel bed channel. *Sedimentology* **23**, 67–85.

Lavis, M. E. and Smith, K. (1972). Reservoir storage and the thermal regimes of rivers, with special reference to the River Lune, Yorkshire. *Sci. Total Environ.* **1**, 81–90.

Leggett, W. C. (1977). The ecology of fish migrations. *Ann. Rev. Ecol. Syst.* **8**, 285–308.

Lehmkuhl, D. M. (1972). Changes in the thermal regime as a cause of reduction of bethnic fauna downstream of a reservoir. *J. Fish. Res. Bd. Can.* **29**, 1329–1332.

Leopold, L. B., Wolman, W. B. and Miller, J. P. (1964). "Fluvial Processes in Geomorphology." W. H. Freeman and Co., San Francisco.

Letterman, R. D. and Mitsch, W. J. (1978). Impact of mine drainage on a mountain stream in Pennsylvania. *Environ Pollut.* **17**, 53–73.

Lillehammer, A. and Saltveit, S. J. (1979). Stream migration in Norway. *In* "The Ecology of Regulated Streams", pp. 201–213, (Ward, J. V. and Stanford, J. A. eds). Plenum Publishing Corporation, New York.

Lindroth, A. (1942). Sauerstoffverbrauch der Fische. I. Verschiendene Entwicklungs— und Altersstadien vom hachs und Hecht. *Z. vergl. Physiol.* **29**, 583–594.

Lister, D. B. and Walker, C. E. (1966). The effect of flow control on freshwater survival of chum, coho and chinook salmon in the Big Qualicum River. *Can. Fish Cult.* **37**, 3–21.

Lloyd, D. G. and Beaver, J. L. (1078). Environmental factors in overseas situations. *Symposium on Engineering and the Environment: Harmony or Conflict.* Institution of Water Engineers and Scientists, London. 6 and 7 December 1978. 6/1-6/19.

Lloyd, R. and Jordan, D. H. M. (1964). Some factors affecting the resistance of rainbow trout (*Salmo gairdnerii* Richardson) to acid waters. *Int. J. Air Wat. Pollut.* **8**, 393–403.

Logan, P. (1979). The biota of pools and riffles: a comparison. Unpublished M.Sc. Thesis. University of Wales.

Lowe, R. L. (1979). Phytobenthic ecology and regulated streams. In "The Ecology of Regulated streams", pp. 25–34. (Ward, J. V. and Stanford, J. A. eds). Plenum Publishing Corporation, New York.

Macan, T. T. (1960). The effects of temperature on *Rhithrogena semicolorata* (Ephemeroptera). *Int. Rev. ges. Hydrobiol.* **45**, 197–201.

Macan, T. T. and Maudsley, R. (1966). The temperature of a moorland fishpond. *Hydrobiologia* **27**, 1–22.

M.A.F.F. (1976). The fisheries implications of water transfers between catchments. Ministry of Agriculture, Fisheries and Food and the National Water Council.

Makkaveyev, N. I. (1970). The impact of large water engineering projects on geomorphic processes in stream valleys. *Geomorfologiya* **2**, 28–34.

Mann, R. H. K. (1976). Observations on the age, growth, reproduction and food of the chub *Squalius cephalus* (L) in the River Stour, Dorset. *J. Fish. Biol.* **8**, 265–288.

McCleave, J. D. (1978). Rhythmic aspects of estuarine migration of hatchery-reared Atlantic salmon (*Salmo salar*) smolts. *J. Fish Biol.* **12**, 559–570.

McGrath, C. J. (1958). Dams as barriers or deterrents to the migration of fish. IUCN 7th Technical Meeting, Athens, Georgia. September 1958. **4**, 81–92.

McIntire, C. D. (1964). Some effects of current velocity on periphyton communities in laboratory streams. *Hydrobiologia* **27**, 559–570.

Mellquist, P. (1979). The use of weirs to reduce adverse effects of regulation of Norwegian rivers. Presented to the 1st International Symposium on Regulated Streams, Erie, Pennsylvania, 18–21 April 1979.

Menendez, R. (1976). Chronic effects of reduced pH on brook trout (*Salvelinus fontinalis*). *J. Fish. Res. Bd. Can.* **33**, 118–123.

Mermel, T. W. (1979). Major dams of the world. *Water Power and Dam Construction* (special issue) November 1979. 95–105.

Merry, D. G. (1979). The riparian vegetation of the River Wye. Report to Nature Conservancy Council. London.

Middleton, J. B. (1967). Control of temperatures of water discharged from a multiple-purpose reservoir. Fishery Resources Symposium American Fisheries Society, University of Georgia, Athens. 37–46.

Mills, D. H. and Shackley, P. E. (1971). Salmon smolt transportation experiments on the Conon River system, Ross-shire. Department of Agriculture and Fisheries for Scotland. *Freshwater and Salmon Fisheries Research* **40**.

Minckley, W. L. (1963). The ecology of a spring stream Doe Run, Meade County, Kentucky. *Wildl. Monogr. Chestertown*, **11**, 124pp.

Moore, C. M. (1969). Effects of small structures on peak flows. In "Effects of Watershed Changes on Streamflow", pp. 101–117, (Moore, C. M. and Morgan, C. W. eds). University of Texas Press, Austin.

Morris, D. L. and Brooker, M. P. (1979). The vertical distribution of macroinvertebrates in the substratum of the upper reaches of the R. Wye, Wales. *Freshwat. Biol.* **9**, 573–583.

Morris, L. A., Langemeier, R. N., Russell, T. R. and Witt, A. (1968). Effects of main stem impoundments and channelization upon the limnology of the Missouri River, Nebraska. *Trans. Am. Fish. Soc.* **97**, 380–388.

Mostafa, M. G. (1955). Riverbed degradation below large capacity reservoirs. *Proc. Am. Soc. Civ. Engrs* **81**, 1–9.

Moyle, P. B. (1976). Some effects of channelization on the fishes and invertebrates of

Rush Creek, Modoc County, California. *Calif. Fish Game* **62**, 179–186.

Müller, K. (1956). Das productionbiologische zusammenspeil zwischen See und Fluss. *Ber. Limn. Flusostn. Freudenthal.* **7**, 1–8.

Müller, K. (1962). Limnological-fishery biology studies in the regulation of streams of the Swedish Lapplands. *Oikos* **13**, 125–154.

Mundie, J. H. (1974). Optimization of the salmonid nursery stream. *J. Fish. Res. Bd. Can.* **28**, 849–860.

Nebeker, A. V. (1971). Effect of high water temperatures on adult emergence of aquatic insects. *Wat. Res.* **5**, 777–783.

Nebeker, A. V., Andros, J. D., McCrady, J. K. and Stevens, D. G. (1978). Survival of steelhead trout (*Salmo gairdneri*) eggs, embryos and fry in air-supersaturated water. *J. Fish. Res. Bd. Can.* **35**, 261–264.

Neel, J. K. (1963). Impact of reservoirs. *In* "Limnology in North America", pp. 575–593, (Frey, D. G. ed.). University of Wisconsin Press, Wisconsin.

Nilsson, B. (1976). The influence of man's activities in rivers on sediment transport. *Nordic Hydrology* **7**, 145–160.

Nilsson, C. (1978). Changes in the aquatic flora along a stretch of the River Umëalven, N. Sweden following hydroelectric exploitation. *Hydrobiologia* **61**, 229–236.

Nisbet, M. (1961). Un example de pollution de riviere par vidage d'une retenue hydroélectrique. *Verh. Internat Verein. Limnol.* **14**, 678–680.

Nishizawa, T. and Yamabe, K. (1970). Changes in downstream temperature caused by the construction of reservoirs. Part I. *Tokyo Univ. Sci. Rep. Sect. C.* **10**, 27–42.

Norris, K. E., Brocato, F. and Calandrino, F. (1960). A survey of fish transportation. Methods and equipment. *Calif. Fish and Game* **46**, 5–33.

Northrup, W. L. (1965). Republican River channel deterioration. Miscellaneous Publications, U.S. Department of Agriculture. Publication 970, Paper 47, 409–424.

Oborne, A. C., Brooker, M. P. and Edwards, R. W. (1980). The chemistry of the River Wye. *J. Hydrol.* **45**, 233–252.

Oswood, M. W. (1979). Abundance patterns of filter-feeding caddisflies (Trichoptera: Hydropsychidae) and seston in a Montana (U.S.A.) lake outlet. *Hydrobiologia* **63**, 177–183.

Parsons, J. W. (1955). The trout fishery of the tailwater below Dale Hollow Reservoir. *Trans. Am. Fish. Soc.* **85**, 75–92.

Pearson, W. D., Kramer, R. H. and Franklin, D. R. (1968). Macroinvertebrates in the Green River below Flaming Gorge Dam, 1964–5 and 1967, *Proc. Utah Acad. Sci. Arts, Lett.* **45**, 148–167.

Peltier, W. H. and Welch, E. B. (1969). Factors affecting growth of rooted aquatics in a river. *Weed Science* **17**, 412–416.

Penáz, N., Kubicek, F., Marvan, P. and Zelinka, M. (1968). Influence of the Vir River Valley Reservoir on the hydrological and ichthyological conditions in the River Svratka. *Acta. Sci. Nat. Brno* **2**, 1–60.

Percival, E. and Whitehead, H. (1929). A quantitative study of some types of stream-bed. *J. Ecol.* **17**, 282–314.

Peters, J. C. (1962). The effects of stream sedimentation on trout embryo survival. *In* "Third Seminar on Biological Problems in Water Pollution", pp. 272–279, (Tarzwell, C. M. ed.).

Peterson, R. H., Spinney, H. C. E. and Steedharan, A. (1977). Development of Atlantic salmon (*Salmo salar*) eggs and alevins under varied temperature regimes. *J. Fish. Res. Bd. Can.* **34**, 31–43.

Petran, M. and Kothe, P. (1978). Influence of bedload transport on the macro-benthos of running waters *Verh. Internat. Verein. Limnol.* **20**, 1867–1872.

Petts, G. E. (1978). The adjustment of river channel capacity downstream of reservoirs. Unpublished Ph.D. Thesis. University of Southampton.

Pfitzer, D. W. (1954). Investigations of waters below storage reservoirs in Tennessee. *Trans. North Am. Wildl. Conf.* **19**, 271–282.

Pfitzer, D. W. (1967). Evaluations of tailwater fishery resources resulting from high dams. In "Reservoir Fishery Resources Symposium", pp. 477–488. American Fisheries Society, University of Georgia, Athens.

Philipson, G. N. (1969). Some factors affecting the net-spinning of the caddisfly *Hydropsyche instabilis*. *Hydrobiologia* **34**, 369–377.

Philipson, G. N. and Moorhouse, B. H. S. (1974). Observations on ventilatory and net-spinning activities of the larvae of the genus *Hydropsyche* Pictet (Trichoptera: Hydropsychidae) under experimental conditions. *Freshwat. Biol.* **4**, 525–533.

Poddubnyi, A. G. and others (1971). The behaviour of salmon (*Salmo salar* L.) in the conditions of the artificial barriers of the Nizhne—Tuluma hydroelectric station. *Informatsionnyi Byulletenl-Biologiya Vnutrennikh Vod.* **10**, 57–61. (Translation: Boston Spar, RTS 8956).

Poole, W. C. and Stewart, K. W. (1976). The vertical distribution of macrobenthos within the substratum of the Brazos River, Texas. *Hydrobiologia* **50**, 151–160.

Powell, G. C. (1958). Evaluation of the effects of a power dam water release pattern upon the downstream fishery. *Colorado Coop. Fish. Unit Quart. Rep.* **4**, 31–37.

Pyefinch, K. A. (1966). Hydroelectric schemes in Scotland—biological problems and effects on salmonid fisheries. In "Man-made Lakes". (Lowe-McConnell, R. H. ed.), pp. 139–147. Symposia of the Institute of Biology No. 15.

Raastad, J. E. (1979). Investigations of the bottom fauna of regulated rivers with special emphasis on black-flies (Diptera, Simuliidae). *Terskelprosjekter Informasjon* **8**.

Rabeni, C. F. and Minshall, G. W. (1977). Factors affecting microdistribution of stream benthic insects. *Oikos* **29**, 33–43.

Radford, D. S. and Hartland-Rowe, R. (1971). A preliminary investigation of the bottom fauna and invertebrate drift in an unregulated and regulated stream in Alberta. *J. appl. Ecol.* **8**, 883–903.

Raleigh, R. F. and Ebel, W. J. (1967). Effect of Browntree Reservoir on migrations of anadromous salmonids. In "Reservoir Fishery Resources Symposium", pp. 415–443. American Fisheries Society. University of Georgia, Athens.

Ramadan, F. M. (1972). Characterization of Nile waters prior to the High Dam. *Wasser und Abwasser-Forschung* **1**, 21–24.

Raymond, H. L. (1969). Effect of John Day Reservoir on the migration rate of juvenile chinook salmon in the Columbia River. *Trans. Am. Fish. Soc.* **97**, 356–359.

Reynolds, A. J. (1976). A decade's investigation of the stability of erodible stream beds. *Nordic Hydrology* **7**, 161–180.

Ridley, J. E. and Steel, J. A. (1975). Ecological aspects of river impoundments. In "River Ecology", pp. 565–587, (Whitton, B. A. ed.). Blackwell Scientific Publications. 565–587.

Ridley, J. E., Cooley, P. and Steel, J. A. P. (1966). Control of thermal stratification in Thames Valley Reservoirs. *Proc. Soc. Wat. Treat. Exam.* **15**, 225–244.

Robinson, G. D., Dunson, W. A., Wright, J. E. and Mamolito, G. E. (1976). Differences in low pH tolerance among strains of brook trout (*Salvelinus fontinalis*). *J. Fish Biol.* **8**, 5–17.

Rogers, A. and Cane, A. (1979). Upstream passage of adult salmon through an unlit tunnel. *J. Inst. Fish. Mgmt.* **10**, 87–92.

Royal, L. A. and Cooper, A. C. (1958). Dams as barriers or deterrents to the migration of fish. IUCN 7th Technical Meeting **4**, 93–100.

Rucker, R. R. (1972). Gas-bubbles disease of salmonids: a critical review. U.S. Department of the Interior Fish and Wildlife Service. Bureau of Sport Fisheries and Wildlife. Technical Paper No. 58. 11pp.

Rupprecht, R. (1975). The dependence of emergence-period in insect larvae on water temperature. *Verh. Internat. Verein. Limnol.* **19**, 1–15.

Rutter, E. J. and Engstrom, LeRoy. (1964). Hydrology of flow control. Part III. Reservoir regulation. *In* "Handbook of Applied Hydrology" (Ven Te Chow, ed.). McGraw-Hill, New York.

Saunders, R. L. and Henderson, E. B. (1969). Survival and growth of Atlantic salmon fry in relation to salinity and diet. *Fish. Res. Bd. Can. Tech. Rep.* **148**.

Schoning, R. W. and Johnson, D. R. (1956). A measured delay in the migration of adult chinook salmon at Bonneville Dam on Columbia River. *Oregon Fish Comm., Contr.* **23**.

Serruya, C. (1969). Problems of sedimentation in the Lake of Geneva. *Verh. Internat. Verein. Limnol.* **17**, 209–218.

Shumway, D. L., Warren, C. E. and Doudoroff, P. (1964). Influence of oxygen concentration and water movement on the growth of steelhead trout and coho embryos. *Trans. Am. Fish. Soc.* **93**, 342–356.

Slack, J. G. (1978). Water quality aspects of bankside storage. *J. Inst. Wat. Engr. Sci.* **32**, 153–167.

Smith, N. (1971). "A History of Dams." Peter Davies, London.

Snieszko, S. F. (1974). The effects of environmental stress on outbreaks of infectious diseases of fishes. *J. Fish Biol.* **6**, 197–208.

Solomon, D. J. (1978). Some observations on salmon smolt migration in a chalk stream. *J. Fish. Biol.* **12**, 571–574.

Soltero, R. A., Wright, J. C. and Horpestad, A. A. (1973). Effects of impoundment on the water quality of the Bighorn River. *Wat. Res.* **7**, 343–354.

Somme, S. (1958). The effects of impoundment on salmon and sea trout rivers. IUCN 7th Technical Meeting, Athens, **4**, 77–80.

Spence, J. A. and Hynes, H. B. N. (1971a). Differences in fish populations upstream and downstream of a mainstream impoundment. *J. Fish. Res. Bd. Can.* **28**, 45–46.

Spence, J. A. and Hynes, H. B. N. (1971b). Differences in the benthos upstream and downstream of an impoundment. *J. Fish. Res. Bd. Can.* **28**, 35–43.

Sprules, W. M. (1947). An ecological investigation of stream insects in Algonquin Park, Ontario. *Univ. Toronto. Stud. biol. Serv.* **56**, 1–81.

Stalnaker, C. B. (1979). The use of habitat structure perferenda for establishing flow regimes necessary for maintenance of fish habitat. *In* "The Ecology of Regulated Streams", pp. 321–337, (Ward, J. V. and Stanford, J. A. eds). Plenum, New York.

Stanford, J. A. and Ward, J. V. (1979). Stream regulation in North America. *In* "The Ecology of Regulated Streams", pp. 215–236, (Ward, J. V. and Stanford, J. A. eds). Plenum, New York.

Steichen, J. M., Garton, J. E. and Rice, C. E. (1979). The effect of lake destratification on water quality. *J. Am. Wat. Wks. Assoc.* **71**, 219–225.

Symons, P. E. K. (1979). Estimated escapement of Atlantic salmon (*Salmo salar*) for maximum smolt production in rivers of different productivity. *J. Fish. Res. Bd. Can.* **36**, 132–140.

Symons, P. E. K. and Heland, M. (1978). Stream habitats and behavioural inter-actions of under yearling and yearling Atlantic salmon (*Salmo salar*). *J. Fish. Res. Bd. Can.* **35**, 175–183.

Symons, J. M., Irwin, W. H. Clark, R. M. and Robeck, G. G. (1967). Management and measurement of DO in impoundments. *Proc. Amer. Soc. Civ. Engrs.* **SA6**, 181–209.

Thompson, R. W. S. (1954). Stratification and overturn in lakes and reservoirs. *J. Inst. Wat. Engrs.* **8**, 19–52.

Toms, I. P., Wood, G. and Owens, M. (1975). Grafham Water: some effects of the impoundment of nutrient rich water. Proceedings Water Research Centre Symposium on The Effects of Storage on Water Quality, University of Reading. March 24–26, 1975. 127–161.

Treharne, W. D. (1963). Pollution problems on the Afon Rheidol. *Wat. Wast. Treat.* March/April, 610–613.

Trojnar, J. R. (1977). Egg hatchability and tolerance of brook trout (*Salvelinus fontinalis*) fry at low pH. *J. Fish. Res. Bd. Can.* **34**, 574–579.

Trotsky, H. M. and Gregory, R. W. (1974). The effects of water flow manipulation below a hydroelectric power dam on the bottom fauna of the upper Kennebec River, Maine. *Trans. Am. Fish. Soc.* **103**, 318–324.

Truesdale, G. and Taylor, G. (1978). Quality implications in reservoirs filled from surface water sources. *Prog. Wat. Tech.* **10**, 289–300.

Turner, D. J. (1971). Dams and ecology. Can they be made compatible? Civil Engineering ASCE. September, 1971, 76–80.

Tyler, P. A. and Buckley, R. T. (1974). Stratification and biogenic meromixis in Tasmanian reservoirs. *Aust. J. Mar. Freshwat. Res.* **25**, 299–313.

Walker, K. F. (1979). Regulated streams in Australia: the Murray-Darling river system. *In* "The Ecology of Regulated Streams", pp. 143–163. (Ward, J. V. and Stanford, J. A. eds). Plenum, New York.

Walters, R. C. S. (1966). Conservation of water by impounding reservoirs, river regulation reservoirs and pumped storage reservoirs. *In* "River Engineering and Water Conservation Works", pp. 393–421, (Thorn, R. B. ed.). Butterworths, London.

Ward, J. V. (1974). A temperature-stressed stream ecosystem below a hypolimnial re-lease mountain reservoir. *Arch. Hydrobiol.* **74**, 247–275.

Ward, J. V. (1975). Downstream fate of zooplankton from a hypolimnial release mountain reservoir. *Verh. Internat. Verein Limnol.* **19**, 1798–1804.

Ward, J. V. (1976a). Effects of thermal constancy and seasonal temperature dis-placement on community structure of stream macroinvertebrates. *In* "Thermal Ecology II", pp. 302–307, (Esch, G. W. and McFarlane, R. W. eds). ERDA Symposium Series (CONF-750425).

Ward, J. V. (1976b). Effects of flow patterns below large dams on stream benthos: a review. *In* "Instream Flow Needs Symposium", Vol. II, pp. 235–253, (Osborn, J. S. and Allman, C. H. eds). American Fisheries Society.

Ward, J. V. and Short, R. A. (1978). Macroinvertebrate community structure of four special lotic habitats in Colorado, U.S.A. *Verh. Internat. Verein. Limnol.* **20**, 1382–1387.

Ward, J. V. and Stanford, J. A. (1979a). Ecological factors controlling stream zoobenthos with emphasis on thermal modification of regulated streams. *In* "The Ecology of Regulated Streams", pp. 35–55, (Ward, J. V. and Stanford, J. A. eds). Plenum, New York.

Ward, J. V. and Stanford, J. A. (1979b). Limnological considerations in reservoir operation: optimization strategies for protection of aquatic biota in the receiving stream. *In* "Proceedings of the Mitigation Symposium". U.S. Department of Agriculture, Collins, Colorado.

Waters, T. F. (1972). The drift of stream insects. *A. Rev. Ent.* **17**, 253–272.

Webster, J. R., Benfield, E. F. and Cairns, J. (1979). Model predictions of effects of impoundment on particulate organic matter transport in a river system. *In* "The Ecology of Regulated Streams", pp. 339–364, (Ward, J. V. and Stanford, J. A. eds). Plenum, New York.

Weitkamp, D. E. and Katz, M. (1977). Dissolved atmospheric gas supersaturation of water and gas bubble disease of fish. The Water Resources Scientific Information Center, Department of the Interior, Washington, D.C. 107pp.

Wetzel, R. G. (1975). "Limnology". W. B. Saunders Company.

Whitcomb, D. (1963). Aquatic weeds and the distribution of freshwater animals. *Proc. Brit. Coarse Fish Conf.* **1**, 40–47.

Wiebe, A. H. (1960). The effects of impoundments upon the biota of the Tennessee River System. *Proceedings 7th Technical Meeting I.U.C.N.* **4**, 101–117.

Wickett, W. P. (1954). Production of chum and pink salmon in a controlled stream. *Prog. Rep. Fish. Res. Bd. Can.* **11**, 933–953.

Williams, D. D. and Hynes, H. B. N. (1974). The occurrence of benthos deep in the substratum of a stream. *Freshwat. Biol.* **4**, 507–524.

Wong, S. L. Clark, B., Kirby, M. and Koscinw, R. F. (1978). Water temperatures fluctuations and seasonal periodicity of *Cladophora* and *Potamogeton* in shallow rivers. *J. Fish. Res. Bd. Can.* **35**, 866–870.

Wright, J. C. (1967). Effects of impoundments on productivity, water chemistry and heat budgets of rivers. *In* "Reservoir Fishery Resources Symposium", pp. 188–199, American Fisheries Society. University of Georgia, Athens.

Wunderlich, W. O. and Elder, R. A. (1967). The mechanics of stratified flow in reservoirs. *In* "Reservoir Fishery Resources Symposium", pp. 56–68. American Fisheries Society, University of Georgia, Athens.

Wurtsbaugh, W. A. and Davis, G. E. (1977). Effects of temperature and ration level on the growth and food conversion efficiency of *Salmo gairdneri* Richardson. *J. Fish. Res. Bd. Can.* **11**, 87–98.

Zalumi, S. G. (1970). The fish fauna of the lower reaches of the Dnieper River. Its present composition and some features of its formation under conditions of regulated and reduced river discharge. *J. Ichthyol.* **10**, 587–590.

The Effects of Discharge on Sediment Dynamics and Consequent Effects on Invertebrates and Salmonids in Upland Rivers

N. J. MILNER* AND J. SCULLION,

*UWIST Field Centre, Newbridge-on-Wye,
Powys, Wales*

P. A. CARLING AND D. T. CRISP

*FBA Teesdale Unit, c/o Northumbrian Water Authority,
Lartington Treatment Plant, Barnard Castle, UK*

*Present address: Welsh Water Authority, Area Laboratory, Highfield, Priestley Road, Caernarfon, Wales

I. INTRODUCTION

The discharge regimes of streams and rivers and their consequent effects upon channel form, bed movements and sediment deposition can have considerable influence on the biology of invertebrate fauna and salmonid fishes. Discharge regime and the related physical effects are themselves modified by catchment management, especially by such activities as river regulation and transfer, drainage works, afforestation and deforestation. Such changes in land and water use are occurring with increasing frequency and it is therefore surprising that relatively few studies concerned with their biological consequences have been undertaken in western Europe. In contrast, there is a large body of published information on the biological effects of discharge and discharge-related factors in North America. This relates particularly to salmon of the genus *Oncorhynchus* in rivers of the western seaboard of Canada and the USA. The application of this work to upland rivers in Britain presents some difficulties because the species concerned are different and the hydrographic regime and gravel bed morphology of Laurentian and Scandinavian streams differ from those in Britain. This is largely a function of latitude and climate, i.e. cool temperate and arctic climates dominate in the upper catchments of the former areas, compared with a temperate climate in Britain. Further, these areas have been de-glaciated only recently and, as a result, there is an abundance of unconsolidated material available for transportation. In Britain less material is available and the terrestrial vegetation cover is denser and more continuous. These factors influence both channel morphology and gravel bed composition.

In recent years there has been increased interest in these topics in Britain and the present review attempts to bring together the relevant, but scattered, literature. The review focuses primarily on information about British species but draws upon information about species from elsewhere where appropriate. For example, in the section on salmonid fishes, most of the literature reviewed refers to the Atlantic salmon (*Salmo salar* L.) and the trout (*Salmo trutta* L.) but major gaps in our knowledge of these species are indicated and are filled, wherever possible, by means of information on the genus *Oncorhynchus* or American species of the genus *Salmo*.

Cooperative studies by sedimentologists, hydrologists and biologists could be a fruitful development. Therefore a section of the review is concerned with physical methodology and the problems of application of physical measurements in a biological context. Study of the hydraulics of upland rivers is a "young" branch of science and is fraught with its own inherent problems and uncertainties. The present account examines these complexities in some detail because they are likely to become increasingly relevant to biological studies in the future.

Every effort has been made to adhere strictly to the given title and for this reason a number of effects commonly associated with discharge regime changes (e.g. changes in temperature regime as a result of impoundment) have not been covered.

II. Hydraulics

A. INTRODUCTION

Successful analysis of the biological impacts of discharge regime in upland rivers requires definition of "upland" rivers and identification and quantification of the relevant differences in discharge regime between "upland" and "lowland" rivers. However, no hydraulic-geomorphic parameters have been devised which serve as general organizers of lotic biotic data (Sedell *et al.*, 1978).

Physical effects likely to be of biological importance are the frequency and magnitude of variations in water level, velocity, bed composition, intragravel flow and movements of bed material and finer materials. These are examined and, as far as possible, their relationships to hydraulic parameters are defined.

B. DISCHARGE REGIME

Though there are broad regional differences in discharge regime, it is not yet possible to define "upland" streams precisely (e.g. Newson, 1981). In general, the occurrence of high and low flow periods is progressively earlier in the year in the upland areas of northern and western Britain; (Ward, 1968; Jones and Peters, 1977), and summer base flow periods are usually longer in duration in upland streams (Smith, 1966, 1969).

These differences may be related to the variable contribution to base flow from storage and surface run-off, the latter being more important in upland areas. Consequently, run-off rates per unit area (Ward, 1968) and the frequency of spates and the rate of change of discharge are greater in upland than in lowland areas (Jones and Peters, 1977).

The base flow index (B.F.I.) (Beran and Gustard, 1977; Anon., 1978) may be interpreted as an index of stream "flashiness". B.F.I. cannot be used to separate neatly upland and lowland streams but is a useful parameter to compare or rank streams. In an analysis of data from

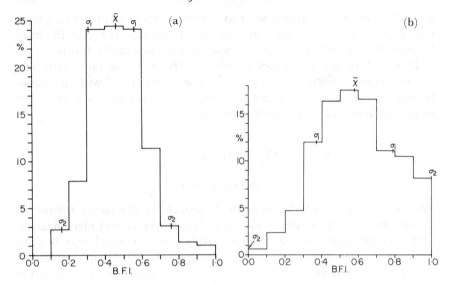

Fig. 1. (a) Base flow indices (B.F.I.) for 291 catchments in "upland" Britain. Data for England and Wales considered west of the Exe-Humber line (excluding the Cheshire Plain). $\bar{x}$ = mean value, σ_1 = one standard deviation, σ_2 = two standard deviations. (b) Base flow indices (B.F.I.) for 342 catchments in "lowland" Britain. Data for England considered east of the Exe-Humber line (but including the Cheshire Plain). $\bar{x}$ = mean value, σ_1 = one standard deviation, σ_2 = two standard deviations.

634 basins in England and Wales (Fig. 1) "upland" streams had a mean B.F.I. of 0·46 and "lowland" streams a mean of 0·58. "Upland" stream data are, commonly, biased towards the lower reaches of larger upland streams, therefore, summary hydrological data (e.g. Ward, 1968; Jones and Peters, 1977) should not be applied uncritically in evaluating flow, sedimentation and biotic interrelationships in small upland water courses.

Important local variations are superimposed on any regional pattern. In upland areas, where precipitation falls principally as snow during the winter, extended base flow periods may extend throughout December until February (Lavis, 1973). Annual peak discharges may then occur in March with the spring thaw. Similarly, the significance of rainfall is mediated by seasonal influences. Frontal rain-storms in winter may have an influence on the flood hydrograph and sediment response different from that of convectional summer storms (Hall, 1967). Recent investigations have demonstrated the importance of local differences in catchment response to rainfall (Hewlett, 1974; Weyman, 1975); the response of the hydrograph varying within a single storm event as source areas change in degree of importance (Betson and Marius, 1969; Kirkby and Weyman, 1974; Dunne *et al.*, 1975; Bevan, 1978).

Despite these problems, knowledge of hydraulic relationships, notably between velocity and discharge and discharge and cross-sectional area, is fundamental to the description of upland channel flow and ultimately to channel morphological response. As an example, isolation of a "significant discharge" (usually bankful discharge) (Andrews, 1980) and its recurrence interval, is important as traditionally this represents an important channel-forming discharge (Wolman and Miller, 1960). In lowland streams this flow is frequently equated with a discharge of recurrence interval one to two years. For upland streams there are few data on recurrence intervals, but the latter may be significantly shorter. Analyses of data from gravel-bedded streams in the UK have indicated that bankful discharge is exceeded on average 2·2 days per year, (Nixon, 1959). This may be related to the increase in frequency and relative magnitude of spates in headwater streams (Jones and Peters, 1977) and to distinctive channel and basin characteristics (Bowen, 1959; Wolf, 1959), though the exact relationship to hydraulics remains unclear (Nixon, 1959; Charlton, 1977; Williams, 1978).

C. PHYSICAL AND HYDRAULIC CHARACTER

Velocity distribution away from the bed may be described as a positive logarithmic function of the vertical distance from the bed. This has been written in the form:

$$\frac{\overline{U}}{U_*} = 5{\cdot}75 \log 11{\cdot}0 \frac{D}{k_s} \tag{1}$$

for relatively wide shallow streams.*

In natural streams water movement is usually turbulent throughout the depth. A laminar sublayer may exist close to the bed. The thickness of this layer decreases as turbulence (current speed) increases (Simons and Sentürk, 1977);

$$\frac{1}{\sigma} = \frac{1}{11{\cdot}6} \frac{U_*}{v} \tag{2}$$

Equation (1) should be satisfactory for use in gravel-bedded streams in the UK when $D/k_s > 10$ (Kamphuis, 1974) and $k_s \simeq 3{\cdot}5d_{90}$ (Charlton *et al.*, 1978). However, where this is not the case no

* For notation in all equations see Appendix I (p. 220).

satisfactory relationships have yet been derived. As a cautionary note, many upland streams in the United Kingdom may fall into this latter category. Streams may be narrow, shallow and have beds consisting of large boulders (Newson and Harrison, 1978) and possibly large organic debris (Keller and Swanson, 1979). As such, these streams are not alluvial, as is common for lowland streams, but owe a degree of their form to the primary geology and geomorphology of the catchment.

Flow varies rapidly both temporally and spatially, leading to complex response of depth, width, resistance coefficients and energy slope. Peterson and Mohanty (1966) on the basis of laboratory investigations divided upland stream flows into three categories: (1) tranquil, (2) rapid, and (3) tumbling, related to the relative influence of drag forces associated with grain, bed-form and channel-bank resistances. Judd and Peterson (1969) extended the analysis to include field data from natural gravel- and boulder-bedded streams.

Despite the significance of changes in channel form, resulting from variable hydraulic characteristics, to engineering and biological investigations, very little theory exists relating channel morphology to processes (Leopold and Langbein, 1962; Henderson, 1963; Yang and Song, 1979; Hey, 1979). Li *et al.* (1976) found that a theoretically correct model for small alluvial gravel-bedded steams fitted field data collected by Brush (1961) and Judd and Peterson (1969). Nevertheless, models are frequently restrictive, often applying only to straight channels with simple flow patterns. Consequently, various investigators include a range of parameters in assessing empirically the relationships between channel morphology and hydraulics (Parker and Anderson, 1977; Parker, 1979a) although the significance of some measures is unclear (Riley, 1976).

Leopold and Maddock (1953) expressed the most important relationships as power functions the exponents of which sum to unity:

$$W = a \, Q^b \tag{3}$$

$$\overline{D} = c \, Q^f \tag{4}$$

$$\overline{U} = k \, Q^m \tag{5}$$

thus: $Q = W \overline{D} \overline{U} = ack \, Q^b \, Q^f \, Q^m$

therefore, $ack = 1 \cdot 0$ and $b + f + m = 1 \cdot 0$.

Although power functions may not be the most appropriate exact model (Richards, 1973, 1976a) the adage that the exponents are functions of energy expenditure (Leopold and Langbein, 1962; Langbein, 1964) has led to their continued application in relationships between upland channel geometry and hydraulics (e.g. Charlton, 1977). Exponents b and m for upland streams (Calkins and Dunne, 1970; Kellerhals, 1970; Bevan *et al.*, 1979) are frequently larger than those for lowland streams (Leopold *et al.*, 1964; Stall and Yang, 1970; Klein, 1976). For example, examination of the data of Bevan *et al.* (1979) gives a value for $\bar{m}$ of 0·68 compared with summary values of $\bar{m}$ for lowland streams of 0·34—0·43 (Leopold *et al.*, 1964). It is frequently necessary to estimate flow parameters in gravel-bedded rivers without recourse to detailed field data. Usually semi-empirical relationships are used and these commonly contain constraints, such as the assumption of a straight channel (Chow, 1967; Graf, 1971; Simons and Sentürk, 1977). Basic to most requirements is the assessment of the relative roughness of the channel bed during high flows (Bray, 1979). Commonly used are the Manning equation:

$$\overline{U} = \frac{1}{n} R^{0·66} S^{0·5}, \tag{6}$$

or the Darcy-Weisbach equation:

$$f = \frac{8 \, AgS}{\rho \overline{U}^2}. \tag{7}$$

Values of Manning's roughness number (n), are commonly chosen subjectively from published tables (e.g. Graf, 1971; Simons and Sentürk, 1977) and type photographs (Barnes, 1967) relating values of "n" to visual characteristics of the channels. A common alternative is to relate the roughness to a grain size representative of the river bed material, especially using variants of the Strickler equation where:

$$n = ad_n^{0·167}. \tag{8}$$

Numerous workers have empirically chosen various d_n sizes to suit their own particular requirements. Bray (1979) discusses variants of the Strickler relationship and other methods of characterizing roughness, for gravel bedded rivers, whilst basic derivations may be found in Graf (1971).

Although, with the selection of a suitable roughness coefficient, fairly accurate estimates of velocity may be achieved for gravel bedded streams (Roberson and Wright, 1973; Bray, 1979) the relative roughness is not a simple linear function of discharge (Culbertson and Dawdy, 1964; Amein and Fang, 1970). Commonly, the relative roughness in gravel-bedded streams decreases with increasing discharge (Martinec,

1973; Bevan *et al.*, 1979), whilst in sand-bedded streams the roughness may initially increase as ripples grow in size with increasing discharge. In addition, complex relationships between roughness and discharge exist in streams with beds consisting of both gravels and sands (Simons *et al.*, 1979), whilst very little is known of the quantitative importance of vegetation in streams in modifying drag coefficients (Kouwen and Unny, 1973).

In order to calculate the stress on the bed, which may influence gravel movement and the behaviour of organisms, values of the Darcy-Weisbach coefficient (f) are commonly introduced into equations of the form:

$$\frac{\overline{U}}{U_*} = \sqrt{\frac{8}{f,}} \tag{9}$$

where excessive values of f would yield unrealistic shear velocities. The importance of correcting for frictional losses other than those on the bed material was recognized by Johnson (1942) and Williams (1970). Corrections based upon proportioning the flow into areas over banks and the area over the active bed are given by Lundgren and Jonsson (1964) and Knight and Macdonald (1979). Corrections based upon dividing the hydraulic radius are given by Einstein and Barbarossa (1952) and specifically for gravel bedded rivers by Hey (1979); these yield the bed resistance alone.

D. BED MATERIALS

The importance, for stream management, of recognizing the mechanisms of gravel supply, transport and subsequent deposition in upland water courses has been acknowledged for a considerable time (Clayton, 1951). Widespread North American literature, refers largely to extensive braided stream deposits (e.g. Fahnestock, 1963; Williams and Rust, 1969; McKenzie and Walker, 1974; Smith, 1974; Jopling and McDonald, 1975) and may not be applicable in Britain where investigations of sediment source, rate of supply and depositional characteristics remain relatively few (Bluck, 1971, 1976; Slaymaker, 1972; Painter *et al.*, 1974; Harvey, 1974; Lewin *et al.*, 1974; Lewin, 1976; Hitchcock, 1977; Harvey, 1977; Walling *et al.*, 1979).

In Britain downstream distribution of materials in gravel-bedded streams is largely manifest as a series of pool and riffle sequences, although braided channels and point bar deposits may locally reflect loci of extensive gravel deposition. Exact mechanisms of pool and riffle formation are unclear but are discussed by Langbein and Leopold

(1968), Keller (1971), Keller and Melhorn (1974), Richards (1976b) and Lisle (1979). Keller (1971) observed that, although during low flows current speed over riffles exceeded current speed in pools, this pattern was reversed during high flows (cf. Andrews, 1979). He used this observation to explain grain size differences, contrasting pools and riffles.

As material is transported and abraded downstream from source areas, mean grain size is reduced, sediment becomes more rounded (Hall, 1967) and better sorted.

Locally, tributaries (Swenson, 1942; Knighton, 1980) and changes of gradient (Leopold *et al.*, 1964) may disrupt this pattern. The significance of gradient in controlling the grain size of deposits remains unclear (Hack, 1957; Wilcock, 1971). In addition, the composition of the material may also change as particles of various geological materials undergo differential wear and solution (Plumley, 1948; Kuenen, 1956).

Within an individual reach of a channel, particle size may vary considerably (Iwamoto *et al.*, 1978; Shirazi *et al.*, 1979) from fine silt and clay (mean grain size, $\bar{x} < 0.063$ mm) to large boulders ($\bar{x} > 256$ mm) and differences may be substantial between pools and riffles. Generally, materials in pools are smaller in grain size (Hynes, 1970). Leopold *et al.* (1964) found a 250–500% increase in median grain size in a traverse from pool to riffle.

River gravels commonly have two size modes. The dominant mode may be regarded as a framework (Pettijohn, 1975) or lattice-work (Moss, 1962) of large cobbles. Such a modal class existing alone, termed an openwork gravel, is rare in nature (Cary, 1951). Usually the void space is swiftly infilled with fine sandy materials which constitute the sub-dominant mode (Moss, 1962). Simultaneous deposition of cobbles and fine sands and silts is rare in fluvial systems (Fraser, 1935). Usually, if the fine matrix has a mean grain size some 15% or less than that of the coarse mode (Pettijohn, 1975), then the former may be interpreted as a secondary addition to the gravel (Plumley, 1948; Dal Cin, 1967). There is little information on mechanisms of silt deposition in gravel beds (Einstein, 1968; Beschta and Jackson, 1979), although there is some evidence that deposition is periodic and related to various channel and catchment characteristics (Adams and Beschta, 1980). This is important because addition of fine material ($\bar{x} < 0.83$ mm) to a gravel bed reduces intergravel flow speeds (Vaux, 1962) by reducing permeability.

Platts *et al.* (1979) reported a linear correlation between river bed porosity (λ) and the geometric mean grain size (d_g) of the bed materials (size range, 10–70 mm). The permeability (β) of the sediments they investigated may be proportional to (d_g)^2, as $\beta^a \propto a_g^{1.92}$. However, the very good correlation achieved between the geometric mean grain size and the percentage of fines (Platts *et al.*, 1979) is to some degree spurious (Benson, 1965) as the two variables are not independent.

No simple relationships relating grain size, porosity and permeability are evident (Fraser, 1935; Pettijohn, 1975) and therefore the universality of empirical correlations is restricted (Platts *et al.*, 1979).

The most obvious feature of deposits in gravel-bedded streams is the frequent presence of a coarse deposit, at the surface, protecting finer materials beneath from being eroded. This armour layer may frequently consist of imbricated clasts where cobbles are flat rather than spherical (Lane and Carlson, 1954). In extreme examples this may constitute a cobble layer overlying sands (Lane and Carlson, 1954) though, more commonly, fine gravel deposits lie below (Milhous, 1972a). This armour layer may be largely absent where sediments are highly mobile and well-graded.

Analysis of grain-size data and the choice of descriptive statistics are not discussed in detail here. Pettijohn (1975) provides an up-to-date and concise summary with many pertinent references. Dyer (1970) provides statistics specifically to describe sand and gravel mixes which might be suitable for bio-hydrological investigations in upland rivers. However, these statistics have not been tested. In general, the use of the modified Wentworth–Lane terminology (Pettijohn, 1975) to describe sediments, and the use of the phi-scale to grade fractions (Krumbein, 1934) has increased in biological investigations (Hynes, 1970; Bohlin, 1977; Peterson, 1978; Shirazi *et al.*, 1979). Sumner (1978), gives a clear explicit derivation of the phi-scale from a nominal scale in millimetres. More important, when comparing results from differing investigators the compatibility of data sets should be considered. Mean grain size, median grain size and geometric mean grain size have all been used together with log-normal, arc-sin and untransformed distributions. Sampling techniques vary considerably between investigators and are reviewed by Kellerhals and Bray (1971) and Gomez (1979). Certain sampling procedures require transformation of data sets before statistical analysis (Kellerhals and Bray, 1971; Potter, 1979).

E. INTRAGRAVEL FLOW

Water flow through gravel river beds, expressed as permeability, may have important biological consequences by regulating supply of oxygen to the intragravel environment (Stuart, 1953a; Wickett, 1954; Coble, 1961; McNeil, 1962; Sheridan, 1962; Silver *et al.*, 1963; Cooper, 1965) and mediating temperature variations, (Hansen, 1975). Permeability may be reduced by siltation (Phillips and Campbell, 1962; McNeil and Ahnell, 1964; Shapley and Bishop, 1965; Shelton and Pollock, 1966; Peters, 1967; Hansen, 1971; Phillips, 1971; Hansen and Alexander, 1975).

Permeability may be expressed as a volume flux through an area including both void space and solids. The rate of flow is proportional to the hydraulic head and to the viscosity:

$$Q = K\frac{A\varDelta H}{vL},\qquad(10)$$

where K is a permeability parameter. This flow is conceptually different from the seepage discharge in which the bulk area includes only the cross-sectional void space.

$$Q = A_1 \overline{U}\qquad(11)$$

The derivation of values of K for gravel mixes has been widely explored and is dependent upon shape, packing and sorting of the sediment (Krumbein and Monk, 1942; Engelhardt and Pitter, 1951). Further, K varies with depth in stratified sediment mixes (Mast and Potter, 1963; Potter and Pettijohn, 1963). In practical applications in biology (Pollard, 1955) K is often given as:

$$K = \frac{\overline{U}}{S}.\qquad(12)$$

Pollard (1955), Terhune (1958), Gangmark and Bakkala (1958) and Turnpenny and Williams (in press) describe methods to evaluate the permeability and apparent velocity in river beds.

Mechanisms of interchange of water between the river and the gravel are poorly understood, although Vaux (1962) and Cooper (1965, 1974) isolated stream bed surface profile, gravel permeability and relative roughness as important elements influencing interchange. Variations in stream discharge have been correlated with changes in intragravel oxygen concentrations (McNeil, 1962; Wickett, 1968) and with changes in intragravel velocities (Williams and Hynes, 1974). Information on the influence of discharge on vertical fluid exchange is limited and conflicting (Martin, 1970; Walters and Rao, 1971; Clayton and Harrison, 1975). An important development is the field measurement of vertical fluxes of interstitial water in sand and gravel beds (Lee and Hynes, 1977) using a seepage meter (Lee, 1977; Lee and Cherry, 1979; Lee et al., 1980). Grondwater flux to the streambed ranged from 0·001 − 9 cm^3 m^{-2} s^{-1} and spatial changes in flux were related qualitatively to streambed composition. Lee and Hynes (1977) noted diurnal correlations between streamflow and groundwater discharge, but concluded from chemical data that exchange of water from the stream flow to the interstitial environment was limited.

If more complete theoretical models of intragravel water quality and exchange mechanisms similar to the water quality models of Wickett (1975) and O'Brien *et al.* (1978) are to be developed and applied to the natural environment, better empirical data are required to scale these models. In this respect, accurate determinations of interstitial seepage velocity would be preferable to calculations of Darcian apparent velocities. Derivation of the latter requires the assumption of laminar flow (Ward, 1964). The most recent hot-film sensors now used in biological sciences (La Barbera and Vogel, 1976) can be used in natural silt-laden flows (Richardson and McQuivey, 1968; McQuivey, 1973a, b). Interstitial seepage velocities related to infauna behaviour have been measured in natural sand beds (Riedl and Machan, 1972), and in gravel beds (Walkotten, pers. comm.) using hot-film sensors.

F. BED DYNAMICS

Fluvial sediment processes may be divided, for convenience, into three phases which are both complex and connected:

(1) Initiation of motion of the individual particles in the sediment body.

(2) Sediment transport.

(3) Cessation of motion of the particles (deposition).

1. *Initiation of motion*

Initiation of motion of an individual particle on a bed surface occurs when the hydrodynamic forces acting upon the grain exceed a critical threshold value. Bull (1979) reviews the concept of thresholds in the earth sciences, but here we may state that the threshold is dependent upon the physical nature of the flow, and of the sediment grain and the grain's relationship to other associated bed materials and the earth's gravitational field. Graf (1971) gives a full exposition of these principles.

Various hydrodynamic criteria have been related to incipient motion of grains, these include depth, velocity and discharge. Although individually these appear attractive (e.g. Hjulstrom, 1935; Neill, 1967; Virmani *et al.*, 1973) they are in fact interdependent and cannot be used independently in a very precise manner. For example Moss *et al.* (1980) and Bagnold (1977) have demonstrated that both depth and velocity are significantly related in a non-linear fashion to the shear at the sediment-water interface.

More recently, the shear stress at the bed, τ_0 (Gessler, 1971) and the available steampower $\omega = \tau_0 \, \overline{U}$ (Bagnold, 1977) have been used widely to characterize initiation of transport, although not without reservations (Yang, 1973). The shear stress may be written:

$$\tau_0 = \gamma RS, \tag{13}$$

which is the classic equation of Du Boy.

Alternatively an entrainment function is frequently expressed as a dimension-less relationship:

$$\frac{\tau_0}{(\rho_s - \rho) \, gd} = f \, (U_* \, d/v). \tag{14}$$

This is the classic Shields' criterion for the initiation of motion (Miller et al., 1977).

Nearly all practical derivations of incipient motion criteria are based on theoretical reasoning and empirical data. Considerable scatter of data arises from the differing, usually subjective, definitions of "incipient motion". Further, there is the difficulty of including the particle resistance factors such as the friction angle, particle shape, packing and the degree of particle exposure (Lane and Carlson, 1954; Laronne and Carson, 1976; Mantz, 1977; Cheetham, 1979).

Similarly, some attempt has been made to distinguish incipient motion curves for both laminar and turbulent flows (Yalin and Karahan, 1979) whilst Walters and Rao (1971) and Clayton and Harrison (1975) indicated that interstitial flow influences incipient motion criteria.

Once material has been set in motion the bed may restabilize owing to sediment sorting and the formation of an armour layer. Although the importance of these armouring processes has been investigated widely in recent years (e.g. Gessler, 1965; Kellerhals, 1967; Little and Mayer, 1972; Bayazit, 1975; Rákóczi, 1975), a loose unarmoured bed is a basic assumption in most models of incipient motion (Church, 1972; Laronne and Carson, 1976).

Graf (1971) and Mantz (1977) reviewed the extensive data on the incipient motion of sand and non-cohesive silts. However, in most upland British streams, gravel, cobbles, and boulders are important constituents of the streambeds. The movement of larger material is of particular interest with regard to overall bed stability. Important contributions to deriving criteria of incipient motion for sediments in these larger size ranges include Neill (1967; 1968), Church (1972); Baker (1973); Baker and Ritter (1975); Mears (1979) and Graf (1979). The effect of lift forces on particles in shallow flows leads to considerable deviation from the classic Shields' curve (Baker and Ritter, 1975) (Fig. 2). Graf (1979) provides expressions for incipient motion of both submerged and partially emergent particles.

Relationships proposed by these authors imply steady-state conditions, although the instantaneous shear stress may be three to ten times the calculated values (Kalinske, 1947; Cheetham, 1979). Further, broad channels are assumed; and where channels are narrow and bank effects (pp. 157–158) are significant, the models may deviate considerably from observed data. Although progress has been made in identifying the effects of unsteady flow near the bed in sand-bedded flows (Lesht, 1979) macro-turbulence intensity and its periodic dissipation in steep upland streams remains largely unexplored (Matthes, 1947).

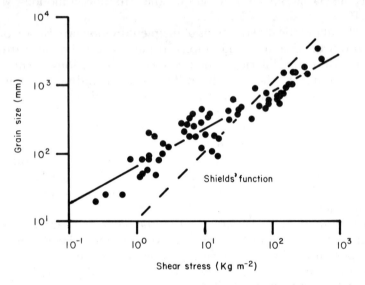

FIG. 2. Particle size and mean shear stress for coarse bedload in natural streams, illustrating the deviation from Shields' laboratory-derived function (after Baker and Ritter, 1975).

Although most models of sediment transport require estimation of bed shear stress or a related parameter, it is possible to avoid its use in sand-bedded streams (Yang, 1979) or to estimate values from extrapolation of linearized sediment transport curves to zero transport (Wilcock, 1971). The problems in applying these methods in upland gravel-bedded streams are two-fold. First, the significance of defining a threshold of motion becomes more important as sediment size increases, because the ratio between available energy and the energy necessary to initiate motion is reduced; consequently, considerable errors may result in sediment transport investigations from the use of ill-defined threshold values. Second, the extrapolation technique requires the assumption of continued linearity of the extrapolated rating curve at very low threshold conditions. It is at these low transport rates that considerable deviation from a linear relationship may occur (Bagnold, 1977).

Because of the problems in choosing threshold values appropriate to field conditions, many field workers obtain limited field data relating grain size to hydrodynamic parameters, in order to choose and qualify relationships for further general application. This is usually achieved with tracers such as tagged cobbles (Butler, 1977). However, radioactive tracers (Ramette and Heuzel, 1962), surrogate cobbles (Cheetham, 1979) and, more recently, acoustic devices (Richards and Milne, 1979) have been used. Alternatively, mobile materials may be sampled directly and the grain size related to the dynamics of the flow (p. 168).

2. Sediment transport

Transport mechanisms. Materials transported in a stream have four immediate sources; (1) atmospheric fall-out, (2) contributions from overland flow, (3) bank erosion, and (4) mobilization of previously deposited material from the streambed. Item (1) largely consists of organic matter such as leaf fall and is discussed on pp. 170–171. Items (2) and (3) are of interest here in that they represent sources of materials that may become part of the bed material.

Once the threshold stress acting on individual grains of the bed has been exceeded sediment transport will occur. The modes of sediment transport can conveniently be defined as traction load, saltation load, suspension load and dissolved load. Traction load and saltation load are frequently grouped together as bedload in fluvial systems, whilst material that is so fine that it does not settle from suspension in natural flow is called wash load. Dissolved load is not considered here as it rarely contributes to the bed material in upland UK streams.

Traction load consists of particles always in contact with the bed; they move by sliding and rolling.

Saltation load makes frequent contacts with the bed, moving by bouncing (Francis, 1973), and usually does not extend more than a few grain diameters from the bed (Bagnold, 1973; Francis, 1973; Moss *et al.*, 1980).

Suspension load is supported by the flow. No contacts which can be attributed to gravitational forces are made between grains and the bed. However, turbulence will result in infrequent contacts between particles and the bed. Generally the criterion for suspension may be written:

$$U_* > V_s. \tag{15}$$

As the flow strengthens and slackens with, for example, the passage of a flood hydrograph, material may be transported in any or all of these modes.

Bedload. Most bedload material in upland streams consists of gravel- and cobble-sized materials. Sampling methods do not usually differentiate the two modes, traction and saltation.

Sampling is achieved by trapping material in samplers lowered into the flow or by collecting materials in traps inserted into the stream bed. Both methods may be costly and time-consuming, and the former may be very inaccurate (Novak, 1957). Sampling and sediment tracing methods have been reviewed by Hubbell (1964), Painter (1972), and Richards and Milne (1979); whilst practical applications of samplers are given by Hollingshead (1971), Nanson (1974) and Emmett (1976) and of traps by Leopold and Emmett (1976).

Sediment transport as bedload is usually expressed as an immersed weight ($0.62 \times$ dry weight) per bed width per second, e.g. $kg\ m^{-1}\ s^{-1}$. Rating curves are expressed as power functions relating transport to parameters such as stream-power measured at a cross-section (Emmett, 1976; Bagnold, 1977). Clearly, transport measured past a station is not directly related to the stress exerted on the bed at that location but rather to conditions immediately upstream of the sampling point. In this respect, in reaches where spatial variations in bed composition and shear stress are significant, more scatter may be expected in data. The commonly drawn distinction between "bedload" and "bed material load" (e.g. Colby, 1961) emphasizes this point. Bed material is not usually completely mobilized until the stress is great enough to move the bed elements forming the framework of the deposit; the bedload, therefore, does not equate with the bed material load until high discharge. As a result similar size distribution curves for bed material and bedload usually reflect extensive bed mobilization. This may be a useful threshold in hydrological-biological investigations. At lower discharges the bedload may not be representative of the bed material (Everts, 1973).

Because of sampling difficulties very little is known of the spatial and temporal distribution of bedload transport throughout a spate, and the frequency and duration of such events throughout the year. Dilution effects and lag effects similar to those described for suspended loads (p. 170) probably occur (Helland-Hansen, 1971; Milhous, 1972b).

Newson (1980a, Fig. 9) recorded bedload movement events for two upland UK streams. Event frequency ranged from two to 13 events per year over a four year period. Annual variation in total load transported km^{-2} of catchment approximated to only one order of magnitude, 10^0 to 10^1 tonnes km^{-2} (Newson, 1980b).

Imeson and Ward (1972) recorded 16 bedload movement events during a year in a small lowland stream and noted that some 98% of material was transported by only five spates.

The depth to which bed material is reworked and the spatial extent of reworking by spates is not known with precision (e.g. Andrews, 1979). Leopold et al. (1964) give data for large rivers where scour may be severe. The maximum scour depth below the preceding bed level for a sand and gravel bed river was 38 m! Data on small headwater streams are very limited; Tagart (1976) reported scour depths of typically 38 mm, with an extreme of 229 mm, associated qualitatively with the largest spate in the water year. McNeil and Ahnell (1964) reported "extensive" scour of salmonid redds but included no measures of scour depth.

Suspended load. A variety of techniques exists to measure suspended load. Most practical methods involve the collection of discrete samples at fixed intervals in time and space. Determinations of the concentration of solids are then made in the laboratory. Alternatively, continuous monitoring techniques are being developed. At present only the use of turbidity meters is practical. These yield surrogate values for concentration and, as they must be calibrated for each individual station, their use is dependent on the assumption of stationarity of processes in the system.

In deep and wide streams there may be considerable concentration gradients laterally and with depth. In these cases, depth-integrated samples are required to represent fully the complete flow. In shallow upland flows, $(D/k_s \simeq 1 \cdot 0)$, concentrations are usually homogeneous in all dimensions. Sampling methods and the inherent errors are discussed and compared by several investigators (Walling and Teed, 1971; Gregory and Walling, 1973; Bennett and Sabol, 1973; Fleming, 1969; Yorke, 1976), whilst an example of information on the lateral and depth variations in suspended sediment concentrations for the R. Tyne, UK is presented by Hall (1964).

Preliminary models of concentration data are normally related to stream discharge, regarded as the prime controlling factor, although more complex models introducing other factors are possible, (Guy, 1964; Walling, 1971).

Simple power functions are used, where:

$$\overline{C} = a \ \overline{Q}^b \text{ and} \tag{16}$$

$$\overline{Q}_s = k \ \overline{Q}^z. \tag{17}$$

The exponent b is commonly equal to 2·0 although it may be as low as 0·80 (Gregory and Walling, 1973), whilst the exponent z is equal to $b + 1$. Equation 16 frequently shows considerable scatter which is attributable to sampling techniques (Rákóczi, 1977), and also to lag and dilution effects in the response of the sediment to the discharge (Costa, 1977). The sediment concentration curve may peak before or after, or may coincide with the hydrograph peak (Heidel, 1966; Al-Ansari *et al.*, 1977). Concentration data may therefore plot as a hysteresis loop against discharge with higher concentrations on the rising stage and lower concentrations on the falling stage for a given discharge (or vice versa).

Scatter may be reduced in the regression by producing individual rating curves for rising and falling stages and for individual seasons. Walling (1977a) identifies eight main subdivisions of the basic rating curve; examples of this differentiation are given by Hall (1967), Gregory and Walling (1973) and Nanson (1974), whilst Walling (1977a, b) explores in some detail the overall assumptions and the accuracy of its application.

Few investigators comment on the dynamics of organic particles in the sediment load. Generally, concentration of organic detritus remains constant in a downstream direction (Fisher and Likens, 1973; Sedell *et al.*, 1978). Fisher and Likens (1973) further noted that the logarithms of the concentration of both coarse (> 1 mm) and fine (< 1 mm) detritus were related to discharge in a positive linear manner. Other investigators have found positive functions, Crisp and Robson (1979) for a wide range of discharge values from a peat catchment found a logistic function and a power law useful after logarithmic transformation, whilst Bormann *et al.* (1974) noted quadratic functions. Differing models probably reflect differences in catchment characteristics as well as sampling variability. Considerable scatter in a relationship for coarse organic material (Fisher and Likens, 1973) was related to irregular releases of organic debris (e.g. branches, leaves) as log jams broke up under high flows (e.g. Maciolek, 1966). Transport of 74% of organic throughput was related for this reason to only two spates in a year.

Similarly Bormann *et al.* (1974) found 87% of throughput occurred during 0·4% of the year.

Evidence relating the dynamics of fine organic fractions (< 53 μm) to flow is conflicting. Some investigations have indicated that concentrations are not related to discharge, and transport of organics is dynamically dissimilar to inorganic sediment transport (Bormann *et al.*, 1974; Sedell *et al.*, 1978; Naiman and Sedell, 1979). In contrast, positive correlations between discharge and concentration of fine organic particles (< 1 mm) have been recorded, as well as hysteresis loops similar to those described for inorganic fractions (Fisher and Likens, 1973; Bilby and Likens, 1979). In particular, Gurtz *et al.* (1980) present time sequences of comparable concentrations of inorganic and organic concentrations that fluctuate in sympathy with each other and with discharge.

Some disagreement between the results of investigations may be related to differing definitions of particle sizes but must also be related to the variable characteristics of the source of material in individual catchments. Maciolek (1966) concluded that most organic debris is allocthonous and that storage of fine organics in the bed was not important. In contrast Nelson and Scott (1962) concluded that only some 66% of material was allocthonous in their experimental catchment, and that storage in the river bed (up to 300 g m^{-2}) was an important source of suspended organics. Similarly, the concentration of organics has been reported higher in winter than summer (Malmquist *et al.*, 1978) and vice versa (Nelson and Scott, 1962). Seki *et al.* (1969) reported that material was finer in winter than summer, whilst Naiman and Sedell (1979), summarizing earlier work, concluded that most organic particles were in the size range 3–20 μm.

Laboratory investigations would be of considerable assistance. Fisher *et al.* (1979) report a small number of data from flume studies. They concluded that organic fragments (leaves, stems, etc.) are resuspended at similar threshold stress values as spheres of the same volume and density. However, they admitted to considerable scatter in their data. In contrast, Shapley and Bishop (1965) from indirect field evidence, concluded that organic fractions had lower settling rates than inorganic particles.

Transport equations. Equations developed to predict sediment transport may be theoretical or semi-empirical models, (Engelund and Fredsøe, 1976). Tanner (1977) reviewed the status of sediment transport investigations. He concluded that most streams were capable of rational analysis and models could be rationalized, even if empirical.

Purely empirical models offer the best solution to accurate short term derivations of transport potential, but require considerable field data. However, semi-empirical models may be used with limited information on the hydraulics and sediment characteristics of the system. Models are divided into (1) total load models (2) suspended load models and (3) bedload models. Widely applied models are reviewed by White *et al.* (1978). Although suspended sediment models may predict observed field relationships, bedload models suffer from the lack of field verification. Most models of bedload transport apply strictly to sand beds and have few empirical data supporting their application to gravel-bedded rivers.

The choice of a model is usually made pragmatically, by fitting limited field data to various models and selecting the most appropriate (e.g. Muir, 1970; Al-Ansari *et al.*, 1978). Widely applied models in gravel-bedded streams include the Schoklitsch, Shulitz, Meyer–Peter and Mueller models (Church, 1972; Bjorn *et al.*, 1974). However, Emmett and Leopold (1977) tested five models against extensive bedload transport data and concluded that none gave very good agreement. The major limitation lay in the necessity to derive a measure of bed roughness based on a "representative" grain size for the bed. This was found to vary with stage, a point already noted (p. 160). Bagnold (1977) used a streampower model wherein the relative depth is allowed to vary and obtained a good fit to observed data from gravel-bedded rivers but this model has not yet been tested elsewhere.

Considerable progress cannot be expected until more detailed and careful measurements are made of bedload transport rates in streams. The recent investigations of Emmett (1976); Leopold and Emmett (1976, 1977) and Newson (1980a) are examples of the kind of data that are required to improve model formulation.

3. *Deposition*

Deposition occurs when the shear velocity is less than the settling velocity of a grain:

$$U_* < V_s. \tag{18}$$

This is a simplification, as the influence of adjacent grains (Maude and Whitmore, 1958) the vertical distribution of sediment (Parker, 1979b) and upward turbulent dispersion (Engelund, 1970) are factors influencing settling rates. An extremely useful review of commonly applied equations, estimating the settling velocities of grains, is given by Graf and Acaroglu (1966).

There is very little information on the periodic deposition of fine sediments along a river reach (Costa, 1977); although deposition is commonly linked to the recession of a flood hydrograph (Wolman *et al.*, 1972). Deposition may counter-balance resuspension of materials, maintaining a quasi-equilibrium state of the active bed (Mackin, 1948; Leopold and Maddock, 1953; Crickmay, 1974).

Trapping of fine sediments in gravel beds has been investigated by Einstein (1968) and Beschta and Jackson (1979). The latter found that sand intrusion rates were a complex function of concentration, velocity and grain size. Field investigations are few (Adams and Beschta, 1980) although the sediment traps described by Welton and Ladle (1979) might be profitably applied. The importance of a laminar sublayer in trapping fine sediments (McCave, 1970) has not been demonstrated in fluvial systems.

III. INVERTEBRATES

A. INTRODUCTION

Benthic invertebrate populations show considerable spatial and temporal variation in abundance, diversity and composition. Many closely inter-related factors influence invertebrate populations in rivers and streams. The most important are considered to be water velocity, the nature of the substratum (including attached vegetation), food availability, temperature, water quality and predation (Macan, 1963; Hynes, 1970). In addition to the general review by Hynes (1970), certain aspects of the streamflow requirements of the benthos have been discussed by other authors in relation to river regulation (Giger, 1973; Ward, 1976a; Brooker, 1981). The present discussion is limited to the effects on invertebrates of physical channel characteristics such as width, depth, gradient, roughness and type of bed, and hydraulic features such as water velocity variation.

B. INVERTEBRATE FAUNA: GENERAL CONSIDERATIONS

Similarity of riffle fauna found in unpolluted rivers and streams has been noted by many authors in the UK (Jones, 1941, 1943, 1948; Hynes, 1961; Maitland, 1964; Morgan and Egglishaw, 1965; Minshall and Kuehne, 1969; Armitage *et al.*, 1974, 1975; Fahy, 1975; Brooker and Morris, 1980). The fauna is usually dominated by insect groups such as Ephemeroptera, Plecoptera, Trichoptera and Diptera, many of which are highly adapted to conditions in fast-flowing water, and exhibit a wide variety of morphological, physiological and behavioural modifications. Several species are confined to the lotic habitat because of inherent current requirements associated with respiratory physiology (Philipson, 1954; Ambühl, 1959, 1961; Jaag and Ambühl, 1964) or feeding (Harrod, 1964; Edington, 1965, 1968; Philipson, 1969). Many stream invertebrates are not exposed to appreciable water velocities because they normally reside in a thin zone above solid surfaces, the laminar sublayer, within which velocities are greatly reduced (p. 157). Other "dead water" zones behind large particles and in the interstices between particles offer additional shelter. Morphological adaptations of invertebrates, such as body flattening and streamlining, enable them to avoid strong currents by occupying these places.

C. BED COMPOSITION

1. *Nature of surface bed materials*
Several studies in Europe (Percival and Whitehead, 1929; Illies, 1958; Thorup, 1966) and North America (Linduska, 1942; Pennak and Van Gerpen, 1947; Sprules, 1947; Cummins, 1964, 1966; Kennedy, 1967; Minshall and Minshall, 1977) suggest that substratum type may be the most important factor determining the distribution of many species. Heterogeneity of substratum particle size is of critical importance in providing varied microhabitats which are able to support an abundant and diverse fauna (Hynes, 1970; Giger, 1973; Ward, 1976a). Qualitative invertebrate surveys (e.g. Percival and Whitehead, 1929; Pennak and Van Gerpen, 1947; Sprules, 1947) indicate a direct relationship between increasing particle size, ranging from sands through to cobbles, and invertebrate abundance and diversity. Some authors (Scott and Rushforth, 1959; Scott, 1960) suggested that invertebrate abundance may be directly related to the surface area of individual particles, particularly their lower and lateral surfaces which afford protection from strong currents. However, this has not been confirmed in other studies (Allen, 1959; Minshall and Minshall, 1977; Rabeni and Minshall, 1977). There is apparently little field information on the influence of particle shape, texture and degree of embeddedness of surface particles on invertebrate distribution and abundance. Some evidence suggests that increased irregularity of shape and texture of particles (large gravels and cobbles) leads to increased abundance and diversity of invertebrates, particularly insects (Hart, 1978). From a laboratory study, Brusven and Prather (1974) reported a progressive reduction in numbers of several species of Ephemeroptera, Plecoptera and Trichoptera on cobble substrata (115 mm, mean diameter) which were increasingly embedded in surrounding sand (1·0

−1·5 mm diameter). They attributed the faunal reductions to a restriction in access to the lower, more sheltered, surfaces of large particles. In contrast, Percival and Whitehead (1929) observed an increase in faunal densities on firmly embedded stones compared with loose stones, which suggests that substratum stability may be an additional important factor.

2. Microdistribution of invertebrates

Even within fairly uniform riffle areas, there is a mosaic pattern of invertebrate distribution which can be related to local patterns and interactions of water velocity, substratum type and food (Needham and Usinger, 1956; Ulfstrand, 1967; Chutter, 1969a; Egglishaw, 1969; Minshall and Minshall, 1977; Gore, 1978). Linduska (1942) described the typical mayfly fauna of a Montana stream in terms of the positions of individual species on boulders relative to local variation in water velocity. Scott (1958) attributed the water velocity preferences of several species of trichopteran larvae to food supply. *Stenophylax* and *Odontocerum* were always found under stones in areas where surface velocities were less than 20 cm s^{-1}. These species feed on detritus which collects in such situations. At surface velocities of over 40 cm s^{-1} *Rhyacophila* occurred under stones where there was a rich supply of small insects on which it feeds and on the upper surfaces of stones where moss provided shelter. Many other invertebrate groups tolerate a wider range of water velocities (Needham and Usinger, 1956; Chutter, 1969a).

Few studies in the UK have investigated variations in faunal abundance and diversity within individual riffles, but several North American studies on natural rivers and experimental channels suggest that both abundance and diversity are greatest in areas of medium water velocity (20–75 cm s^{-1}, usually determined during low flows) over cobble substrata (6–25 cm diameter) at depths of 15–40 cm (Surber, 1951; Ruggles, 1966; Kennedy, 1967; Gore, 1978). Where algae and moss predominate, usually during the summer months, many authors have reported a considerable increase in abundance of various groups of invertebrates (Percival and Whitehead, 1929; Idyll, 1943; Frost, 1942; Hynes, 1961). Occasionally, heavy algal growths such as those often found downstream of impoundments may reduce the densities of certain invertebrate groups such as the Simuliidae and some ephemeropteran nymphs (*Rhithrogena*) by decreasing the surface area of exposed stones used for attachment sites (Ward, 1976b). Mosses usually provide shelter and attachment sites for other groups such as eliminthid larvae (Coleoptera), chironomids and crustaceans which may feed on trapped detrital material (Percival and Whitehead, 1929; Frost, 1942; Hynes, 1961). Heavy algal growths frequently support high densities of herbivorous naidids (Oligochaeta) and certain chironomids (Von Mitis, 1938; Oliff *et al.*, 1965).

Organic detritus may also influence the microdistribution of the benthos. Many studies have indicated that most, if not all, running-water habitats in temperate regions which have not been significantly altered by man are predominantly heterotrophic (Cummins, 1966; Hynes, 1970; Mundie, 1974; Boling *et al.*, 1975). Thus the maintenance of invertebrate communities depends upon the import of organic matter. Numerous studies have emphasized the importance of autumn-shed leaves (Hynes, 1963; Egglishaw, 1964; Kaushik and Hynes, 1968, 1971; Cummins, 1973; Cummins *et al.*, 1973; Bärlocher and Kendrick, 1974; Petersen and Cummins, 1974; Willoughby, 1974; Berrie, 1975). However, many upland streams in Britain rise on open peat moorland with few or no riparian trees; in such situations the import of fine peat particles provides a major proportion of organic matter input (Crisp, 1966; Brennan *et al.*, 1978; Crisp and Robson, 1979). The microdistribution of several insect groups such as the plecopterans, *Leuctra inermis*, *Amphinemura sulcicollis*, the ephemeropterans *Baetis*, *Rhithrogena* and certain Chironomidae have been positively correlated with the distribution of coarse and fine organic detritus chiefly arising from terrestrial leaf material (Egglishaw, 1964, 1968, 1969). Such inter-relationships have been confirmed for certain Ephemeroptera, Plecoptera, Trichoptera and Chironomidae in North American studies on the colonization of artificial substrata (Barber and Kevern, 1973; Rabeni and Minshall, 1977). Studies on the upper reaches of the R. Tyne (England) have revealed close associations between the micro-distributions of various species of chironomids and the amount of deposited peat material used as a food source and for tube-building (Brennan *et al.*, 1978; McLachlan *et al.*, 1978; Brennan and McLachlan, 1979; Walentowicz and McLachlan, 1980).

3. *Macrodistribution of invertebrates: riffles and pools*
The most biologically important physical differences between riffles and pools appear to be their water velocities, depths and substratum types. Most studies in the UK (Egglishaw and Mackay, 1967; Hughes, 1975; Scullion, 1977) and North America (Lymen and Dendy, 1943; Briggs, 1948; O'Connell and Campbell, 1953; Minshall and Minshall, 1977) suggest that riffles support higher densities of macroinvertebrates than pools. Differences in invertebrate community structure within riffles and pools have also been observed; with higher densities of burrowing

forms such as oligochaetes and certain tipulids and chironomids in pools and higher densities of ephemeropterans (Baetidae and Ecdyonuridae), trichoptera (Hydropsychidae) and dipterans (Simuliidae) in riffles (O'Connell and Campbell, 1953; Egglishaw and Mackay, 1967; Minshall and Minshall, 1977; Scullion, 1977). However, few species appear to be restricted to either riffle or pool zones within the same river system.

Substratum differences may be more important in determining invertebrate distribution within riffles and pools than water velocity or depth variation. The introduction of unsilted artificial substrata (32–64 mm diameter) into pools resulted in an increase in the abundance of certain species of Ephemeroptera, Plecoptera and Trichoptera, whose densities were normally low in pools compared with riffles (Minshall and Minshall, 1977). Similar increases in faunal abundance on artificial substrata in pools have been reported in other North American studies (Wene and Wickliff, 1940; Rabeni and Minshall, 1977).

4. *Vertical distribution of invertebrates*

A substantial proportion of benthic riffle fauna has been found to extend down into gravel beds at least to depths of 20–50 cm in North America (Coleman and Hynes, 1970; Mundie, 1971; Radford and Hartland-Rowe, 1971a; Williams and Hynes, 1974; Gilpin and Brusven, 1976; Poole and Stewart, 1976) and in Europe (Schwoerbel, 1961; Morris and Brooker, 1979). Some groups such as Simuliidae and certain of the Ephemeroptera are associated with surface materials (Coleman and Hynes, 1970; Williams and Hynes, 1974; Poole and Stewart, 1976; Morris and Brooker, 1979) while groups such as Oligochaeta, Diptera and Coleoptera often form a considerable proportion of the interstitial fauna (Gilpin and Brusven, 1976; Poole and Stewart, 1976; Morris and Brooker, 1979).

The vertical distribution of invertebrates within stream beds has been little studied in relation to the chemical and physical environment. Factors likely to be important include particle size, shape, texture and composition, the degree of compaction and amount of void space which, in turn, may affect the distribution of water velocity, oxygen concentration and fine organic detritus. Williams and Hynes (1974) employed a coring device to collect invertebrates at depth and a freeze-coring technique (Stocker and Williams, 1972) to collect undisturbed substratum samples. No significant correlations were found between invertebrate abundance at depths down to 70 cm and mean particle size or heterogeneity of particle size, intragravel flow or temperature. The distribution of fine organic material and the oxygen concentration were, however, directly related to faunal abundance. Other workers have reported associations between the densities of interstitial fauna and the amount of fine organic matter (Schwoerbel, 1961) and oxygen concentrations (Poole and Stewart, 1976). Poole and Stewart (1976) also reported that increased proportions of fine sands and silts below 20 cm within a river bed apparently restricted the vertical distribution of insect groups such as Ephemeroptera and Coleoptera.

D. BED DYNAMICS

1. *Erosion (flow increase)*
Behavioural reactions. As discharges increase invertebrates can avoid local increases in water velocity by repositioning on stones, moving into slower-flowing areas or burrowing into the substratum. Species highly adapted to fast-flowing conditions such as *Simulium* frequently seek shelter on stable substrata during high flows and thus are rarely swept downstream (Zahar, 1951). Other species associated with slower-flowing reaches have been observed to move towards more sheltered banks as flow increases, these include the amphipod *Gammarus* (Hynes, 1954) and ephemeropterans such as *Leptophlebia*, *Habrophlebia* and *Caenis* (Pleskot, 1953). Whilst the occurrence of invertebrates deep within the bed has been well documented, little is known about the effects of discharge on their vertical distribution. A few North American studies suggest that some taxa may rapidly move downwards in response to increasing discharge. Williams and Hynes (1974) reported that peak numbers of invertebrates shifted from a depth of 10 cm to a depth of 30 cm 24 h after a spate, but samples taken 48 h after another spate did not differ from samples taken in low flows. Poole and Stewart (1976) observed that after high flows in the impounded River Brazos, densities of the net-spinning trichopterans *Cheumatopsyche* and *Hydropsyche* and the mayfly nymph, *Neocoroterpes* were reduced at the substratum surface but increased at depths of 10–20 cm.

Invertebrate drift. Downstream invertebrate drift has been studied intensively, principally in relation to its quantity, composition, diel variation and value as a source of food for fish. Most attention has been given to low or moderate discharges where increases in discharge have been related to increases in drift rates (Chapman, 1966b; Elliott, 1967; Waters, 1969; Everest and Chapman, 1972). There have been fewer studies on the quantitative and qualitative changes in invertebrate drift which may occur during spates, but this "catastrophic drift" has been reported in a general manner (Harrison and Elsworth, 1958; Maitland, 1964; Anderson and Lehmkuhl, 1968: Pearson and Franklin, 1968; Pearson and Kramer, 1972; MacKay and Kalff, 1973).

In only two studies in the United Kingdom have increases in both total numbers and densities of drifting invertebrates been associated with increases in flow. Brooker and Hemsworth (1978) observed a seven-fold increase in the total numbers of drifting invertebrates as a result of about a three-fold increase in discharge associated with a reservoir release in the River Wye, mid-Wales, over a three day period. The normal diel variation in drift rate, with increased numbers at night, reported under natural discharges (Waters, 1962; Elliott, 1968; Elliott and Minshall, 1968) was enhanced by increased discharge in groups such as the Ephemeroptera and Coleoptera, whereas the chironomid larva, *Rheotanytarsus*, responded immediately to flow increases as has been reported for other chironomids under natural flows (Anderson and Lehmkuhl, 1968). In a detailed investigation of total invertebrate drift in a small (1 m wide) moorland stream in Cumbria, Crisp and Robson (1979) found that maximum concentrations and transport of invertebrates such as Simuliidae, Plecoptera and Trichoptera preceded peak discharge. Maximum concentration of the terrestrial component of the drift coincided with maximum flows.

Bed stability. Spates can be associated with substantial bedload movement (pp. 168–169) and several authors have reported a considerable reduction in invertebrate abundance following such mechanical disturbance of bed materials (Beauchamp, 1932; Surber, 1937; Tarzwell, 1937; Briggs, 1948; Jones, 1951; Elliott, 1967; Kennedy, 1967; Herricks and Cairns, 1972; Wise, 1980). However, the apparent reductions in faunal abundance immediately after floods may be explained, in part, by the sheltering of certain components of the benthos at greater depths within the river bed (Poole and Stewart, 1976). Recovery of the fauna from the effects of exceptional spates is normally quite rapid (< 1 year) and in many rivers which experience considerable disturbance of bed materials by regular seasonal spates, there is a corresponding seasonal change in invertebrate abundance, with fewest animals occurring after floods (Harrison, 1958; Harrison and Elsworth, 1958; Gaufin, 1959; Oliff, 1960, 1963).

There is evidence that the scouring effects of large spates on invertebrates are not uniform across the width or along the length of a river. Petran and Kothé (1978) reported reductions in fauna where bedload movement was most prevalent, usually in midstream. Allen (1951,

1959) also reported a greater reduction in numbers of invertebrates on the bed surface in the faster-flowing centre of a river than in more sheltered areas near the banks. Downstream increases in faunal abundance following severe spates were observed by Minckley (1963) who attributed it to the scouring of invertebrates from upstream areas where densities were reduced.

Several studies have shown a differential effect of floods on various invertebrate groups (Sprules, 1947; Jones, 1951; Harker, 1953; Minckley, 1963; Brooker and Hemsworth, 1978) and on closely related species (Harker, 1953). Groups such as the Ephemeroptera, Plecoptera and Chironomidae appear to be more susceptible to the scouring action of spates than other, perhaps, more firmly attached groups such as Trichoptera and Simuliidae. Harker (1953) reported that two closely related species of ecdyonurid mayflies were affected differentially by spates, *Ecdyonurus torrentis* being more tolerant than *Heptagenia lateralis*. The effects seemed to vary with size in *Ecdyonurus torrentis*; larger specimens were frequently crushed during floods while smaller individuals survived. Little quantitative information is available on the differential effects of spates on invertebrates inhabiting different types of bed materials. Maitland (1964) reported that following high river flows there was a much greater reduction in the numbers of burrowing forms such as Tubificidae (Oligochaeta) and Chironomidae in sand substrata than in an upstream stony riffle area. Mechanical disturbance of bed materials during spates may cause a change in species composition of the benthos. In gravel beds of Bere stream the lumbriculid oligochaete, *Stylodrilus heringianus* was replaced by tubificids washed from upstream soft sediments after flood disturbance (Ladle, 1971).

Downstream of some river impoundment schemes, the erosive effects of high variable discharges have resulted in washout of fine particles and a general reduction in invertebrate abundance (Blanz *et al.*, 1969; Radford and Hartland-Rowe, 1971b; McGary and Harp, 1972; Trotsky and Gregory, 1974; Ward and Short, 1978). The effect of these modified discharge regimes on the benthos appear to be similar to, but more severe than, the action of natural floods: the normal riffle insect fauna comprising Ephemeroptera, Plecoptera, Trichoptera and Coleoptera is greatly reduced. However, within these taxa, certain tolerant species are able to survive immediately downstream of dams. Radford and Hartland-Rowe (1971b) found that three species of mayfly, especially adapted to torrential conditions, comprised the bulk of the fauna under water velocities exceeding 2 m s^{-1}. The burrowing stonefly nymph, *Alloperla*, which normally inhabits the hyporheic zone has frequently been reported under similar flow regimes (Radford and

Hartland-Rowe, 1971b; Trotsky and Gregory, 1974; Ward and Short, 1978).

Stability of coarse bed materials (large gravels and cobbles) may be attained over a long period immediately downstream of hydroelectric impoundments and results in an increase in abundance of a few tolerant species such as certain Chironomidae and amphipod and isopod crustaceans (Blanz *et al.*, 1969; Hoffman and Kilambi, 1970). Radford and Hartland-Rowe (1971b) obtained similar results and concluded that the negligible effect of spates on the invertebrate fauna in a regulated river, compared with a considerable reduction in the fauna in an unregulated tributary, resulted from the increased bed stability afforded by long-term exposure to numerous reservoir releases. Several studies have shown a general reduction in both macrophytes and invertebrate fauna, or at least components of them, in unstable upland streams as compared with less "flashy" streams, although differences in temperature and water quality may also be important limiting factors (Jones, 1948; Badcock, 1953; Egglishaw and McKay, 1967).

2. Transport

The transport phase of bed dynamics is concerned with suspensions composed of organic (both living and dead) and inorganic particles.

Invertebrates. Various aspects of invertebrate drift have been discussed earlier in relation to erosion. The present discussion deals with rates of downstream displacement of drifting invertebrates. This phenomenon has rarely been investigated in upland rivers and streams (Mackay, 1970; Elliott, 1971; Hemsworth and Brooker, 1979). The proportion of benthos drifting at any one time is generally small (0·5%) under low or moderate flows in upland rivers (Elliott, 1965; Ulfstrand, 1968; Bishop and Hynes, 1969; Radford and Hartland-Rowe, 1971a; Armitage, 1977; Hemsworth and Brooker, 1979). Assuming that such drifting organisms travel at the average water velocity, estimates of downstream displacement of the total benthos in the upper reaches of the River Wye ranged from 2·9–55·9 m d^{-1} (Hemsworth and Brooker, 1979). Considerably smaller estimates of downstream displacement (0·6–1·9 m d^{-1}) were recorded by Crisp and Gledhill (1970) in a lowland chalk stream. Estimates such as these are only approximate as they do not take into account within-site variability in drift rates, discharge variations, water velocity variations between reaches (riffles and pools), behavioural reactions of drifting invertebrates, the usual sampling errors of benthic populations and the vertical distribution of invertebrates within the river bed. Nevertheless they probably reflect the order of mean displacement rates which occur in upland rivers. Evaluation of such movements in terms of distributional stability of populations has been limited to studies on the crustacean, *Gammarus* (Lehmann, 1967) and a few species of Ephemeroptera, Plecoptera and Trichoptera (Hemsworth and Brooker, 1979). For most species studied, the estimated net downstream displacement of cohorts in their aquatic phase is less than 10 km per generation period. Since no account was taken of spates it is possible that catastrophic drift (pp. 178–179) may be responsible for larger downstream shifts of some invertebrates (Crisp and Robson, 1979).

Organic detritus. Much of the hydrological literature makes no attempt to separate organic and inorganic components of the suspended solid load, though this distinction is of considerable biological importance (pp. 170–171). Many common stream insects such as the Simuliidae and certain Trichoptera have developed highly specialized filtering mechanisms to feed on seston. This is the subject of a detailed review (Wallace and Merritt, 1980). Filter-feeder abundance is strongly influenced by both the quality and quantity of seston: high densities of simuliids and hydropsychids have been found in lake outlets and below some impoundments, followed by a dramatic downstream decrease associated with declining seston quantity as drifting phytoplankton and zooplankton are rapidly removed (Cushing, 1963; Maciolek and Tunzi, 1968; Spence and Hynes, 1971; Ward and Short, 1978; Jonsson and Sandelund, 1979). The relative nutritive value of various types of drifting organic matter has yet to be established, although microalgae are considered to be of high quality because of the protein and lipid content (Naiman and Sedell, 1979). Some authors have concluded that peat particles are nutritionally poor, partly because of their high organic carbon and low organic nitrogen content (Toscano and McLachlan, 1980; Walentowicz and McLachlan, 1980). The absence of certain detritivorous insect larvae downstream of irrigation and hydroelectric power dams has been related to the erosion of allochtononous material from the river bed by frequent periodic flooding (Radford and Hartland-Rowe, 1971b; Ward and Short, 1978).

Mineral particles. High suspended solid concentrations of short duration are a characteristic feature of spates in upland rivers. Although the abrasive action of suspended mineral particles has been related to reductions in macrophyte or faunal abundance (Harrison and Elsworth, 1958; Radford and Hartland-Rowe, 1971b; Brennan *et al.*, 1978), its direct effect has not been quantified and is frequently confounded by concomitant increases in water velocity and bedload movement. Prolonged high concentrations of inert particulate suspensions may have an adverse effect on mosses (Lewis, 1973) but direct effects of suspensions on invertebrates have not been established.

3. Deposition (flow reduction)

Siltation. There is considerable evidence from pollution studies that the deposition of heavy loads of fine (< 2 mm diameter) inorganic particles on a river bed generally results in a reduction in total faunal abundance (e.g. reviews by Cordone and Kelly, 1961; Gebhardt, 1969; Iwamoto *et al.*, 1978). Members of groups such as the Ephemeroptera, Plecoptera, Trichoptera and Coleoptera appear to be more sensitive than others such as Oligochaeta and Chironomidae (European Inland Fisheries Advisory Commission, 1964; Nuttall, 1972; Scullion and Edwards, 1980). Deposition rates, even in pools, are rarely as high as those reported in most pollution studies; though siltation of pools may sometimes be sufficient to reduce plant growth and faunal abundance and alter invertebrate community structure (pp. 176–177). Similar results have been reported from field and laboratory studies on the effects of "light" siltation by fine (< 0·2 mm) inorganic materials on the fauna in riffle areas in which the interstices between larger particles (gravels and cobbles) were not completely silted (Hamilton, 1961; Herbert *et al.*, 1961; Chutter, 1969b; Cummins and Lauff, 1969; Gammon, 1970; Rabeni and Minshall, 1977; Rosenberg and Wiens, 1978). Much more information is required on the deposition and redistribution of fine particles, in terms of both organic and mineral components, and the effects of flow regime variations in terms of flood magnitude, frequency and timing. Without this information, it is difficult to assess biological effects, except in the most general terms.

Hydraulic aspects. In addition to particle deposition, flow reduction may exert a broader influence on invertebrate populations. In unregulated rivers, reductions in discharge tend to be more gradual than increases and base flows are normally reached over a long period. Rapid discharge reductions in experimental channels (Minshall and Winger, 1968), or as a result of reservoir manipulations associated with irrigation (Gore, 1977) and hydroelectric power generation (Radford and Hartland-Rowe, 1971b) have resulted in an increase in the drift rate and density of insect groups such as Ephemeroptera, Plecoptera and Diptera (Simuliidae). Laboratory studies have shown that an abrupt cessation in water flow initiates swimming reactions in certain members of these groups (Elliott, 1967; Madsen, 1968, 1969). In addition, the stranding and subsequent dessication of major insect groups has been observed during rapid flow reductions below impoundments (Kroger, 1973; Gore, 1977).

Periods of prolonged low summer discharges are a characteristic feature of upland rivers and streams in temperate climates and appear to have few harmful effects on the invertebrate fauna. Increased faunal densities have been recorded in rivers subjected to considerable discharge reductions as a result of exceptional natural droughts, even though wetted areas were not substantially reduced as compared with normal summer base flows (Brooker, 1978). During extended droughts, loss of submerged area may result in a decrease in area of invertebrate production. In extreme drought conditions the elimination of groups such as the Ephemeroptera, Plecoptera and Trichoptera, by stranding and dessication, have been observed in headwaters of small streams (Engelhardt, 1951; Hynes, 1958, 1961; Larimore *et al.*, 1959). Other groups such as Platyhelminthes, Oligochaeta, Ostracoda, Copepoda, Hydracarina and certain chironomids are able to avoid dessication by burrowing into the river bed. Other studies (Gilpin and Brusven, 1976) showed that less dramatic flow reductions had little effect on the vertical distribution of invertebrates.

IV. Salmonid Biology

A. INTRODUCTION

The two main European salmonid species, the Atlantic salmon (*Salmo salar* L.) and the trout (*Salmo trutta* L.) in both its freshwater and anadromous forms, typically use the gravel beds of upland rivers for egg deposition. The interactions between discharge, substratum composition, and sediment movements can have an important influence upon spawning site choice, survival of intragravel stages, emergence of swim-up fry and the growth and survival of older stages.

There is considerable literature on these topics. Most of it is of North American origin and much of it refers to siltation effects in association with logging. This literature has been reviewed by Cordone and Kelley (1961); Gibbons and Salo (1973); Giger (1973); Parkinson and Slaney (1975) and Iwamoto *et al.* (1978). Other information, with special reference to stream regulation, has been brought together by Fraser (1972, 1975).

B. SPAWNING, REDD FORMATION AND REDD STRUCTURE

River-spawning salmonids typically deposit their eggs in gravel beds which are commonly situated in the upper reaches of river systems. Spawning behaviour, redd formation and structure have been described in detail for Atlantic salmon (Belding, 1934; White, 1942; Jones and Ball, 1954; Jones and King, 1949, 1950) and brown trout (Greeley, 1932; Hobbs, 1937; Stuart, 1953b, 1954; Jones and Ball, 1954). The general observation that salmonids spawn in the same areas year after year suggests that spawning gravel beds have characteristic properties which the fish are able to detect, but the cues which stimulate spawning in a particular area are not fully understood and may vary between species. Various mechanisms have been proposed including tactile and visual responses to the gravel surface structure (Jones, 1959), accelerating water velocity (White, 1942), groundwater seepage (Benson, 1953; Hansen, 1975), downward movement of stream water into gravel (Stuart, 1953a, 1954) and oxygen concentration (Hansen, 1975).

The reaction of spawners to physical cues usually results in redd formation in areas of low silt content and high intragravel flows (White, 1942; Stuart, 1953a, b, 1954). The selection is not always precise and fish may spawn in unsuitable silted gravel (Hobbs, 1937; White, 1942), particularly if over-crowding occurs on the better spawning beds (Peterson, 1978). Stuart, (1953b) states that in small rocky streams brown trout may spawn in small patches of gravel between larger boulders which, although suitable for egg incubation at the time of excavation, subsequently become scoured away by floods and ice, or dry up. Redd excavation may result in significant coarsening of the gravel (Platts et al., 1979) as fine particles, which may contain a high proportion of organic material (McNeil and Ahnell, 1964), are carried away by the current (Hobbs, 1937; Royce, 1959; Peters, 1962; Cooper, 1965, 1974). A certain minimum current is necessary to initiate motion of particles (p. 164) in addition to minimum velocities which may be necessary to enable fish to orientate to the gravel bed and dig efficiently (McCart, 1969).

There are few observations on the velocity and depth requirements of Atlantic salmon and brown trout. Jones and King (1950) observed that Atlantic salmon in an experimental stream spawned in depths ranging from 15 to 61 cm; although selected surface velocities varied less, with optimal values between 30 and 45 cm s^{-1}: spawning was not observed at velocities < 5–8 cm s^{-1} (Jones and King, 1950). Reiser (1976) found that brown trout utilized velocities ranging from

22–38 cm s^{-1} (measured at 0·6 of water depth) and depths from 14–28 cm as measured over the pit of the redd. Fraser (1975) has summarized depth and velocity criteria from many North American salmonids and, while interspecific differences in current selection have been observed (Arnold, 1974; Fraser, 1975), much of the variation is probably an expression of size variation in the fish, larger fish being able to stem faster currents by virtue of their greater swimming speed (Brett, 1965).

Particle sizes utilized for redd formation also vary according to fish size (Burner, 1951). Several detailed surveys have been carried out on the physical characteristics of Pacific salmonid spawning gravel (Burner, 1951; Fraser, 1975), but less information is available for Atlantic salmon and brown trout. White (1942) observed that the former can spawn in a mixture of sand and fine gravel containing little coarse gravel, but not in coarse gravel alone. Apparently normal behaviour was observed in Atlantic salmon spawning in lake shore gravel (ranging from 6–51 mm, with admixtures of coarse silt) in an experimental stream tank (Jones and Ball, 1954). Peterson (1978) made a detailed survey of gravel composition in Atlantic salmon spawning areas of New Brunswick streams. A wide range of composition was used, with particles > 150 mm diameter contributing 12–81% by weight of the samples. Few systematic data are available for brown trout although there are qualitative descriptions (Hobbs, 1937; Stuart, 1953b; Solomon and Templeton, 1976). Reiser (1976) concluded that for brown trout in Wyoming streams 70% by weight of the gravel should fall within the range 6–76 mm.

Somme (1960) suggested that salmonids may be unable to distinguish gravel which will dry out during the incubation period from that which remains permanently submerged and that losses from drying out in a Norwegian river were greater for sea trout than salmon due to the greater excavation depth of the latter. The depths at which salmonids deposit their eggs increase with fish size (Greeley, 1932; White, 1942). Depths of 15–30 cm have been reported for Atlantic salmon (Belding, 1934; White 1942) and brown trout eggs have been found at depths of 8–22 cm (Hardy, 1963).

Redd superimposition may be a major source of mortality in Pacific salmonids that spawn at high densities (Neave, 1958; McNeil, 1964, 1967), though Parkinson and Slaney (1975) considered that anadromous game fishes did not spawn in dense enough populations for redd superimposition to have a serious effect on egg survival. It has been regarded as a significant cause of egg mortality in brown trout (Hobbs, 1937; Horton, 1961) and sea trout (Alm, 1950; Harris, 1970), although systematic quantitative data are lacking. Clearly, there is a need for more information on spatial requirements of European salmonid species on spawning grounds and for descriptions of basic parameters such as utilized depths, velocities and bed composition.

C. EGG AND ALEVIN DEVELOPMENT

The successful development of salmonid eggs is directly affected by the physical environment in the streambed, which is determined mainly by the structure and composition of bed materials. Two proximate factors influencing egg survival can be identified: bed stability and intragravel water movements which control oxygen supply and removal of toxic metabolic wastes. Emergence of alevins, the final stage of intragravel development, can also be affected by the size distribution and compaction of gravel particles.

Water temperature is not considered in this review. It is, however, an important factor determining the rate of development and, hence, the time for which the eggs and alevins are present in the gravel and vulnerable to washout and siltation.

1. Bed stability

Gravel movements and disturbance by shifting debris are known to cause washout of eggs and physical damage from crushing and smothering by deposition from unstable upstream areas. These processes have mainly been documented for Pacific salmon spawning beds on North American west coast rivers where losses can be considerable, being up to 90% for pink and chum salmon (McNeil, 1966). The importance of egg washout may vary considerably according to the physiography of different streams and stream sections and the significance of this mortality is less clear for Atlantic salmon and brown trout particularly in UK rivers on the Atlantic seaboard which, in general, are less steep than North American west coast rivers and less subject to rapid run-off from melting snow and ice (Allen, 1969). Stuart (1953b) found that, although superficial surface movement occurred, no major disturbance of brown trout redds was detectable even during severe floods in small Scottish streams. Hobbs (1937) reported that 27% of planted egg boxes were washed away in a New Zealand stream, but considered that erosion was not a major factor controlling fry production because brown trout normally selected areas of stable gravel. In contrast, Harris (1970) reported that on average 27% (range 0–58%) of brown trout and sea trout redds were washed away in tributaries of the Afon Dyfi, West Wales, and concluded that erosion was a major cause of egg loss. Elliott (1976) found that numbers of brown trout eggs drifting downstream were related to water velocity and possibly density of eggs in the gravel. He concluded that the effect of velocity was logarithmic and might be fairly constant in streams of similar size.

The relationship between current velocity and gravel movement has been investigated in the context of sediment transport but, although limited data are available on the depths of egg deposition, the relationships between water velocity and depth of erosion are complex and, at present, unpredictable (p. 169). Similarly, the recurrence intervals of discharges causing egg loss through erosion are unknown, although they are likely to be shorter in upland streams than in the lower parts of catchments (p. 155).

2. *Intragravel water movements*
Oxygen consumption by salmonid eggs varies during development. Total consumption per egg (Hayes *et al.*, 1951) and critical concentration (the level below which oxygen demand of the embryo is reduced) increase during embryonic development and then decrease sharply after hatching. For Atlantic salmon critical dissolved oxygen concentrations have been estimated for: early eggs -0.76 mg 1^{-1} (Lindroth, 1942); eyed ova -3.1 mg 1^{-1} (Hayes *et al.*, 1951); pre-hatching -5.8 mg 1^{-1} (Lindroth, 1942); hatching—10.0 mg 1^{-1} (Lindroth, 1942) and 7.1 mg 1^{-1} (Hayes *et al.*, 1951). Oxygen consumption data for Atlantic salmon have been summarized by Hamor and Garside (1975). Considerable variation was associated with differences in age, egg size, temperature, previous history, illumination and method of measurement. The sublethal effects of hypoxia include reduced growth rate and efficiency of yolk utilization (Garside, 1966; Hamor and Garside, 1977), premature hatching and reduced size at hatching (Garside, 1966) and morphological changes (Garside, 1959). Delayed effects of low oxygen concentration have been observed. Mason (1969) found that fry subject to hypoxia prior to emergence were less successful than normally-incubated fry in competition for food and space.

Oxygen supply to embryos in gravels is determined by the velocity of intragravel water and its oxygen content. Even at high oxygen concentrations, low velocities may be limiting because toxic waste products of metabolism, free carbon dioxide and ammonia, are not removed sufficiently quickly (Wickett, 1962; Shumway *et al.*, 1964; McNeil, 1966). In an experimental study of the combined effects of oxygen, water velocity and temperature on Atlantic salmon embryos, Hamor and Garside (1976) found that survival during embryogeny and hatching was limited primarily by oxygen supply and secondarily by water velocity; an increase in both increased survival. Temperature ranked third in effect, lower temperature (5°C compared with 10°C) reduced the rate of development but increased survival to hatching.

Several features of gravel composition and structure influence oxygen supply, chiefly by controlling water movement through the gravel. Three variables have received attention: particle size distribution, permeability and apparent velocity.

The proportion of fine particles appears to be an important feature in spawning gravel. They generally form only a small ($\leqslant 20\%$) component of the bed material and small changes in that proportion may have major effects on egg survival by reducing void space and the rate of water percolation through the gravel (Cooper, 1974). Many workers have demonstrated that an increase in fine particles reduces survival of salmonid embryos (Harrison, 1923; Hobbs, 1937; Shelton, 1955; Wickett, 1958; McNeil and Ahnell, 1964; Hall and Lantz, 1969; Parkinson and Slaney, 1975; Turnpenny and Williams, 1980). McNeil and Ahnell, (1964) concluded that productive redds of pink and chum salmon should contain $\leqslant 5\%$ of particles smaller than 0·83 mm. Wickett (1962) suggested that all particles smaller than the egg size of pink and chum salmon ($\simeq 5$ mm) were potentially harmful, small particles ($< 0·3$ mm) reducing waterflow through coarse materials more than larger particles (Cooper, 1974). In an early attempt at standardization, Warner (1953) described criteria for "good", "fair" and "poor" gravels with sand/silt contents of $\leqslant 10\%$, 11–19% and $\geqslant 20\%$ respectively.

Vertical stratification of the composition of bed material has received little attention. The formation of an armouring layer of large particles on the surface may result in a different particle size distribution compared with that obtaining deeper down (Peterson, 1978; Shirazi et al., 1979). Peterson (1978), in an analysis of Atlantic salmon spawning gravels, found that percentage sand ($< 2·2$ mm) increased with depth and reached 20–30% by weight at depths of 15–25 cm. In contrast, Platts et al. (1979) did not find significant changes in vertical size distribution of chinook salmon redd gravel sampled by freeze coring. In the strata at depths of 0–15·2, 15·2–30·5 and 30·5–45·7 cm the geometric mean particle sizes were 39·2, 20·1 and 35·2 mm respectively; this may have resulted from destratification caused by the redd excavation and infilling processes.

Available evidence indicates that there can be considerable variation in particle size distribution even within a spawning area, but the effect of this on selection of spawning sites by adult fish and on survival of embryos has not been reliably quantified in spite of the "criteria" put forward by many workers. There is a need for basic data on the range of gravel size compositions utilized by Atlantic salmon and brown trout. In the UK such data are very few.

Particle size analysis has value as a method of classifying spawning bed materials because a change in composition, particularly amongst the finer fractions, may have some effect on the intragravel environment. However, the method has limitations because of the complex way in which different size fractions interact to modify intragravel flows and because, for a given grading curve, permeability can vary with particle shape and compaction (Cooper, 1965) and hydraulic head. An additional difficulty is imposed by the variation in threshold particle diameters employed by different workers to define the upper limit of fines: sizes of 1–5 mm have been quoted.

Several workers have demonstrated significant negative correlations between the proportions of fine particles and gravel permeability (Terhune, 1958; McNeil and Ahnell, 1964; Cooper, 1965; Wickett, 1968; Peterson, 1978) and Wickett (1958, 1962, 1968) demonstrated correlations between permeability, egg survival and optimum number of spawners for pink and chum salmon.

With the exception of Turnpenny and Williams (1980), no work of this nature has been undertaken in rivers of the British Isles. For Atlantic salmon in New Brunswick streams Peterson (1978) found that permeabilities measured in spawning gravels were generally < 2000 cm h^{-1}, but values (< 1600 cm h^{-1}) which caused very low survival ($< 1\%$) in pink salmon (Wickett, 1958) allowed Atlantic salmon survival of $0.2–14\%$ (mean—3.6%). He suggested that in the streams he studied the lower limits of permeability permitting salmon survival were between 500 and 1000 cm h^{-1}. Turnpenny and Williams (1980) showed that the threshold permeability for rainbow trout survival in streams in South Wales was 40 cm h^{-1} and in natural gravels permeability ranged up to 6000 cm h^{-1}. Permeability in brown trout redds has been found to range from 160 to 6000 cm h^{-1} (mean —2400 cm h^{-1}) (Reiser, 1976).

From the limited data available, Pacific salmon appear to utilize gravels of far higher permeability than those reported for other species and regions. This may reflect the bed dynamics and discharge regimes characteristic of rivers on the West coast of North America, but, it is also possible that *Oncorhynchus* species are unable to survive at the lower permeabilities tolerated by the genus *Salmo*. Brown trout may be able to tolerate lower permeabilities than Atlantic salmon, but whether this is a result of smaller egg size or lower relative oxygen requirements remains equivocal.

The rate at which oxygen is delivered to eggs depends on oxygen concentration and seepage velocity which, for a given stream profile and hydraulic gradient, is proportional to permeability. If hydraulic head varies then permeability will not necessarily describe the potential oxygen availability to eggs.

Although the effects of reduced oxygen concentration on salmonid eggs have been demonstrated in the laboratory, few studies have measured oxygen within gravel as an index of spawning bed quality. McNeil (1962) and Sheridan (1962) found considerable spatial and temporal variation in dissolved oxygen concentration in spawning beds. After the initial disturbance of redd excavation, increasing sedimentation generally results in a gradual decrease in concentration of dissolved oxygen (Peters, 1962; Turnpenny and Williams, 1980). Oxygen measurements need to be complemented by some estimate of intragravel flow in order to determine the rate at which oxygen is made available to eggs (Hobbs, 1937). Wickett (1954) developed standpipe techniques for measuring dissolved oxygen and apparent velocity and then combined these studies with data on oxygen consumption of salmonid eggs to calculate limiting intragravel flows for the varying oxygen demand of eggs and dissolved oxygen concentrations of water. This approach has been repeated by several workers with some improvement of measuring technique (Pollard, 1955; Terhune, 1958; Turnpenny and Williams, in press).

Direct relationships have been demonstrated between the proportion of fines in gravel mixtures and apparent velocity (Peters, 1962; Cooper, 1974), and between apparent velocity and egg survival (Wickett, 1954; Andrew and Geen, 1960; Peters, 1962; Cooper, 1974; Turnpenny and Williams, 1980). In a study on rainbow trout eggs Turnpenny and Williams (1980) found that at the prevailing oxygen concentrations the threshold apparent velocity for survival was $\simeq$ 2 cm h^{-1}. Similar thresholds were measured by Terhune (1958). Peterson (1978) calculated that minimal velocities for Atlantic salmon, based on correlations between *in situ* measurements of velocity and survival, were 4·5–12·0 cm h^{-1} and Peters (1962) noted that wide temporal variation in apparent velocity can occur, recording a decrease of 90–5 cm h^{-1} corresponding to a dissolved oxygen decrease from 8·1–6·4 mg l^{-1}.

Clearly, no one variable completely describes the suitability of gravel as a medium for development of salmonid eggs, and the requirements of the eggs themselves are not precisely known. Assuming adequate dissolved oxygen content of the perfusing water, the most critical factor appears to be the proportion of "fine" particles although its effect on intragravel water movements will depend on other factors such as remaining gravel composition, bed profile and hydraulic head.

The entrapment of hatched alevins in the gravel has not been examined specifically for Atlantic salmon or brown trout, but studies on Pacific salmon have demonstrated that this can be a significant cause of mortality and that the rate of entrapment increases with the proportion of fine particles (Shelton, 1955; Phillips, 1964; Cooper, 1965; Hall and Lantz, 1969; Phillips and Koski, 1969; Hausle and Coble, 1976; Platts *et al.*, 1979).

Although some feeding occurs in the gravel the limited food availability and restricted movements of alevins result in rapid mortality after yolk reserves have been utilized if emergence is delayed (Dill, 1967; Hurley and Brannon, 1969).

Modification of gravel composition by altered discharge regimes may have important consequences for egg and alevin development. Naturally-occurring low discharges, causing sedimentation and low intragravel flows have been regarded as an important source of egg mortality in Pacific salmon (Wickett, 1958; McNeil, 1966). The scouring action of spates is believed to remove much of the deposited fine material (Stuart, 1953a, b; Eustis and Hillen, 1954; McNeil and Ahnell, 1964) and the elimination of flood events by regulation can result in accumulation of fines (Eustis and Hillen, 1954). However, the erosive effects of floods may also be deleterious and the elimination of winter floods by flow stabilization has resulted in increases in survival of Pacific salmon eggs previously subjected to erosion and washout at high flows (Wickett, 1952; Gangmark and Bakkala, 1960; Lister and Walker, 1966).

D. JUVENILE AND RESIDENT ADULT STAGES

There is evidence that the major population regulatory processes take effect during the juvenile post-emergent phase and that they operate through density dependent mortality (Ricker, 1954; Larkin, 1956; Le Cren, 1965, 1973; Gee *et al.*, 1978). This can be related to space and food-limited territoriality (Chapman, 1966a) and to modification of territory size as a result of changes in discharge and bed composition (Kalleberg, 1958). Flow and substratum modifications may, therefore, have direct impact on salmonid stocks by influencing regulatory processes of the early stages. Juvenile salmonid biology and territorial behaviour has been described in detail for Atlantic salmon (Lindroth, 1955; Kalleberg, 1958; Keeneleyside and Yamamoto, 1962), sea trout (Lindroth, 1955) and brown trout (Kalleberg, 1958; Hartman, 1963).

The surface structure of the streambed influences survival from the time of emergence from the gravel, providing shelter from water currents and predators, visual isolation and support. Marr (1963) suggested that brown trout alevins lack static stability in the vertical plane and require grooved surfaces for support during periods of inactivity. Fry reared on rough surfaces have higher growth and survival rates than fry reared on smooth surfaces (Brannon, 1965; Marr, 1966). Photonegative responses in salmon fry ensure close association with the substratum during the first days after emergence.

Le Cren (1973) described the early post-emergence behaviour of brown trout fry. Initially they rested on the streambed close to each other, but, after a few days, became more active, moved off the bottom to feed and became more aggressive. As territoriality developed they dispersed evenly downstream of the redd to a distance of $\simeq$ 100 m within a week. Territory size was about 0·05 m² and preferred habitat was slow-flowing gravelly areas with 2–10 cm depth of water.

As they grow salmonids require larger bed material and greater water depth (Schuck, 1945; Allen, 1969; Everest and Chapman, 1972; Jones, 1975; Bohlin, 1977; Symons and Heland, 1978). MacCrimmon (1954) considered that crevices between stones provide the chief source of shelter for Atlantic salmon parr. Increased visual isolation and reduced territory size afforded by increasing particle size was thought to account for a higher density of salmon fry in riffles (Kalleberg, 1958) and Le Cren (1973) considered that a similar mechanism could maintain high densities of brown trout fry. Interspecific differences can be detected and Atlantic salmon juveniles are typically found in more riffly areas than brown trout (Jones, 1975) which competitively dominate salmon of similar size (Lindroth, 1955; Kalleberg, 1958; Le Cren, 1965; Egglishaw, 1967). Symons and Heland (1978) found that the greatest densities of both 0+ and 1+ juvenile Atlantic salmon occurred where velocities, measured 5 cm above the streambed, averaged 50–65 cm s⁻¹. However, experiments in laboratory streams showed that 0+ fish primarily utilized shallow (10–15 cm), pebbly (1·6–6·4 cm diameter) riffle areas and as they grew increasingly made use of deeper (> 30 cm) riffles with boulder (> 25·6 cm diameter) cover. 1+ fish were dominant to 0+ fish in competition for space as the younger fish approached a size (> 6·6 cm) commensurate with a move to deeper water. They concluded that pebbly riffles without boulders offered a prime nursery habitat because competition from older salmon was eliminated.

Deep runs and pools were utilized by larger juveniles and adults of brown trout, particular in winter months when they were provided with cover (including undercut banks and bankside vegetation) and reduced velocity (Hartman, 1963). The numerous examples of habitat improvement and of increased standing crop of trout being encouraged by the addition of suitable cover, testify to its importance in determining the carrying capacity of streams (Shetter *et al.*, 1946; Boussu, 1954; Saunders and Smith, 1962; Hunt, 1969).

Clearly, for production of all ages of stream-dwelling salmonid populations there should be provision for a range of discharge, substratum and channel morphology requirements. Wetted area alone is not an adequate indicator of stream rearing potential and due account should be taken of the spatial requirements at all levels of the population age structure.

The effect of discharge on alteration of salmonid habitats has not been fully investigated. Siltation of crevices between bed material and infilling of pools can lead to loss of habitat for juvenile (Macrimmon, 1954; Bjorn *et al.*, 1974) and adult resident salmonids (Saunders and Smith, 1965) and reduction in availability of angling pools (Graesser, 1979). Such effects have been found to derive from prolonged low flows, increased suspended solids load and deposition of debris and gravel following floods.

V. General Conclusions

A. HYDRAULICS

Although considerable progress has been made in developing the statistical analysis of hydrological data, time series of hydrological data in the UK are relatively brief (Anon., 1975). Consequently accurate quantitative analyses of flow regime, especially for upland streams. are scarce (Pardé, 1939; Ward, 1968; Anon., 1975). Existing schemes to categorize streams biologically, rely on a spatial and correlative basis (Jones and Peters, 1977), yet analysis of stream regimes in a biological context will require seasonal indices to define and quantify temporal hydro-biological interaction. Presently, no such indices have been developed (Sedell *et al.*, 1978).

Detailed investigations of the hydraulics of upland channels are few (e.g. Judd and Peterson, 1969). As complete theoretical models of channel flow are unobtainable at present, good empirical, but versatile models should be satisfactory for practical application (e.g. Bevan *et al.*, 1979). Unfortunately, although novel methods of measuring various factors such as "significant roughness" have yielded useful equations to predict flow parameters, considerable field data, laboriously collected, are required (Judd and Peterson, 1969; Bathurst, 1978). Methods of classifying upland streams hydraulically are required but are unlikely to be developed satisfactorily until a large quantity of hydraulic data from a wide variety of streams (with various catchment and climatic characteristics) have been collated.

Despite the importance of gravel-bed stability to salmonid spawning, very little is known of the dynamics of gravel movement through a specified channel reach. Recent investigations of sediment sources, dynamics of transport of materials and mechanisms and spatial and temporal patterns of deposition have shown that there are complex relationships between these factors. Despite the progress which has been made in understanding these interactions (e.g. Klingeman and Emmett, in press; Andrews, 1979; Adams and Beschta, 1980), there remains a need for further effort by multi-disciplinary teams on instrumented catchments. In particular, attention should be paid to the exchange of fluid and sediment at the sediment–water interface.

Prediction of the movement of coarser materials is imprecise in narrow, sinuous, cobble-bedded streams, especially where beds may be compacted. Models are urgently needed that will predict the frequency, extent and depth of scour in biologically-significant gravel beds. Further, the relationship between bedload and bed materials needs to be defined. For example, although 25% of the bed material of the East Fork River is gravel, less than 9% of the bedload is gravel (Andrews, 1979).

In order to give continuous, accurate records suitable for detailed analysis, advances in instrumentation to measure bedload, suspended load and changes in channel geometry will be needed.

Organic particles transported through streams systems vary considerably in size and density. Consequently, considerable data are required to derive models of organic detritus transport. At present, a conflict of evidence (p. 171) reflects differences in catchment characteristics, sampling techniques, definition of grain-size class intervals and the primary requirements of individual investigations. Laboratory studies of the transport of organic debris are rare (e.g. Fisher *et al.*, 1979). These would supply useful data and should not prove unduly difficult. A priority in field investigations should be, the assessment of (a) sampling methods, (b) spatial and temporal sampling intervals, and (c) grain size class intervals. A distillation of the available data in the light of these assessments, could provide guidelines in developing uniform techniques. Comparability of techniques between investigators would allow direct and incontrovertible comparison of results.

B. INVERTEBRATES

Knowledge of the effects of flow regime upon benthic invertebrates is sparse and fragmentary. This reflects our limited understanding of the complex of interrelated processes which link changes in the fauna to river flow patterns and the dynamics of bed materials.

Preliminary studies (Gore, 1978) have attempted to define both minimum and optimum river flows within riffle areas, in terms of certain hydraulic variables (water velocity and depth) required to maintain the most abundant and diverse fauna. Application of such information to detailed hydrological surveys within individual river systems, may provide a useful predictive tool for assessing the impact of flow modification schemes. However, this approach is not suitable for more general application to river systems in different geographical areas. The latter can only be achieved by gaining a deeper understanding of the physical and hydraulic processes to which the invertebrate fauna responds.

C. SALMONIDS

Giger (1973) and Fraser (1975) have reviewed in detail many of the approaches and methodologies developed in North America to establish criteria of streamflow and habitat required by salmonids.

Substratum and flow requirements for spawning have received the most attention (Warner, 1953; Savage, 1962; Pitney, 1969; Hutchinson and Aney, 1964; Baxter, 1961; Thompson, 1972; Fraser, 1975). The basic approach has been an empirical one in which various physical characteristics of a river or spawning area have been correlated with the spawning performance of the salmonids. Similar approaches, but incorporating estimated food production and shelter requirements, have been used to suggest criteria for juveniles and resident adults (Kelley *et al.*, 1960; Thompson, 1972; Tennant, 1976).

Such empirically-derived models present problems in extrapolation to flow regimes or geographical areas other than those for which they were derived and, in general, they give little insight into the detail of the processes involved. It is arguable that, in most of the available approaches, the judgement of individual workers plays an important part in predicting the effects of flow modifications (Giger, 1973).

More objective assessments require a great deal of further study, particularly into the functional responses amongst flow factors and sediment dynamics, as related to erosive processes and intragravel flows. Experimental field studies of the kind pioneered by Wickett (1952,

1954, 1958, 1962) are needed and more data are required on the range of composition and stability of bed materials utilized by spawning salmonids in British rivers. Erosive and depositional processes vary spatially within and between catchments, but the distribution and success of spawning in relation to these processes has not been investigated. Such information is necessary to determine the consequences to fish stocks of loss or improvement of spawning areas and to facilitate management decisions.

Acknowledgements

This review was prepared as a part of research programmes at the University of Wales Institute of Technology (funded by the Department of the Environment) and by the Freshwater Biological Association (supported by the Department of the Environment, the Natural Environment Research Council and the Northumbrian Water Authority). The authors are indebted to Professor R. W. Edwards and Dr M. Newson for comments on the manuscript and to Mrs D. Jones who typed it. Raw data on Base Flow Index were provided by the Institute of Hydrology.

References

Adams, J. N. and Beschta, R. L. (1980). Gravel composition in Oregon Coastal Streams. *Can J, Fish. Aquat. Sci.* **37**, 1514–1521.

Al-Ansari, N. A., Al-Jabbari, M. H. and McManus, J. (1977). The effect of farming upon solid transport in the River Almond, Scotland. Proc. Paris Symp. July 1977. Erosion and solid matter transport in inland waters. *Int. Ass. Hydraul. Res. Publ.* **122**, 118–125.

Al-Ansari, N. A., Al-Jabbari, M. H. and McManus, J. (1978). Sediment discharge in the River Almond, Scotland. *Iraqi J. Sci.* **19**, 145–157.

Allen, K. R. (1951). The Horokiwi stream. A study of a trout population. *N.Z. Mar. Dept., Fish. Bull.* **10**, 1–238.

Allen, K. R. (1959). The distribution of stream bottom fauna. *Proc. N.Z. Ecol. Soc.* **6**, 5–8.

Allen, K. R. (1969). Limitations on the production of salmonid populations in streams. *In* "Symposium on salmon and trout in streams". (Northcote, T. G., ed.) H.R. pp. 3–18. MacMillan Lectures in Fisheries. Univ. of British Columbia, Vancouver.

Alm, G. (1950). The sea-trout population in the Ava stream. *Rept. Inst. Freshwat. Res. Drottningholm* **31**, 26–56.

Ambühl, H. von (1959). Die Bedeutung der Strömung als ökologischer Faktor. *Schweitz. Z. Hydrol.* **21**, 133–164.

Ambühl, H. von (1961). Die Strömung als physiologischer und ökologischer Faktor. Experimentelle Untersuchungen an Bachtieren. *Verh. Int. Verein. theor. angew. Limnol.* **14**, 390–395.

Amein, M. and Fang, C. S. (1970). Implicit flood routing in natural channels. *J. Hydraul. Div. Am. Soc. civ. Engrs.* **96**, (HY 12), 2481–2500.

Anderson, N. H. and Lehmkuhl, D. M. (1968). Catastrophic drift of insects in a woodland stream. *Ecology* **49**, 198–206.

Andrew, F. S. and Geen, G. H. (1960). Sockeye and pink salmon production in relation to proposed dams in the Fraser River System. *Int. Pac, Salmon Fish. Comm. Bull.* **11**.

Andrews E. D. (1979). Scour and fill in a stream channel, East Fork River, western Wyoming. *Geol. Surv. Prof. Pap.* **1117**.

Andrews, E. D. (1980). Effective and bankful discharges of streams in the Yampa River Basin, Colorado and Wyoming. *J. Hydrol.* **46**, 311–330.

Anon. (1975). "Flood studies report—Hydrological studies." Vol. 1. N.E.R.C. London.

Anon. (1978). "Low Flow Study Report" No. 1. April. Institute of Hydrology, Wallingford, England.

Armitage, P. D. (1977). Invertebrate drift in the regulated River Tees, and an unregulated tributary Maize Beck, below Cow Green Reservoir. *Freshwat. Biol.* **7**, 167–183.

Armitage, P. D., MacHale, A. M. and Crisp, D. C. (1974). A survey of stream invertebrates in the Cow Green basin (Upper Teesdale) before inundation. *Freshwat. Biol.* **4**, 369–398.

Armitage, P. D., MacHale, A. M. and Crisp, D. C. (1975). A survey of the invertebrates of four streams in the Moor House National Nature Reserve in northern England. *Freshwat. Biol.* **5**, 479–495.

Arnold, G. P. (1974). Rheotropism in fishes. *Biol. Revs.* **49**, 515–576.

Badcock. R. M. (1953). Comparative studies in the population of streams. *Rep. Inst. Freshwat. Res. Drottningholm* **35**, 38–50.

Bagnold, R. A. (1973). The nature of saltation and of "bedload" transport in water. *Proc. R. Soc. Lond. A.* **332**, 473–504.

Bagnold. R. A. (1977). Bedload transport by natural rivers. *Water Resour. Res.* **13**, 303–312.

Baker, V. R. (1973). Paleohydrology and sedimentology of Lake Missoula flooding in Eastern Washington. *Geol. Soc. Am. Spec. Pap.* **144**, 1–79.

Baker, V. R. and Ritter, D. F. (1975). Competence of rivers to transport coarse bed load material. *Bull. geol. Soc. Am.* **86**, 975–978.

Barber, W. E. and Kevern, N. R. (1973). Ecological factors influencing macroinvertebrate standing crop distribution. *Hydrobiologia* **43**, 53–75.

Bärlocher, F. and Kendrick, B. (1974). Dynamics of the fungal population on leaves in a stream. *J. Ecol.* **62**, 761–791.

Barnes, H. H. (1967). Roughness characteristics in natural channels. *U.S. Geol. Surv. Wat. Supply Pap.* **1849**.

Bathurst, J. C. (1978). Flow resistance of large-scale roughness. *J. Hydraul. Div. Am. Soc. civ. Engrs.* **104**, (HY 12), 1587–1603.

Baxter, G. (1961). Environmental effects of dams and impoundments. *Ann. Rev. Ecol. Syst.* **8**, 255–283.

Bayazit, M. (1975). Simulation of armour coat formation and destruction. *Int. Ass. Hydraul. Res. Congr. 16th v.*, São Paulo, Brazil **2**, 73–80.

Beauchamp, R. S. A. (1932). Some ecological factors and their influence on competition between stream and lake-dwelling triclads. *J. Anim. Ecol.* **1**, 175–190.

Belding, D. L. (1934). The spawning habits of the Atlantic salmon. *Trans. Am. Fish. Soc.* **64**, 211–218.

Bennett, J. P. and Sabol, G. V. (1973). Investigation of sediment transport curves constructed using periodic and aperiodic samples. Int. Ass. Hydraul. Res.

Int. Symp. on River Mechanics. 9–12 January, Bangkok. 49–60.

Benson, M. A. (1965). Spurious correlation in hydraulics and hydrology. *Hydraul. Div. Am. Soc. civ. Engrs.* **91** (HY 4), 35–42.

Benson, N. G. (1953). The importance of groundwater to trout populations in the Pidgeon River, Michigan. Trans. Eighteenth N. Amer. Wildl. Conf. 269–281.

Beran, M. S. and Gustard, A. (1977). A study into the low-flow characteristics of British rivers. *J. Hydrol.* **35**, 147–157.

Berrie, A. D. (1975). Detritus, micro-organisms and animals in freshwater. *In* "The Role of Terrestrial and Aquatic Organisms in Decomposition Processes." (Anderson, J. M. and Macfadyen, A., eds), pp. 323–338. The 17th Symposium of the British Ecological Society, 15–18 April, 1975.

Beschta, R. L. and Jackson, W. L. (1979). The intrusion of fine sediments into a stable gravel bed. *J. Fish. Res. Board Can.* **36**, 204–210.

Betson, R. P. and Marius, J. P. (1969). Source areas of storm run-off. *Wat. Resour. Res.* **5**, 574–582.

Bevan, K. J. (1978). The hydrological response of headwater and sideslope areas. *Hydrol. Sci. Bull.* **23**, 419–437.

Bevan, K., Gilman, K. and Newson, M. (1979). Flow and flow routing in upland channel networks. *Hydrol. Sci. Bull.* **24**, 303–325.

Bilby, R. E. and Likens, G. E. (1979). Effect of hydrological fluctuation on the transport of fine particulate organic carbon in a small stream. *Limnol. Oceanogr.* **24**, 69–75.

Bishop, J. E. and Hynes, H. B. N. (1969). Downstream drift of invertebrate fauna in a stream ecosystem. *Arch. Hydrobiol.* **66**, 56–90.

Bjorn, T. C., Brusven, M. A., Molnan, M., Watts, F. J. and Wallace, R. L. (1974). Sediment in streams and its effects on aquatic life. Completion Report Project B-025-IDA. July 1972–Oct. 1974. Univ. Idaho, Moscow, Idaho, USA.

Blanz, R. E., Hoffman, C. E., Kilambi, R. V. and Liston, C. R. (1969). Benthic macroinvertebrates in cold tailwaters and natural streams in the state of Arkansas. *Proc. Am. Conf. S.E. Fish Game Comm.* **23**, 281–292.

Bluck, B. J. (1971). Sedimentation in the meandering River Endrick. *Scot. J. Geol.* **7**, 93–138.

Bluck, B. J. (1976). Sedimentation in some Scottish rivers of low sinuosity. *Trans. R. Soc. Edinburgh* **69**, 425–456.

Bohlin, T. (1977). Habitat selection and intercohort competition of juvenile sea-trout, *Salmo trutta. Oikos* **29**, 112–117.

Boling, R. H., Goodman, E. D., Van Sickle, J. A., Zimmer, J. O., Cummins, K. W., Peterson, R. C. and Reice, S. R. (1975). Toward a model of detritus processing in a woodland stream. *Ecology* **56**, 141–151.

Bormann, F. H., Likens, G. E., Siccama, J. G., Pierce, R. S. and Eaton, J. S. (1974). The export of nutrients and recovery of stable conditions following deforestation at Hubbard Brook. *Ecol. Monogr.* **44**, 225–277.

Boussu, M. F. (1954). Relationship between trout populations and cover on a small stream. *J. Wildl. Mgmt.* **18**, 227–239.

Bowen, H. C. (1959). Discussion of "A study of the bankful discharges of rivers in England and Wales" by M. Nixon. *Proc. Inst. Civ. Engrs.* **14**, 396–397.

Brannon, E. L. (1965). The influence of physical factors on the development and weight of sockeye salmon embryos and alevins. *Int. Pac. Salmon Fish. Comm. Prog. Rep.* **13**.

Bray, D. I. (1979). Estimating average velocity in gravel bed rivers. *J. Hydraul.*

Div. Am. Soc. civ. Engrs. **105**, (HY 9), 1103–1122.

Brennan, A. and McLachlan, A. J. (1979). Tubes and tube-building in a lotic chironomid (Diptera) community. *Hydrobiologia*, **67**, 173–178.

Brennan, A., McLachlan, A. J. and Wotton, R. S. (1978). Particulate material and midge larvae (Chironomidae: Diptera) in an upland river. *Hydrobiologia* **59**, 67–73.

Brett, J. R. (1965). The relation of size to rate of oxygen consumption and sustained swimming speed of sockeye salmon (*Oncorhynchus nerka*). *J. Fish. Res. Board. Can.* **22**, 1491–1501.

Briggs, J. C. (1948). The quantitative effects of a dam upon the bottom fauna of a small California stream. *Trans. Am. Fish. Soc.* **78**, 70–81.

Brooker, M. P. (1978). After the drought. *Trout and Salmon Mag.* **Feb.**, 43.

Brooker, M. P. (1981). The impact of impoundments on the downstream fisheries and general ecology of rivers. *Applied Ecology* **6**, 91–152.

Brooker, M. P. and Hemsworth, R. J. (1978). The effect of the release of an artificial discharge of water on invertebrate drift in the R. Wye, Wales. *Hydrobiologia* **59**, 155–163.

Brooker, M. P. and Morris, D. L. (1980). A survey of the macroinvertebrates of the River Wye. *Freshwat. Biol.* **10**, 437–458.

Brush, L. M. (1961). Drainage basins, channels and flow characteristics of selected streams in central Pennsylvania. *Prof. Pap. U.S. geol. Surv.* 282-F.

Brusven, M. A. and Prather, K. V. (1974). Influence of stream sediments on distribution of macrobenthos. *J. Ent. Soc. of British Columbia* **71**, 25–32.

Bull, W. B. (1979). Threshold of critical power in streams. *Bull. Geol. Soc. Am.* Part I. **90**, 453–464.

Burner, C. J. (1951). Characteristics of spawning nests of Columbia River salmon. *Bull. U.S. Wildl. Serv. Fish.* **52**, 97–110.

Butler, P. R. (1977). Movement of cobbles in a gravel-bed stream during a flood season. *Bull. Geol. Soc. Am.* **88**, 1072–1074.

Calkins, D. and Dunne, T. (1970). A salt tracing method for measuring channel velocity in small mountain streams. *J. Hydrol.* **11**, 379–392.

Cary, A. S. (1951). Origin and significance of open-work gravel. *Trans. Am. Soc. civ. Engrs.* **116**, 1296–1308.

Chapman, D. W. (1966a). Food and space as regulators of salmonid populations in streams. *Amer. Naturalist* **100**, 345–357.

Chapman, D. W. (1966b). The relative contributions of aquatic and terrestrial primary producers to the trophic relations of stream organisms. *Spec. Publ. Pymatuning Lab. Field Biol.* **4**, 116–130.

Charlton, F. G. (1977). An appraisal of available data on gravel rivers. Report No. Int. 151, August 2nd impression. Hydraulic Research Station, Wallingford, England.

Charlton, F. G., Brown, P. M. and Benson, R. W. (1978). The hydraulic geometry of some gravel rivers in Britain. Report No. Int. 180 July. Hydraulic Research Station, Wallingford, England.

Cheetham, G. H. (1979). Flow competence in relation to stream channel form and braiding. *Bull. Geol. Soc. Am.* **90**, 877–886.

Chow, V. T. (1967). "Open Channel Hydraulics." 2nd Ed. McGraw-Hill, New York.

Church, M. (1972). Baffin Island Sandurs:—A study of Arctic fluvial processes. *Bull. Geol. Surv. Can.* **216**, 208 pp.

Chutter, F. M. (1969a). The distribution of some stream invertebrates in relation to current speed. *Int. Revue ges. Hydrobiol.* **54**, 413–422.

Chutter, F. M. (1969b). Effects of silt and sand on the invertebrate fauna of streams and rivers. *Hydrobiologia* **34**, 57–76.

Clayton, C. L. (1951). The problem of gravel in highland water courses. *J. Inst. Wat. Engrs.* **July**, 400–406.

Clayton, L. and Harrison, S. S. (1975). Effects of ground-water seepage on fluvial processes. Research Project Technical Completion Report to the Office of Water Research & Technology. Dept. of the Interior, Washington, D.C.

Coble, D. W. (1961). The influence of water exchange and dissolved oxygen in redds on survival of steelhead trout embryos. *Trans. Am. Fish. Soc.* **90**, 469–474.

Colby, B. R. (1961). Effects of depth of flow on discharge of bed material. *U.S. Geol. Surv. Wat. Supply Pap.* **1498**-D.

Coleman, M. J. and Hynes, H. B. N. (1970). The vertical distribution of the invertebrate fauna in the bed of a stream. *Limnol. Oceanogr.* **15**, 31–40.

Cooper, A. C. (1965). The effect of transported stream sediments on the survival of sockeye and pink salmon eggs and alevins. *Int. Pac. Sal. Fish. Comm. Bull.* **18**.

Cooper, A. C. (1974). Physical aspects of sediment in relation to survival of salmon eggs. *Symposium on Stream Ecology, Cent. Cont. Educ. Univ. B.C. F.P.* **2407**, 1–13.

Cordone. A. J. and Kelly, D. W. (1961). The influences of inorganic sediment on the aquatic life of streams. *Calif. Fish Game.* **47**, 189–228.

Costa, J. E. (1977). Sediment concentration and duration in stream channels. *J. Soil Wat. Conserv.* **32**, 168–170.

Crickmay, C. H. (1974). "The Work of the River." Macmillan, London.

Crisp, D. T. (1966). Input and output of minerals for an area of Pennine moorland: the importance of precipitation, drainage, peat erosion and animals. *J. Appl. Ecol.* **3**, 327–348.

Crisp, D. T. and Gledhill, T. (1970). A quantitative description of the recovery of the fauna in a muddy reach of a mill stream in Southern England after draining and dredging. *Arch. Hydrobiol.* **67**, 502–541.

Crisp, D. T. and Robson, S. (1979). Some effects of discharge upon the transport of animals and peat in a north Pennine headstream. *J. Appl. Ecol.* **16**, 721–736.

Culbertson, J. K. and Dawdy, D. R. (1964). A study of fluvial characteristics, and hydraulic variables, Middle Rio Grande, New Mexico. *Wat. Supply. Pap. U.S. geol. Surv.* **1498**-F.

Cummins, K. W. (1964). Factors limiting the micro-distribution of larvae of the caddis-flies *Pycnopsyche lepida* (Hagen) and *Pycnopsyche guttifer* (Walker) in a Michigan stream. *Ecol. Monogr.* **34**, 271–295.

Cummins, K. W. (1966). A review and future problems in benthic ecology. *In* "Organism–Substrate Relationships in Streams." (Cummins, K. W., Tryon, C. A. and Hartman, R. T., eds), pp. 2–51. Spec. Publ. Pymatuning Lab. of Ecology, Univ. Pittsburgh.

Cummins, K. W. (1973). Trophic relations in aquatic insects. *Ann. Rev. Ent.* **18**, 183–206.

Cummins, K. W. and Lauff, G. H. (1969). The influence of substrate particle size on the microdistribution of stream macrobenthos. *Hydrobiologia* **34**, 145–181.

Cummins, K. W., Petersen, R. C., Howard, F. O., Wuychech, J. C. and Holt, V. I. (1973). Utilization of leaf litter by stream detritivores. *Ecology*, **54**, 336–345.

Cushing, C. E. Jr. (1963). Filter-feeding insect distribution and planktonic food in the Montreal River. *Trans. Am. Fish. Soc.* **92**, 216–219.

Dal Cin, R. (1967). Le ghiaie del Piave. *Mem. Mus. Tridentino Sci. Nat.* **16**, (3) 1–117.

Dill, L. M. (1967). Studies on the early feeding of sockeye salmon alevins. *Can. Fish. Cult.* **39**, 23–34.

Dunne, T., Moore, T. R. and Taylor, C. H. (1975). Recognition and prediction of run-off producing zones in humid regions. *Hydrol. Sci. Bull.* **20**, 305–327.

Dyer, K. R. (1970). Grain-size parameters for sandy gravels. *J. Sedim. Petrol.* **40**, 616–620.

Edington, J. M. (1965). The effect of water flow on populations of net-spinning Trichoptera. *Mitt. int. Verein. theor. angew. Limnol.* **13**, 40–48.

Edington, J. M. (1968). Habitat preferences in net-spinning caddis larvae with special reference to the influence of water velocity. *J. Anim. Ecol.* **37**, 675–692.

Egglishaw, H. J. (1964). The distributional relationships between the bottom fauna and plant detritus in streams. *J. Anim. Ecol.* **33**, 463–476.

Egglishaw, H. J. (1967). The food, growth and population structure of salmon and trout in two streams in the Scottish highlands. *Freshwat. Sal. Fish. Res.* **38**, 1–32.

Egglishaw, H. J. (1968). The quantitative relationship between bottom fauna and plant detritus in streams of different calcium concentrations. *J. Appl. Ecol.* **5**, 731–740.

Egglishaw, H. J. (1969). The distribution of benthic invertebrates on substrata in fast flowing streams. *J. Anim. Ecol.* **38**, 19–33.

Egglishaw, H. J. and MacKay, D. W. (1967). A survey of the bottom fauna of streams in Scottish Highlands. 3. Seasonal changes in the fauna of three streams. *Hydrobiologia* **30**, 305–334.

Einstein, H. A. (1968). Deposition of suspended particles in a gravel bed. *J. Hydraul. Div. Am. Soc. civ. Engrs.* **94**, (HY 45), 1197–1205.

Einstein, H. A. and Barbarossa, N. L. (1952). River channel roughness. *Trans. Am. Soc. civ. Engrs.* **117**, 1121–1146.

Elliott, J. M. (1965). Invertebrate drift in a mountain stream in Norway. *Norsk Ent. Tidsskr.* **13**, 97–99.

Elliott, J. M. (1967). Invertebrate drift in a Dartmoor stream. *Arch. Hydrobiol.* **63**, 202–237.

Elliott, J. M. (1968). The daily activity pattern of mayfly nymphs. *J. Zool. London.* **155**, 201–221.

Elliott, J. M. (1971). The distances travelled by drifting invertebrates in a Lake District stream. *Oecologia* **6**, 350–379.

Elliott, J. M. (1976). The energetics of feeding, metabolism and growth of brown trout (*Salmo trutta* L.) in relation to body weight, water temperature and ration size. *J. Anim. Ecol.* **45**, 923–948.

Elliott, J. M. and Minshall, G. W. (1968). The invertebrate drift in the River Duddon, English Lake District. *Oikos,* **19**, 39–52.

Emmett. W. W. (1976). Bedload transport in two large, gravel-bed rivers. Idaho and Washington. Proc. 3rd Fed. Inter-Agency Sedimentation Conf. Denver, Colorado. March 22–26. 4–101, 104–114.

Emmett, W. W. and Leopold, L. B. (1977). A comparison of observed sediment transport rates with rates computed using existing formulae. Geomorphology in Arid Regions: Ed. Doehring, D. O. Proc. 8th Annual Geomorphology Symp. p. 187.

Engelhardt, W. (1951). Faunistisch—ökologische Untersuchungen über Wasserinsekten an den südlichen Zuflüssen des Ammersees. *Mitt. münch. ent. Ges.* **41**, 1–135.

Engelhardt, W. and Pitter, H. (1951). Über die Zusammenhange zwischen Porosität,

Permeabilitat, und Korngrösse bei sand und Sandsteinen. *Heidelberger Beitr. Mun. Pet.* **2**, 477–491.

Engelund, F. (1970). Instability of erodible beds. *J. Fluid Mech.* **42**, 225–244

Engelund, F. and Fredsøe, J. (1976). A sediment transport model for straight alluvial channels. *Nord. Hydrol.* **7**, 293–306.

European Inland Fisheries Advisory Commission (E.I.F.A.C.), (1964). Water quality criteria for European freshwater fish. Report on finely divided solids in inland fisheries. E.I.F.A.C. Tech. pap. 1.

Eustis, A. B. and Hillen, R. H. (1954). Stream sediment removed by controlled reservoir release. *Progve. Fish Cult.* **16**, 30–35.

Everest, F. H. and Chapman, D. W. (1972). Habitat selection and spatial interaction of juvenile chinook salmon and steelhead trout in two Idaho streams. *J. Fish. Res. Board Can.* **29**, 91–100.

Everts, C. H. (1973). Particle over-passing on flat granular boundaries. *J. Watways Harb. Coastal Engng. Div. Am. Soc. civ. Engrs.* **99**, (WW 4), 425–438.

Fahnestock, R. K. (1963). Morphology and hydrology of a glacial stream—White River, Mount Ranier, Washington. Prof. Pap. U.S. geol. Surv. 422-A. A1–A70.

Fahy, E. (1975). Quantitative aspects of the distribution of invertebrates in the benthos of a small stream system in western Ireland. *Freshwat. Biol.* **5**, 167–182.

Fisher, S. G. and Likens, G. E. (1973). Energy flow in Bear Brook, New Hampshire: an integrated approach to stream ecosystem metabolism. *Ecol. Monogr.* **43**, 421–439.

Fisher, J. S., Pickral, J. and Odum, W. (1979). Organic detritus particles: Initiation of motion criteria. *Limnol. Oceanog.* **24**, 529–532.

Fleming, G. (1969). The Clyde basin: hydrology and sediment transport. Ph.D. Thesis. University of Strathclyde.

Francis, J. R. D. (1973). Experiments on the motion of solitary grains along the bed of a water stream. *Proc. R. Soc. London A.* **332**, 443–471.

Fraser, H. J. (1935). Experimental study of the porosity and permeability of clastic sediments. *J. Geol.* **43**, 910–1010.

Fraser, J. C. (1972). Regulated stream discharge for fish and other aquatic resources—an annotated bibliography. *FAO Fisheries Technical Paper*, No. 112.

Fraser, J. C. (1975). Determining discharges for fluvial resources. *FAO Fisheries Tech Pap.* No. 143.

Frost, W. E. (1942). River Liffey Survey IV. The fauna of the submerged "mosses" in an acid and an alkaline water. *Proc. R. Ir. Acad.* **47B**, 293–369.

Gammon, J. R. (1970). The effect of inorganic sediment on stream biota. Envir. Protect. Agency Water Pollut. control. Research Series No. 18050DWC. Gov. Printing Office, D.C.

Gangmark, H. A. and Bakkala, R. G. (1958). Plastic standpipe for sampling streambed environment of salmon spawn. U.S. Fish. Wildl. Serv. Spec. Scientific Rept. 261.

Gangmark, H. A. and Bakkala, R. G. (1960). A comparative study of unstable and stable (artificial channel) spawning streams for incubating King Salmon at Mill Creek. *Calif. Fish. Game* **46**, 151–164.

Garside. E. T. (1959). Some effects of oxygen in relation to temperature on the development of lake trout embryos. *Can. J. Zool.* **37**, 689–698.

Garside, E. T. (1966). Effects of oxygen in relation to temperature on the development of embryos of brook trout and rainbow trout. *J. Fish. Res. Board Can.* **23**, 1037–1134.

Gaufin. A. R. C. (1959). Production of the bottom fauna in the Provo River. *Utah St. Coll. J. Sci.* **33**, 395–419.

Gebhardt, G. A. (1969). The influence of stream disturbance activity on aquatic organisms—a review. Unpubl. report, Bureau of Land Mgmt. Salem, Oregon.

Gee, A. S., Milner, N. J. and Hemsworth, R. J. H. (1978). The effect of density on mortality in juvenile Atlantic salmon (*Salmo salar*). *J. Anim. Ecol.* **47**, 497–505.

Gessler, J. (1965). Der Geschiebetribbeginn bei Mischungen untersucht an natürlichen Abpflasterung—serscheinungen in Kanälen. Mitt. VWE (Zurich) No. 69. (Translation T-5 by E. A. Prych (English)). W. M. Keck Laboratory of Hydr. & Water Resources, Calif. Inst. Tech. Pasadena, California. Revised edit. Oct. 1968.

Gessler, J. (1971). Critical shear stress for sediment mixtures. Proc. 14th Cong. Internat. Assoc. for Hydr. Res. Vol. 13, Paper No. CI Paris 1971, 1–8.

Gibbons. D. R. and Salo, E. O. (1973). An annotated bibiliography of the effects of logging on fish of the Western United States and Canada. USDA For. Serv. Gen. Tech. Pap. PNW–10.

Giger, R. D. (1973). Stream flow requirements of salmonids. Oregon Wildl. Comm. Federal Aid Progress Rept., Proj. No. AFS–62–1. 1–117.

Gilpin, B. R. and Brusven, M. A. (1976). Subsurface sampler for determining vertical distribution of stream bed benthos. *Progve. Fish Cult.* **38**, 192–194.

Gomez, B. (1979). An evaluation of bed material sampling technique for use in mixed sand and gravel-bed streams. Discussion Paper Jan. 1979 Dept. of Geography, University of Southampton, U.K.

Gore, J. A. (1977). Reservoir manipulations and benthic invertebrates in a prairie river. *Hydrobiologia* **55**, 113–123.

Gore, J. A. (1978). A technique for predicting in-stream flow requirements of benthic macro-invertebrates. *Freshwat. Biol.* **8**, 141–151.

Graesser, N. W. (1979). How land improvements can damage Scottish salmon fisheries. *Salm. Trout Mag.* **215**, 39–43.

Graf, W. H. (1971). "Hydraulics of Sediment Transport." McGraw-Hill, New York.

Graf, W. H. and Acaroglu, E. R. (1966). Settling velocities of natural grains. *Bull. internat. Ass. Scient. Hydrol.* XIe Annees No. **4**. 27–43.

Graf, W. L. (1979). Rapids in canyon rivers. *J. Geol.* **87**, 533–551.

Greeley, J. R. (1932). The spawning habits of brook, brown and rainbow trout and the problem of egg predators. *Trans. Am. Fish. Soc.* **62**, 239–248.

Gregory, K. J. and Walling, D. E. (1973). "Drainage Basin Form and Process." Edward Arnold, London.

Gurtz, M. E., Webster, J. R. and Wallace, J. B. (1980). Seston dynamics in Southern Appalachian streams: Effects of clear-cutting. *Can. J. Fish. Aquat. Sci.* **37**, 624–631.

Guy, M. P. (1964). An analysis of some storm period variables affecting stream sediment transport. *Prof. Pap. U.S. geol. Surv.* **462**-E.

Hack, J. T. (1957). Studies of longitudinal stream profiles in Virginia and Maryland. *Prof. Pap. U.S. geol. Surv.* **294**-B, 45–97.

Hall, D. G. (1964). The sediment hydraulics of the River Tyne. Ph.D. Thesis. University of Durham.

Hall, D. G. (1967). The pattern of sediment movement in the R. Tyne. *Internat. Ass. Scient. Hydrol.* **75**, 117–142.

Hall. J. D. and Lantz, R. L. (1969). Effects of logging on the habitat of coho salmon and cutthroat trout in coastal streams. *In* "Symposium on Salmon and Trout in Streams", (Northcote, T. G., ed.) pp. 335–376. H. R. MacMillan Lectures in Fisheries. Univ. of British Columbia, Vancouver.

Hamilton, J. D. (1961). The effect of sand-pit washings on a stream fauna. *Verh. Int. Verein. theor. angew. Limnol.* **14**, 435–439.

Hamor, T. and Garside, E. T. (1975). Regulation of oxygen consumption by incident illumination in embryonated ova of Atlantic salmon (*Salmo salar* L.). *Comp. Biochem. Physiol.* **52A**, 277–280.

Hamor, T. and Garside, E. T. (1976). Developmental rates of Atlantic salmon, *Salmo salar* L., in response to various levels of temperature, dissolved oxygen and water exchange. *Can. J. Zool.* **54**, 1912–1917.

Hamor, T. and Garside, E. T. (1977). Size relations and yolk utilization in embryonated ova and alevins of Atlantic salmon *Salmo salar* L. in various combinations of temperature and dissolved oxygen. *Can. J. Zool.* **55**, 1892–1898.

Hansen, E. A. (1971). Sediment in a Michigan trout stream, its source, movement, and some effects on fish habitat. U.S. Dep. Agric. For. Serv. Res. Pap. NC-59.

Hansen, E. A. (1975). Some effects of groundwater on brown trout redds. *Trans. Am. Fish. Soc.* **104**, 100–110.

Hansen. E. A. and Alexander, G. R. (1975). Effect of an artificially increased sand bedload on stream morphology and its implications on fish habitat. Rept. to N.C. Forest. Experimental Station, Instit. Forest Geometrics, Rhinelander, Wisconsin. 3/65–3/76.

Hardy, C. J. (1963). An examination of eleven stranded redds of brown trout (*Salmo trutta*) excavated in the Selwyn River during July and August, 1960. *N.Z. J. Sci.* **6**, 107–119.

Harker, J. E. (1953). An investigation of the distribution of the mayfly fauna of a Lancashire stream. *J. Anim. Ecol.* **22**, 1–13.

Harris, G. S. (1970). Some aspects of the biology of Welsh sea trout (*S. trutta trutta* L.). Ph.D. Thesis. University of Liverpool.

Harrison, A. D. (1958). Hydrobiological studies on the Great Berg River, Western Cape Province. Part 2. Quantitative studies on sandy bottoms, notes on tributaries and further information on the fauna, arranged systematically. *Trans. R. Soc. S. Africa.* **35**, 227–276.

Harrison, A. D. and Elsworth, J. F. (1958). Hydrobiological studies on the Great Berg River, Western Cape Province. *Trans. R. Soc. S. Africa.* **35**, 125–329.

Harrison, C. W. (1923). Planting eyed salmon and trout eggs. *Trans. Am. Fish. Soc.* **53**, 191–200.

Harrod. J. (1964). The distribution of invertebrates on submerged aquatic plants in a chalk stream. *J. Anim. Ecol.* **33**, 335–348.

Hart, D. D. (1978). Diversity in stream insects: regulation by rock size and microspatial complexity. *Verh. Internat. Verein. Limnol.* **20**, 1376–1381.

Hartman, G. F. C. (1963). Observations on behaviour of juvenile brown trout in a stream aquarium during winter and spring. *J. Fish. Res. Board Can.* **20**, 769–787.

Harvey, A. M. (1974). Gully erosion and sediment yield in the Howgill Fells, Westmorland. *In* "Fluvial Processes in Instrumented Watersheds" Gregory, K. J. and Walling, D. E., eds), vol. 6, pp. 45–58. Inst. Br. Geogrs. Spec. Publ.

Harvey, A. M. (1977). Event frequency in sediment production and channel change. pp. 301–310. *In* "River Channel Changes." Gregory, K. J., ed.) John Wiley, Chichester.

Hausle, D. A. and Coble, D. W. (1976). Influence of sand in redds on survival and emergence of brook trout (*Salvelinus fontinalis*). *Trans. Am. Fish. Soc.* **105**, 57–63.

Hayes, F. R., Wilmot, I. R. and Livingstone, D. A. (1951). The oxygen consumption of the salmon egg in relation to development and activity. *J. Exp. Zool.* **116**, 377–395.

Heidel, S. G. (1966). The progressive lag of sediment concentration with flood waves. *Trans. Am. geophys. Un.* **37**, 56–66.

Helland-Hansen, E. A. (1971). Time as a parameter in the study of incipient motion of gravel. 18th Annual Pacific Northwest Regional Meeting A.G.U. Corvallis Oregon. Oct. 14–15. Abstract p. 31. Water Resour. Abstr. 1973. Water Resources Scientific Information Centre, U.S. Dept. Interior, Washington, D. C. 20240.

Hemsworth, R. J. and Brooker, M. P. (1979). The rate of downstream displacement of macroinvertebrates in the upper Wye, Wales. *Holarctic Ecology* **2**, 130–136.

Henderson, F. M. (1963). Stability of alluvial channels. *Trans. Am. Soc. civ. Engrs.* **128**, 657–686.

Herbert, D. W. M., Alabaster, J. S., Dart, M. C. and Lloyd, R. (1961). The effect of china-clay wastes on trout streams. *Int. J. Air and Wat. Pollut.* **5**, 56–74.

Herricks, E. F. and Cairms, J. (1972). The recovery of a stream macrobenthic communities from the effects of acid mine drainage. Presented to the 4th Symposium of coal mine drainage research. Bituminous Coal Res. Inc. Monroeville, Pennsylvania. 370–398.

Hewlett, J. D. (1974). Comments on letter relating to "Role of subsurface flow in generating surface run-off". 2. "Upstream source areas" by R. Allen Freeze. *Water Resour. Res.* **10**, 605–607.

Hey, R. D. (1979). Flow resistance in gravel bed rivers. *J. Hydraul. Div. Am. Soc. civ. Engrs.* **105** (HY 4), 365–379.

Hitchcock, D. (1977). Channel pattern changes in divided reaches: An example in the coarse bed material of the Forest of Bowland. 207–217. *In* "River Channel Changes." Gregory, K. J., (ed.). John Wiley, Chichester.

Hjulström, F. (1935). Studies in the morphological activity of rivers as illustrated by the River Fyris. *Bull. Uppsala Univ. Geol. Inst.* **25**, 221–528.

Hobbs, D. F. (1937). Natural reproduction of quinnat salmon, brown and rainbow trout in certain New Zealand waters. *New Zealand Mar. Dep. Fish Bull.* **6**, 1–104.

Hoffman, C. E. and Kilambi, R. V. (1970). Environmental changes produced by cold-water outlets from three Arkansas reservoirs. Water Resour. Res. Center Publ. No. 5, Univ. Arkansas, Fayetteville, Arkansas.

Hollingshead, A. B. (1971). Sediment transport measurements in gravel river. *J. Hydraul. Div. Am. Soc. civ. Engrs.* **97** (HY 11). 1817–1834.

Horton, P. A. (1961). The bionomics of brown trout in a Dartmoor stream. *J. Anim. Ecol.* **30**, 311–338.

Hubbell, D. W. (1964). Apparatus and techniques for measuring bedload. *Wat. Suppl. Pap. U.S. Geol. Surv.* **1748**, 1–74.

Hughes, B. D. (1975). A study of a polluted river. Ph.D. Thesis. University of Wales.

Hunt, R. L. (1969). Effects of habitat alteration on production, standing crops and yield of brook trout in Lawrence creek, Wisconsin. *In* "Symposium on salmon and Trout in streams." (Northcote, T. G., ed.), pp. 281–312. H. R. MacMillan Lectures in Fisheries. Univ. of British Columbia, Vancouver.

Hurley, D. A. and Brannon, E. L. (1969). Effect of feeding before and after yolk absorption on the growth of sockeye salmon. Int. Pac. Salmon Fish. Comm. Prog. Rpt. 21.

Hutchinson, J. M. and Aney, W. W. (1964). The fish and wildlife resources of the lower Willamette Basin, Oregon, and their water use requirements. Oregon State Game Commission, Basin Investigation Section, Federal Aid Project F-69-R-1. Progress Report.

Hynes, H. B. N. (1954). The ecology of *Gammarus duebeni* Lilljeborg and its occurrence in fresh water in western Britain. *J. Anim. Ecol.* **23**, 38–84.

Hynes, H. B. N. (1958). The effect of drought on the fauna of a small mountain stream in Wales. *Verh. int. Verein. theor. angew. Limnol.* **13**, 826–833.

Hynes, H. B. N. (1961). The invertebrate fauna of a Welsh mountain stream. *Arch. Hydrobiol.* **57**, 344–388.

Hynes, H. B. N. (1963). Imported organic matter and secondary productivity in streams. *Proc. XVI Int. Congr. Zool. Washington* **4**, 324–329.

Hynes, H. B. N. (1970). "The Ecology of Running Waters." Liverpool University Press, UK.

Idyll, C. (1943). Bottom fauna of portions of the Cowichan River, B.C. *J. Fish. Res. Board Can.* **6**, 133–139.

Illies, J. (1958). Die Barbenregion mitteleuropäischer Fliessgewässer. *Verh. int. Verein. theor. angew. Limnol.* **13**, 834–844.

Imeson, A. C. and Ward, R. C. (1972). The output of a lowland catchment. *J. Hydrol.* **17**, 145–159.

Iwamoto, R. N., Salo, E. O., Madej, M. A. and McComas, R. L. (1978). Sediment and water quality: a review of the literature including a suggested approach for water quality criteria. U.S. Environ. Protection Agency, Seattle, Washington. EPA 910/9-78-048.

Jaag, O. and Ambühl, H. (1964). The effect of the current on the composition of biocoenoses in flowing water streams. *Adv. Wat. Poll. Res.* **1**, 31–49.

Johnson, J. W. (1942). The importance of considering sidewall friction in bedload investigations. *Civ. Engng.* **12**, 329–331.

Jones, A. N. (1975). Preliminary study of fish segregation in salmon spawning streams. *J. Fish Biol.* **7**, 95–104.

Jones, H. R. and Peters, J. C. (1977). Physical and biological typing of unpolluted rivers. *In* "Biological Monitoring of Inland Fisheries." (Alabaster, J. S., ed.), pp. 39–48. Applied Science.

Jones, J. R. E. (1941). The fauna of the River Dovey, West Wales. *J. Anim. Ecol.* **10**, 12–24.

Jones, J. R. E. (1943). The fauna of the River Teifi, West Wales. *J. Anim. Ecol.* **12**, 115–123.

Jones, J. R. E. (1948). The fauna of four streams in the "Black Mountain" district of South Wales. *J. Anim. Ecol.* **17**, 51–65.

Jones, J. R. E. (1951). An ecological study of the River Towy. *J. Anim. Ecol.* **20**, 68–86.

Jones, J. W. (1959). "The Salmon." Willmer Bros. & Hanem Ltd., London.

Jones, J. W. and Ball, J. N. (1954). The spawning behaviour of brown trout and salmon. *J. Anim. Behav.* **2**, 103–114.

Jones, J. W. and King, G. M. (1949). Experimental observations on the spawning behaviour of the Atlantic salmon (*Salmo salar* Linn.). *Proc. Zool. Soc. Lond.* **119**, 33–48.

Jones, J. W. and King, G. M. (1950). Further experimental observations on the spawning behaviour of the Atlantic salmon (*Salmo salar* Linn.). *Proc. Zool. Soc. Lond.* **120**, 317–323.

Jonsson, B. and Sandelund, O. T. (1979). Environmental factors and life histories of isolated river stocks of brown trout (*Salmo trutta* m. fario) in Søre Osa river system, Norway. *Env. Biol. Fish.* **4**, 43–54.

Jopling, A. V. and McDonald, B. C. (eds.) (1975). Glaciofluvial and Glaciolacustrine Sedimentation. Soc. Econ. Paleont. Mineral Spec. Publ. 23.

Judd, H. E. and Peterson, D. F. (1969). Hydraulics of large bed element channels.

Rept. No. PRWG 17-6, August Utah Wat. Res. Lab. Utah State University, Logan, Utah, USA.

Kalinske, A. A. (1947). Movement of sediment as bed-load in rivers. *Trans. geophys. Un.* **28**, 615-620.

Kalleberg, H. (1958). Observations in a stream tank of territoriality and competition in juvenile salmon and trout. *Inst. Freshwat. Res. Drott. Swed. Rept.* **39**, 55-98.

Kamphuis, J. W. (1974). Determination of sand roughness for fixed beds. *J. Hydraul. Res.* **12**, 193-203.

Kaushik, N. K. and Hynes, H. B. N. (1968). Experimental study on the role of autumn-shed leaves in aquatic environments. *J. Ecol.* **56**, 229-243.

Kaushik, N. K. and Hynes, H. B. N. (1971). The fate of the dead leaves that fall into streams. *Arch. Hydrobiol.* **68**, 465-515.

Keenleyside, M. H. A. and Yamamoto, F. T. (1962). Territorial behaviour of juvenile Atlantic salmon (*Salmo salar* L.). *Behaviour* **19**, 139-169.

Keller, E. A. (1971). Areal sorting of bed-load material: the hypothesis of velocity reversal. *Bull. Geol. Soc. Am.* **82**, 753-756.

Keller, E. A. and Melhorn, W. N. (1974). Form and fluvial processes in alluvial stream channels. Purdue Univ. Wat. Resour. Res. Centre, Tech. Rept. 47.

Keller, E. A. and Swanson, F. J. (1979). Effect of large organic material on channel form and fluvial processes. *Earth Surf. Processes* **4**, 361-386.

Kellerhals, R. (1967). Stable channels with gravel paved beds. *Proc. J. Watways Harb. Coast. Engnr. Am. Soc. civ. Engrs. Div.* **93**, (WW1), 63-84.

Kellerhals, R. (1970). Runoff routing through steep natural channels. *J. Hydraul. Div. Am. Soc. civ. Engrs.* **96**, (HY11), 2201-2218.

Kellerhals, R. and Bray, D. (1971). Sampling procedures for coarse fluvial sediments. *J. Hydr. Div. Am. Soc. civ. Engrs.* **97** (HY 8), 1165-1180.

Kelley, D. W., Delisle, G. E. and Cordone, A. J. (1960). A method to determine the volume of flow required by trout below dams—a proposal for investigation. Calif. Dep. Fish Game, Region II, Inland Fisheries.

Kennedy, D. W. (1967). Seasonal abundance of aquatic invertebrates and their utilisation by hatchery-reared rainbow trout. U.S. Bureau of Sport Fish and Wildl. Tech. Paper 12.

Kirkby, M. J. and Weyman, D. R. (1974). Measurement of contributing area in very small drainage basins. Seminar Paper Series B, No. 3. Dept. Geography, Univ. Bristol.

Klein, M. (1976). The influence of drainage area in producing thresholds for the hydrological regime and channel characteristics of natural rivers. Working Paper No. 147, School of Geography Univ. of Leeds.

Klingeman, P. C. and Emmett, W. W. (in press). Field progress in describing sediment transport. *In* "Engineering Problems in the Management of Gravel-bed Rivers." (R. D. Hey ed) Proc. of a workshop held at Gregynog Hall, Newtown, UK June 23-28 1980.

Knight, D. W. and Macdonald, J. A. (1979). Open channel flow with varying bed roughness. *J. Hydraul. Div. Am. Soc. civ. Engrs.* **105** (HY 9), 1167-1183.

Knighton, A. D. (1980). Longitudinal changes in size and sorting of streambed material in four English rivers. *Geol. Soc. Am. Bull. Part 1.* **91**, 55-62.

Kouwen, N. and Unny, T. E. (1973). Flexible roughness in open channels. *J. Hydraul. Div. Am. Soc. civ. Engrs.* **99**, 713-728.

Kroger, R. L. (1973). Biological effects of fluctuating water levels in the Snake River, Grand Teton National Park, Wyoming. *Amer. Midl. Nat.* **89**, 478-481.

Krumbein, W. C. (1934). Size frequency distributions of sediments. *J. Sedim. Petrol.* **4**, 65–77.

Krumbein, W. C. and Monk, G. D. (1942). Permeability as a function of the size parameters of unconsolidated sand. Am. Inst. Min. Metall. Eng., Tech. Publ. 1492.

Kuenen, P. H. (1956). Experimental abrasion of pebbles. 2. Rolling by current. *J. Geol.* **64**, 336–368.

La Barbera, M. and Vogel, S. (1976). An inexpensive thermistor flow meter for aquatic biology. *Limnol. Oceanogr.* **21**, 750–755.

Ladle, M. (1971). The biology of Oligochaeta from Dorset chalk streams. *Freshwat. Biol.* **1**, 83–97.

Lane, E. W. and Carlson, E. J. (1954). Some observations on the effect of particle size on movement of coarse sediments. *Trans. Am. Geophys. Un.* **35**, 453–462.

Langbein, W. B. (1964). Geometry of river channels. *J. Hydraul. Div. Am. Soc. civ. Engrs.* (**HY 2**), 301–311.

Langbein, W. B. and Leopold, L. B. (1968). River channel bars and dunes—theory of kinematic waves. Prof. Pap. U.S. Geol. Surv. 422–L.

Larimore, R. W., Childers, W. F. and Heckrotte, C. (1959). Destruction and re-establishment of stream fish and invertebrates affected by drought. *Trans. Am. Fish. Soc.* **88**, 261–285.

Larkin, P. A. (1956). Interspecific competition and population control, in freshwater fish. *J. Fish. Res. Board Can.* **13**, 327–343.

Laronne, J. B. and Carson, M. A. (1976). Inter-relationships between bed morphology and bed material transport for a small gravel-bed channel. *Sedimentology* **23**, 67–85.

Lavis, M. E. (1973). Thermal characteristics and low flow hydrology of a Pennine stream. Ph.D. Thesis. University of Durham.

Le Cren, E. D. (1965). Some factors regulating the size of populations of freshwater fish. *Mitt. Int. Verein. Limnol.* **13**, 88–105.

Le Cren, E. D. (1973). The population dynamics of young trout (*Salmo trutta*) in relation to density and territorial behaviour. *Rapports et Proces—Verbaux des Reunions, Conseil International pour L'exploration de la Mer*, **164**, 241–246.

Lee, D. R. (1977). A device for measuring seepage flux in lakes and estuaries. *Limnol. Oceanogr.* **22**, 140–147.

Lee, D. R. and Cherry, J. A. (1979). A field exercise on groundwater flow using seepage meters and mini piezometers. *J. Geol. Educ.* **27**, 6–10.

Lee, D. R., Cherry, J. A. and Pickens, J. F. (1980). Groundwater transport of a salt tracer through a sandy lake bed. *Limnol. Oceanogr.* **25**, 45–61.

Lee, D. R. and Hynes, H. B. N. (1977). Identification of groundwater discharge zones in a reach of Hillman creek in Southern Ontario. *Wat. Poll. Res. Can.* **13**, 121–133.

Lehmann, U. (1967). Drift und Populationsdynamik von *Gammarus pulex fossarum* Koch. *Z. Morph. Ökol.*, **60**, 227–285.

Leopold, L. B. and Emmett, W. W. (1976). Bedload measurements in East Fork River, Wyoming. *Proc. natn. Acad. Sci. USA* **73**, 1000–1004.

Leopold, L. B. and Emmett, W. W. (1977). 1976 Bedload measurements in East Fork River, Wyoming. *Proc. natn. Acad. Sci. USA* **74**, 2644–2648.

Leopold, L. B. and Langbein, W. B. (1962). The concept of entropy in landscape evolution. Prof. Pap. U.S. Geol. Surv. 500A.

Leopold, L. B. and Maddock, T. (1953). The hydraulic geometry of stream channels and some physiological implications. Prof. Pap. U.S. Geol. Surv. 252.

Leopold, L. B., Wolman, M. G. and Miller, J. P. (1964). "Fluvial Processes in Geomorphology", pp. 1–522. San Francisco.

210 N. J. MILNER *ET AL.*

Lesht, B. M. (1979). Relationship between sediment resuspension and the statistical frequency distribution of bottom shear stress. *Mar. Geol.* **32**, M19–27.

Lewin, J. (1976). Initiation of bedforms and meanders in coarse-grained sediment. *Bull. Geol. Soc. Am.* **87**, 281–285.

Lewin, J., Cryer, R. and Harrison, D. I. (1974). Sources for sediments and solutes in mid-Wales. *In* "Fluvial processes in instrumented watersheds." (Gregory, K. J. and Walling, D. E., eds) vol. 6, pp. 73–86. Inst. Br. Geogrs. Spec. Publ.

Lewis, K. (1973). The effect of suspended coal particles on the life forms of the aquatic moss *Eurhynchium riparoides*. 1. The gametophyte plant. *Freshwat. Biol.* **3**, 251–257.

Li, R. M., Simons, D. B. and Stevens, M. A. (1976). Morphology of cobble streams in small watersheds. *J. Hydraul. Div. Am. Soc. civ. Engrs.* **102** (HY 8), 1101–1117.

Lindroth, A. (1955). Distribution, territorial behaviour and movements of sea trout fry in the River Indalsalven. *Rept. Inst. Freshwat. Res. Drottningholm.* **36**, 104–119.

Lindroth, A. C. (1942). Sauerstofferbrauch der Fische. I. Verschiendene Entwicklungs —und Altersstadien com hachs und Hecht. *A. vergl. Physiol.* **29**, 583–594.

Linduska, J. P. (1942). Bottom type as a factor influencing the local distribution of mayfly nymphs. *Can. Ent.* **74**, 26–30.

Lisle, T. (1979). A sorting mechanism for a riffle-pool sequence. *Bull. geol. Soc. Am.* **90**, 616–617.

Lister, D. B. and Walker, C. E. (1966). The effect of flow control on freshwater survival of chum, coho and chinook salmon in Big Qualicum River. *Can. Fish. Cult.* **40**, 41–49.

Little, W. C. and Mayer, P. G. (1972). The role of sediment graduation on channel armouring. Report No. ERC–0672 May 1972. Environmental Resources Center, Georgia, Instit. Technol. Atlanta. 1–104.

Lundgren, H. and Jonsson, I. G. (1964). Shear and velocity distribution in shallow channels. *J. Hydraul. Div. Am. Soc. civ. Engrs.* **90** (HY 1), 1–21.

Lymen, F. E. and Dendy, J. S. (1943). A pre-impoundment bottom-fauna study of Cherokee Reservoir area (Tenessee). *Trans. Am. Fish. Soc.* **73**, 194–208.

Macan, T. T. (1963). "Freshwater Ecology". Longmans, London.

MacCrimmon, H. R. C. (1954). Stream studies on planted Atlantic salmon. *J. Fish. Res. Board Can.* **11**, 362–403.

Maciolek, J. A. (1966). Abundance and character of microseston in a California mountain stream. *Verh. Internat. Verein. Limnol.* **16**, 639–645.

Maciolek, J. A. and Tunzi, M. G. (1968). Microseston dynamics in a simple Sierra Nevada lake-stream ecosystem. *Ecology*, **49**, 60–75.

Mackay, C. L. (1970). A theory concerning the distance travelled by animals entering the drift of a stream. *J. Fish. Res. Board Can.* **27**, 359–370.

Mackay, C. L. and Kalff, J. (1973). Ecology of two related species of caddis fly larvae in the organic substrates of a woodland stream. *Ecology* **54**, 499–511.

Mackin, J. H. (1948). Concept of the graded river. *Bull. geol. Soc. Amer.* **59**, 463–512.

Madsen, B. L. (1968). The distribution of nymphs of *Brachyptera risi* (Morton) and *Nemoura flexuosa* Aub. (Plecoptera) in relation to oxygen. *Oikos* **19**, 304–310.

Madsen, B. L. (1969). Reactions of *Bachyptera risi* (Morton) (Plecoptera) nymphs to water current. *Oikos* **20**, 95–100.

Maitland, P. S. (1964). Quantitative studies on the invertebrate fauna of sandy and stony substrates in the River Endrick, Scotland. *Proc. R. Soc. Edinb.* **68**, 277–301.

Malmqvist, B., Nilsson, L. M. and Svensson, B. S. (1978). Dynamics of detritus in a small stream in southern Sweden and its influence on the distribution of the bottom communities. *Oikos* **31**, 3–16.

Mantz, P. A. (1977). Incipient transport of fine grains and flakes by fluids—extended Shields' diagram. *J. Hydraul. Div. Am. Soc. civ. Engrs.* **103** (HY 6), 601–615.

Marr, D. H. A. (1963). The influence of surface contour on the behaviour of trout alevins *Salmo trutta* L. *Anim. Behav.* **11**, 412.

Marr, D. H. A. (1966). Influence of temperature on the efficiency of growth of salmonid embryos. *Nature* **212**, 957–959.

Martin, C. S. (1970). Effect of a porous sand bed on incipient sediment motion. *Water Resour. Res.* **6**, 1162–1174.

Martinec, J. (1973). Extrapolation of stage-discharge relations for flood computations. Internat. Soc. Hydraul. Res., Internat. Symp. on River Mechanics. 9–12 Jan. 1973. Bangkok, Thailand. 25–35.

Mason, J. C. (1969). Hypoxial stress prior to emergence and competition among the coho salmon fry. *J. Fish. Res. Board Can.* **26**, 63–91.

Mast, R. F. and Potter, P. E. (1963). Sedimentary structures, sand shape, fabrics and permeability. Pt. 2. *J. Geol.* **71**, 548–565.

Matthes, G. H. (1947). Macroturbulence in natural stream flow. *Trans. Am. Geophys. Un.* **28**, 255–262.

Maude, A. D. and Whitmore, R. L. (1958). A generalised theory of sedimentation. *Br. J. appl. Phys.* **9**, 477–482.

McCart, P. (1969). Digging behaviour of *Oncorhynchus nerka* spawning in streams at Babine Lake, British Columbia. *In* "Symposium on Salmon and Trout in streams". (Northcote, T. G., ed.), pp. 39–51. H. R. MacMillan Lectures in Fisheries. Univ. British Columbia, Vancouver.

McCave, I. N. (1970). Deposition of fine grained suspended sediment from tidal currents. *J. geophys. Res.* **75**, 4151–4159.

McGary, J. L. and Harp, G. L. (1972). The benthic macroinvertebrate community of the Greer's Ferry reservoir cold tail-water, Little Red River, Arkansas. *Proc. S.E. Assoc. Game Fish Comm.* **26**, 490–500.

McKenzie, L. S. and Walker, H. J. (1974). Morphology of an arctic river bar. *Coastal Studies Inst. Tech. Rept.* **172**, 32 pp.

McLachlan, A. J., Brennan, A. and Wotton, R. S. (1978). Particle size and chironomid (Diptera) food in an upland stream. *Oikos* **31**, 247–252.

McNeil, W. J. (1962). Variations in the dissolved oxygen content of intragravel water in four spawning streams in S.E. Alaska. Spec. Sci. Rept., Fisheries, Fish & Wildl. Serv. U.S. No. 402.

McNeil, W. J. (1964). Environmental factors affecting survival of young salmon in spawning beds and their possible relation to logging. *U.S. Fish Wildl. Serv. Bur. Comm. Fish., Rep.* 64–1.

McNeil, W. J. (1966). Effect of the spawning bed environment on reproduction of pink and chum salmon. *Bull. U.S. Fish Wildl. Serv. Fish* **65**, 495–523.

McNeil, W. J. (1967). Randomness in distribution of pink salmon redds. *J. Fish. Res. Board Can.* **24**, 1629–1634.

McNeil, W. J. and Ahnell, W. H. (1964). Success of pink salmon spawning relative to size of spawning bed materials. Spec. Sci. Rept.—Fisheries, U.S. Fish. & Wildl. Serv. 469.

McQuivey, R. S. (1973a). Principles and measuring techniques of turbulence characteristics in open channel flows. Prof. Pap. U.S. Geol. Surv. 802 A.

McQuivey, R. S. (1973b). Summary of turbulence data from rivers, conveyance channels and laboratory flumes. Prof. Pap. U.S. Geol. Surv. 802 B.

Mears, A. I. (1979). Flooding and sediment transport in a small alpine drainage basin in Colorado. *Geology* **7**, 53–57.

Milhous, R. T. (1972a). Discussion of Sediment transport measurements in gravel river. *J. Hydraul. Div. Am. Soc. civ. Engrs.* **98**, (HY 11), 2043–2046.

Milhous, R. T. (1972b). Three-dimensional and time aspects of bed material movement in a gravel-bottomed stream. 13th Annual Meeting Oregon Acad. Sci., Portland. Feb. 25–26 1972. Abstract p. 31. Water Resour. Abstr. 1973. Water Resources Scientific Information Center—U.S. Dept. Interior, Washington, D.C. 20240.

Miller, M. C., McCave, I. N. and Komar, P. D. (1977). Threshold of sediment motion under unidirectional currents. *Sedimentology* **24**, 507–527.

Minckley, W. L. (1963). The ecology of spring stream Doe Run, Meade County, Kentucky. *Wildl. Monogr. Chestertown* **11**, 1–124.

Minshall, G. W. and Kuehne, R. A. (1969). An ecological study of invertebrates of the Duddon, an English mountain stream. *Arch. Hydrobiol.* **66**, 169–191.

Minshall, G. W. and Minshall, J. N. (1977). Microdistribution of benthic invertebrates in a Rocky Mountain (USA) stream. *Hydrobiologia* **55**, 231–249.

Minshall, G. W. and Winger, P. V. (1968). The effect of reduction in stream flow on invertebrate drift. *Ecology* **49**, 580–582.

Morgan, N. C. and Egglishaw, H. J. (1965). A survey of the bottom fauna of streams in the Scottish Highlands. Part 1. Composition of the fauna. *Hydrobiologia* **25**, 181–211.

Morris, D. L. and Brooker, M. P. (1979). The vertical distribution of macro-invertebrates in the substratum of the upper reaches of the R. Wye, Wales. *Freshwat. Biol.* **9**, 573–583.

Moss, A. J. (1962). The physical nature of common sandy and pebbly deposits. Pt. 1. *Am. J. Sci.* **260**, 337–373.

Moss, A. J., Walker, P. H. and Hutka, J. (1980). Movement of coarse, sandy detritus by shallow water: an experimental study. *Sediment. Geol.* **25**, 43–66.

Muir, T. C. (1970). Bedload discharge of the River Tyne, England. *Bull. int. Ass. scient. Hydrol.* **15**, 35–39.

Mundie, J. H. (1971). Sampling benthos and substrate materials, down to 50 microns in size, in shallow streams. *J. Fish. Res. Board Can.* **28**, 849–860.

Mundie, J. H. (1974). Optimisation of the salmonid nursery stream. *J. Fish. Res. Board Can.* **31**, 1827–1837.

Naiman, R. J. and Sedell, J. R. (1979). Characterisation of particulate organic matter transported by some Cascade mountain streams. *J. Fish. Res. Board Can.* **36**, 17–31.

Nanson, G. C. (1974). Bedload and suspended load transport in a small steep mountain stream. *Am. J. Sci.* **274**, 471–486.

Neave, F. (1958). Stream ecology and production of anadromous fish. *In* "The investigations of fish-power problems", (Larkin, P. A., ed.), pp. 43–48. H. R. MacMillan Lectures in Fisheries. Univ. British Columbia, Vancouver.

Needham, P. R. and Usinger, R. L. (1956). Variability in the macrofauna of a single riffle in Prosser Creek, California, as indicated by the Surber sampler. *Hilgardia* **24**, 383–409.

Neill, C. R. (1967). Mean velocity criterion for scour of coarse uniform bed-material. In 12th Congress Int. Ass. Hydraul. Res. Fort Collins. 46–54.

Neill, C. R. (1968). A re-examination of the beginning of movement for coarse granular bed materials. Rept. No. Int. 68. Hydraulics Research Station, Wallingford, England. 37 pp.

Nelson, D. J. and Scott, D. C. (1962). Role of detritus in the productivity of a rock-outcrop community in a Piedmont stream. *Limnol. Oceanog.* **7**, 396–413.

Newson, M. (1980a). The geomorphological effectiveness of floods—A contribution stimulated by two recent events in mid-Wales. *Earth Surf. Processes.* **5**, 1–16.

Newson, M. (1980b). The erosion of drainage ditches and its effect on bedload yields in mid-Wales. *Earth Surf. Processes* **5**, 275–290.

Newson, M. (in press). Mountain Streams in:– British Rivers. (Lewin, J., ed.). Allen & Unwin, London.

Newson, M. D. and Harrison, J. G. (1978).. Channel studies in the Plynlimon experimental catchments. Rept. No. 47, Institute of Hydrology, Wallingford, England. 61 pp.

Nixon, M. (1959). A study of the bankful discharges of the rivers of England and Wales. *Proc. Inst. civ. Engrs.* 12 Feb. 157–174.

Novak, P. (1957). Bedload meters—development of a new type and determination of their efficiency with the aid of scale models. Trans. Int. Ass. Hydraul. Res. 7th General Meeting. Lisbon, pp. A91–A99.

Nuttall, P. M. (1972). The effects of sand deposition upon the macro-invertebrate fauna of the River Camel, Cornwall. *Freshwat. Biol.* **2**, 181–186.

O'Brien, R. N., Visaisouk, S., Raine, R. and Alderdice, D. F. (1978). Natural Convection: A mechanism for transporting oxygen to incubating salmon eggs. *J. Fish. Res. Board Can.* **35**, 1316–1321.

O'Connell, T. R. and Campbell, R. S. (1953). The benthos of the Black River and Clearwater Lake, Missouri. *Univ. Mo. Stud.* **26**, 25–41.

Oliff, W. D. (1960). Hydrobiological studies on the Tugela River system. Part 1. The main Tugela River. *Hydrobiologia* **16**, 281–285.

Oliff, W. D. (1963). Hydrobiological studies on the Tugela River system. Part 3. The Buffalo River. *Hydrobiologia* **21**, 355–379.

Oliff, W. D., Kemp, P. H. and King, J. L. (1965). Hydrobiological studies on the Tugela River system. Part 4. The Sundays River. *Hydrobiologia* **26**, 189–202.

Painter, R. B. (1972). The measurement of bedload movement in rivers. *Wat. & Wat. Engrs.* **76**, 291–294.

Painter, R. B., Blyth, K., Mosedale, J. C. and Kelly, M. (1974). The effect of afforestation on erosion processes and sediment yield. *In* "Effects of Man on the Interface of the Hydrological Cycle with the Physical Environment," pp. 62–67. Symp. Proc. of the Paris Symp., Sept. 1974. Int. Ass. Scient. Hydrol. Publ.

Pardé, M. (1939). Hydrologie fluviale des Iles Britanniques. *Annls. Geogr.* **48**, 369–384.

Parker, G. (1979a). Hydraulic geometry of active gravel rivers. *J. Hydraul. Div. Am. Soc. civ. Engrs.* **105** (HY 9), 1185–1205.

Parker, G. (1979b). Self-formed straight rivers with equilibrium banks and mobile bed. Part 1. The sand-silt river. *J. Fluid Mech.* **89**, 109–125.

Parker, G. and Anderson, A. G. (1977). Basic principles of river hydraulics. *J. Hydraul. Div. Am. Soc. civ. Engrs.* **103** (HY 9), 1077–1087.

Parkinson, E. A. and Slaney, P. A. (1975). A review of enhancement techniques applicable to anadromous game fishes. Fisheries Management Report No. 66. British Columbia Fish & Wildlife Branch, August 1975.

Pearson, W. D. and Franklin, D. R. (1968). Some factors affecting the drift rates of *Baetis* and Simuliidae in a large river. *Ecology* **49**, 75–81.

Pearson, W. D. and Kramer, R. H. (1972). Drift and production of two aquatic insects in a mountain stream. *Ecol. Monogr.* **24**, 365–385.

Pennak, R. W. and Van Gerpen, E. D. (1947). Bottom fauna production and physical nature of the substrate in a northern Colorado trout stream. *Ecology* **28**, 42–48.

Percival, E. and Whitehead, H. (1929). A quantitative study of some types of streambed. *J. Ecol.* **17**, 282–314.

Peters, J. C. (1962). The effects of stream sedimentation on trout embryo survival. *In* Trans. 3rd Seminar on biological problems in water pollution. (Tarzwell, C. M., ed.). pp. 272–279.

Peters, J. C. (1967). Effects on a trout stream of sediment from agricultural practices. *J. Wildl. Management* **31**, 805–812.

Peterson, R. C. and Cummins, K. W. (1974). Leaf processing in a woodland stream. *Freshwat. Biol.* **4**, 343–368.

Peterson, D. F. and Mohanty, P. K. (1966). Flume studies of flow in steep rough channels. *J. Hydraul. Div. Am. Soc. civ. Engrs.* **86**, (HY 9), 55–76.

Peterson, R. H. (1978). Physical characteristics of Atlantic salmon spawning gravel in some New Brunswick streams. *Fish. Mar. Serv. Tech. Rept.* **785**.

Pettijohn, E. J. (1975). "Sedimentary Rocks." 3rd Edit. Harper International, New York.

Petran, M. and Kothé, P. (1978). Influence of bedload transport on the macrobenthos of running waters. *Verh. Internat. Verein. Limnol.* **20**, 1867–1872.

Philipson, G. N. (1954). The effect of water flow and oxygen concentration on six species of caddis fly (Trichoptera) larvae. *Proc. Zool. Soc. Lond.* **124**, 547–564.

Philipson, G. N. (1969). Some factors affecting the net-spinning of the caddis fly *Hydropsyche instabilis*. *Hydrobiologia*, **34**, 369–377.

Phillips, R. W. (1964). The influence of gravel size on survival to emergence of coho salmon and steelhead trout. Proc. of the 15th Northwest Fish Culture Conference, Oregon State Univ. Press.

Phillips, R. W. (1971). Effect of sediment on the gravel environment and fish production. *In* "Forest Land Use and Stream Environment." (Krygier, J. K. and Hall, J. D., eds), pp. 64–74. Oregon State Univ. Corvallis, Oregon.

Phillips, R. W. and Campbell, H. J. (1962). The embryonic survival of coho salmon and steelhead trout as influenced by some environmental conditions in gravel beds. 14th Ann. Rep. of Pac. Mar. Fish. Comm. 60–73.

Phillips, R. W. and Koski, K. V. (1969). A fry trap method for estimating salmonid survival from egg deposition to fry emergence. *J. Fish. Res. Board Can.* **26**, 133–141.

Pitney, W. E. (1969). Determination of stream flows for fish life. Oregon State Game Commission.

Platts, W. S., Shirazi, M. A. and Lewis, D. H. (1979). Sediment particle sizes used by salmon for spawning with methods for evaluation. Rept. EPA–60013–79–043. Corvallis Environmental Research Laboratory, Office of Research & Development, U.S. Environmental Protection Agency, Corvallis, Oregon, 97330.

Pleskot, G. (1953). Zur Ökologie der Leptophlebiiden (Ins., Ephemeroptera). *Österr. zool. Z.* **4**, 45–107.

Plumley, W. J. (1948). Black Hills terrace gravels: a study in sediment transport. *J. Geol.* **56**, 526–577.

Pollard, R. A. (1955). Measuring seepage through salmon spawning gravel. *J. Fish. Res. Board Can.* **12**, 706–741.

Poole, W. C. and Stewart, K. W. (1976). The vertical distribution of macrobenthos within the substratum of the Brazos River, Texas. *Hydrobiologia* **50**, 151–160.

Potter, K. W. (1979). Derivation of the probability density function of certain biased samples of coarse riverbed material. *Water Resour. Res.* **15**, 21–22.

Potter, P. E. and Pettijohn, F. J. (1963). "Paleocurrents and Basin Analysis." Springer, New York.

Rabeni, C. F. and Minshall, G. W. (1977). Factors affecting microdistribution of stream benthic insects. *Oikos* **29**, 33–43.

Radford, D. S. and Hartland-Rowe, R. (1971a). Subsurface and surface sampling of benthic invertebrates in two streams. *Limnol. Oceanogr.* **16**, 114–120.

Radford, D. S. and Hartland-Rowe, R. (1971b). A preliminary investigation of the bottom fauna and invertebrate drift in an unregulated and a regulated stream in Alberta. *J. Appl. Ecol.* **8**, 883–903.

Rákóczi, L. (1975). Influence of grain-size composition on the incipient motion and self-pavement of bed materials. Int. Ass. Hydraul. Res. Congr. 16th 2. Sao Paulo, Brazil.

Rákóczi, L. (1977). The significance of infrequent, high suspended sediment concentrations in the estimation of annual sediment transport. *In* Proc. Paris Symp. July 1977. Erosion and solid matter transport in inland waters. Int. Ass. Scient. Hydrol. Publ. 122, 19–25.

Ramette and Heuzel, (1962). A study of pebble movements in the Rhone by means of radioactive tracers. (In French, English Abstr.). *La Houille Blanche No. Speciale A.*, 389–398.

Reiser, D. W. (1976). Determination of physical and hydraulic preferences of brown trout and brook trout in the selection of spawning locations. M.Sc. Thesis. Wyoming University.

Richards, K. S. (1973). Hydraulic geometry and channel roughness—a non-linear system. *Am. J. Sci.* **273**, 877–896.

Richards, K. S. (1976a). Complex width-discharge relations in natural river sections. *Bull. geol. Soc. Am.* **87**, 199–206.

Richards, K. S. (1976b). The morphology of riffle and pool sequences. *Earth Surf. Processes.* **1**, 71–88.

Richards, K. S. and Milne, L. M. (1979). Problems in the calibration of an acoustic device for the observation of bedload transport. *Earth Surf. Processes* **4**, 335–346.

Richardson, E. V. and McQuivey, R. S. (1968). Measurement of turbulence in water. *J. Hydraul. Div. Am. Soc. civ. Engrs.* **94**, (HY 2), 411–430.

Ricker, W. E. (1954). Stock and recruitment. *J. Fish. Res. Board Can.* **11**, 559–623.

Riedl, R. J. and Machan, R. (1972). Hydrodynamic patterns in lotic intertidal sands and their bioclimatical implications. *Mar. Biol.* **13**, 179–209.

Riley, S. J. (1976). Aspects of bankful geometry in a distributary system of eastern Australia. *Hydrol. Sci. Bull.* **XXI**, 4, 545–560.

Roberson, J. A. and Wright, S. J. (1973). Analysis of flow in channels with gravel beds. Proc. 21st Annual Hydraul. Div. Am. Soc. civ. Engrs. Speciality Conf. Montana Univ. Bozeman. Aug. 1973. 63–72.

Rosenberg, D. M. and Wiens, A. P. (1978). Effects of sediment addition on macrobenthic invertebrates in a northern Canadian river. *Water Res.* **12**, 753–763.

Royce, W. F. (1959). On the possibilities of improving salmon spawning areas. Trans. 24th N. Amer. Wildl. Conf. Wildl. Mgmt. Inst. 356–366.

Ruggles, C. P. (1966). Depth and velocity as a factor in stream rearing and production of juvenile coho salmon. *Can. Fish Cult.* **38**, 37–53.

Saunders, J. W. and Smith, M. W. (1962). Physical alteration of stream habitat to improve trout production. *Trans. Am. Fish. Soc.* **11**, 185–188.

Saunders, J. W. and Smith, M. W. (1965). Changes in a stream population of trout associated with increased silt. *J. Fish. Res. Board Can.* **22**, 395–404.

Savage, J. L. (1962). Methods used in salmon spawning area surveys. Mins. of 2nd Ann. Meeting Cal. Dept. of Fish, Game and wat. projs. Poll. Conf. 20–39.

Schuck, H. A. (1945). Survival, population density, growth and movement of the wild brown trout in Crystal Creek. *Trans. Am. Fish. Soc.* **73**, 209–230.

Schwoerbel, J. (1961). Über die Lebensbedingungen und die Besiedlung des hyporheischen Lebensraumes. *Arch. Hydrobiol.* **25**, 182–214.

Scott, D. (1958). Ecological studies on the Trichoptera of the River Dean, Cheshire. *Arch. Hydrobiol.* **54**, 340–392.

Scott, D. (1960). Cover on river bottoms. *Nature, Lond.* **188**, 76–77.

Scott, D. & Rushforth, J. M. (1959). Cover on river bottoms. *Nature, Lond.* **183**, 836–837.

Scullion, J. (1977). The biology of the Taff Bargoed. Ph.D. Thesis. University of Wales.

Scullion, J. and Edwards, R. W. (1980). The effect of coal industry pollutants on the macro-invertebrate fauna of a small river in the South Wales coalfield. *Freshwat. Biol.* **10**, 141–162.

Sedell, J. R., Naiman, R. J., Cummins, K. W., Minshall, G. W. and Vanniote, R. L. (1978). Transport of particulate organic material in streams as a function of physical processes. *Verh. Internat. Verein. Limnol.* **20**, 1366–1375.

Seki, H., Stephens, K. V. and Parson, T. R. (1969). The contribution of allocthonous bacteria and organic materials from a small river to a semi-enclosed area. *Arch. Hydrobiol.* **66**, 37–47.

Shapley, S. P. and Bishop, D. M. (1965). Sedimentation in a salmon stream. *J. Fish. Res. Board Can.* **22**, 919–928.

Shelton, J. M. (1955). The hatching of chinook salmon eggs under simulated stream conditions. *Progr. Fish. Cult.* **17**, 20–35.

Shelton, J. M. and Pollock, R. D. (1966). Siltation and egg survival in incubation channels. *Trans. Am. Fish. Soc.* **95**, 183–187.

Sheridan, W. L. (1962). Waterflow through a salmon spawning riffle in south East Alaska. Spec. Scient. Rept., Fisheries, Fish and Wildl. Serv., U.S. No. 407.

Shetter, D. S., Clarke, O. H. and Hazzard, A. S. (1946). The effects of deflectors in a section of Michigan trout stream. *Trans. Am. Fish. Soc.* **76**, 248–278.

Shirazi, M. A., Lewis, D. H. and Seim, W. K. (1979). Monitoring spawning gravel in managed forested watersheds. Rept. EPA–600/3–79–014. Corvallis Environmental Research Laboratory Office of Research and Development. U.S. Environmental Protection Agency. Corvallis, Oregon, 97330.

Shumway, D. L., Warren, C. E. and Doudoroff, P. (1964). Influence of oxygen concentration and water movement on the growth of steelhead trout and coho salmon embryos. *Trans. Am. Fish. Soc.* **93**, 342–356.

Silver, S. J., Warren, C. E. and Doudoroff, P. (1963). Dissolved oxygen requirements of developing steelhead trout and chinook salmon embryos at different water temperatures. *Trans. Am. Fisheries Soc.* **92**, 327–343.

Simons, D. B., Al-Shaikh-Ali, K. S. and Ruh-Ming Li (1979). Flow resistance in cobble and boulder river beds. *J. Hydraul. Div. Am. Soc. civ. Engrs.* **105**, (HY 5), 477–488.

Simons, D. B. and Sentürk, F. (1977). Sediment transport technology. "Water Resour. Publ." Fort Collins, Colorado.

Slaymaker, H. O. (1972). Patterns of present sub-aerial erosion and landforms in mid-Wales. *Trans. Inst. Br. Geog.* **55**, 47–69.

Smith, K. (1966). Percolation, groundwater discharge and streamflow in the Nidd Valley. *Jour. Inst. wat. Engrs.* **20**, 459.

Smith, K. (1969). The baseflow contribution and runoff in two upland catchments. *Water Wat. Engn.* **Jan.** 18–20.

Smith, N. D. (1974). Sedimentology and bar formation in the upper Kicking Horse River, a braided outwash stream. *J. Geol.* **82**, 205–253.

Solomon, D. J. and Templeton, R. G. (1976). Movements of brown trout *Salmo trutta* in a chalk stream. *J. Fish Biol.* **9**, 411–424.

Somme, S. (1960). The effects of impoundment on salmon and sea trout rivers. Int. Union for Conserv. Nat. Res. 7th Tech. Meeting. 1958. 77–80.

Spence, J. A. and Hynes, H. B. N. (1971). Difference in the benthos upstream and downstream of an impoundment. *J. Fish. Res. Board Can.* **28**, 35–43.

Sprules, W. M. (1947). An ecological investigation of stream insects in Algonquin Park, Ontario. *Univ. Toronto Stud. biol. Ser.* **56**, 1–81.

Stall, J. B. and Yang, C. T. (1970). Hydraulic geometry of 12 selected stream systems of the United States. Research Rept. No. 32. Water Resour. Center, Univ. of Illinois.

Stuart, T. A. (1953a). Water currents through permeable gravels and their significance to spawning salmonids. *Nature, Lond.* **172**, 407–408.

Stuart, T. A. (1953b). Spawning migration, reproduction and young stages of loch trout (*Salmo trutta* L.). *Freshwat. Salm. Fish. Res.* **5**, 1–39.

Stuart, T. A. (1954). Spawning sites of trout. *Nature, Lond.* **173**, 354.

Stocker, Z. S. J. and Williams, D. D. (1972). A freezing core method for describing the vertical distribution of sediments in a stream bed. *Limnol. Oceanogr.* **17**, 136–138.

Sumner, G. N. (1978). "Mathematics for Physical Geographers." Edward Arnold, London.

Surber, E. W. (1937). Rainbow trout and bottom fauna production in one mile of stream. *Trans. Am. Fish. Soc.* **66**, 193–202.

Surber, E. W. (1951). Bottom fauna and temperature conditions in relation to trout management in St. Mary's River, Augusta County, Virginia. *Va. J. Sci.* **2**, 190–202.

Swenson, F. A. (1942). Sedimentation near junction of Maquoketa and Mississippi Rivers. *J. sedim. Pet.* **12**, 3–9.

Symons, P. E. K. & Heland, M. (1978). Stream habitats and behavioural interactions of under yearling and yearling Atlantic salmon (*Salmo salar*). *J. Fish. Res. Board Can.* **35**, 175–183.

Tanner, W. F. (1977). Estimators of alluvial bedload transport. *Earth Surf. Processes.* **2**, 417–420.

Tagart, J. V. (1976). The survival from egg deposition to emergence of coho salmon in the Clearwater River, Jefferson County, Washington. M.Sc. Thesis. University of Washington.

Tarzwell, C. M. (1937). Experimental evidence on the values of trout stream improvement in Michigan. *Trans. Am. Fish. Soc.* **66**, 177–187.

Tennant, D. L. (1976). Instream flow regimes for fish, wildlife and recreation environmental resources. *Fisheries* **1**, 6–10.

Terhune, L. D. B. (1958). The Mark IV groundwater stand-pipe for measuring seepage through salmon spawning gravel. *J. Fish. Res. Board Can.* **15**, 1027–1063.

Thompson, K. E. (1972). Determining stream flows for fish life. Transactions and proceedings of instream flow requirements workshop. Sponsored by Pacific North-west River Basins Commission, Vancouver & Washington, 31–46.

Thorup, J. (1966). Substrate type and its value as a basis for the delimitation of bottom fauna communities in running waters. *Spec. Publs. Pymatuning Lab. Field Biol.* **4**, 59–74.

Toscano, R. J. and McLachlan, A. J. (1980). Chironomids and particles: micro-organisms and chironomid distribution in a peaty upland river. *In* "Chironomidae. Ecology, Systematics, Cytology and Physiology." (Murray, D. A., ed.), pp. 171–178. Pergamon Press, Oxford.

Trotsky, H. M. and Gregory, R. W. (1974). The effects of waterflow manipulation below a hydro-electric power dam on the bottom fauna of the Upper Kennebec River, Maine. *Trans. Am. Fish. Soc.* **103**, 318–324.

Turnpenny, A. W. H. and Williams, R. (1980). Effects of sedimentation on the gravels of an industrial river system. *J. Fish Biol.* **17**, 681–693.

Turnpenny, A. W. H. and Williams, R. (in press). A modified standpipe technique for measuring the oxygen supply for spawning beds. *J. Fish Biol.*

Ulfstrand, S. (1967). Microdistribution of benthic species (Ephemeroptera, Plecoptera, Trichoptera, Diptera: Simuliidae) in Lapland streams. *Oikos* **18**, 293–310.

Ulfstrand, S. (1968). Benthic animal communities in Lapland streams. *Oikos* (suppl.) **10**, 1–120.

Vaux, W. G. (1962). Interchange of stream and intragravel water in a salmon spawning riffle. U.S. Fish. & Wildl. Serv. Spec. Sci. Rept.—Fisheries. 405.

Virmani, J. K., Peterson, D. F. and Watters, G. Z. (1973). Discharge, slope, bed element relations in streams. Proc. 21st Annual Hydraul. Div. Am. Soc. civ. Engrs. Speciality Conf. Montana Univ. Bozeman. Aug. 1973. 73–84.

Von Mitis, H. (1938). Die Ybbs als Typus eines ostalpinen Kalkalpenfluss. Eine vorläufige Mitteilung. *Int. Revue ges. Hydrobiol. Hydrogr.* **37**, 425–444.

Walentowicz, A. T. and McLachlan, A. J. (1980). Chironomids and particles: a field experiment with peat in an upland stream. *In* "Chironomidae. Ecology, Systematics, Cytology and Physiology." (Murray, D. A., ed.), pp. 179–185. Pergamon Press, Oxford.

Wallace, J. B. and Merritt, R. W. (1980). Filter-feeding ecology of aquatic insects. *A. Rev. Ent.* **25**, 103–132.

Walling, D. E. (1971). Instrumented catchments in south-east Devon. Some relation-ships between catchment characteristics and catchment response. Ph.D. Thesis. Exeter University.

Walling, D. E. (1977a). Limitations of the rating curve technique for estimating suspended sediment loads, with particular reference to British rivers. Proc. Paris Symp. July 1977. Erosion and solid matter transport in inland waters. *Int. Ass. Scient. Hydrol. Publ.* **122**, 34–48.

Walling, D. E. (1977b). Assessing the accuracy of suspended sediment rating curves for a small basin. *Water Resour. Res.* **13**, 531–538.

Walling, D. E., Peart, M. R., Oldfield, F. and Thompson, R. (1979). Suspended sediment sources identified by magnetic measurements. *Nature* **281**, 110–113.

Walling, D. E. and Teed, A. (1971). A simple pumping sampler for research into suspended sediment transport in small catchments. *J. Hydrol.* **13**, 325–337.

Walters, G. Z. and Rao, M. Y. P. (1971). Hydrodynamic effects of seepage on bed particles. *J. Hydraul. Div. Am. Soc. civ. Engrs.* **97**, (HY 5) 421–439.

Ward, J. C. (1964). Turbulent flow in porous media. *J. Hydraul. Div. Am. Soc. civ. Engrs.* **(HY 5)**, 1–12.

Ward, J. V. (1976a). Effects of flow patterns below large dams on stream benthos: a review. *In* "Instream Flow Needs Symposium." Vol. II, (Osborn, J. F. and Allman, C. H., eds), pp. 235–253. Amer. Fish. Soc.

Ward, J. V. (1976b). Comparative Limnology of differentially regulated sections of a Colorado mountain river. *Arch. Hydrobiol.* **78**, 319–342.

Ward, J. V. and Short, R. A. (1978). Macroinvertebrate community structure of four special lotic habitats in Colorado, USA. *Verh. Internat. Verein. Limnol.* **20**, 1382–1387.

Ward, R. C. (1968). Some run-off characteristics of British rivers. *J. Hydrol.* **6**, 358–372.

Warner, G. H. (1953). The relationship between flow and available salmon spawning gravel on the Feather River below Nimbus Dam. Adm. Rep. Calif. Dep. Fish. Game. 1953.

Waters, T. F. (1962). Diurnal periodicity in the drift of stream invertebrates. *Ecology* **43**, 316–320.

Waters, T. F. (1969). Invertebrate drift—ecology and significance to stream fishes. *In* "Symposium on Salmon and trout in Streams." (Northcote, T. G., ed.), pp. 121–134. Univ. British Columbia, Vancouver.

Welton, J. S. and Ladle, M. (1979). Two sediment trap designs for use in small rivers and streams. *Limnol. Oceanogr.* **24**, 588–592.

Wene, G. and Wickliff, E. C. (1940). Modification of stream bottom and its effects on the insect fauna. *Can. Ent.* **72**, 131–135.

Weyman, D. R. (1975). "Run-off Processes and Stream Flow Modelling." Oxford University Press.

White, H. C. (1942). Atlantic salmon redds and artificial spawning beds. *J. Fish. Res. Board Can.* **6**, 37–44.

White, W. R., Milli, H. and Crabbe, A. D. (1978). Sediment transport: an appraisal of available methods: Vols. 1 and 2. Report No. Int. 119. Nov. 1973. Second Impression March 1978. Hydraulic Research Station, Wallingford, England.

Wickett, W. P. (1952). Production of chum and pink salmon in a controlled stream. *Progr. Rep. Fish. Res. Board Can.* **93**, 7–9.

Wickett, W. P. (1954). The oxygen supply to salmon eggs in spawning beds. *J. Fish. Res. Board Can.* **11**, 933–953.

Wickett, W. P. (1958). Review of certain environmental factors affecting the production of pink and chum salmon. *J. Fish. Res. Board Can.* **15**, 1103–1126.

Wickett, W. P. (1962). Environmental variability and reproduction potentials of pink salmon in British Columbia. *In* "Symposium on Pink Salmon." (Willimovsky, N. J., ed.), pp. 73–86. H. R. MacMillan Lectures in Fisheries. Univ. of British Columbia, Vancouver.

Wickett, W. P. (1968). Practical results from chum salmon research. Fisheries Res. Bd. Can., Pacific Oceanographic Group Biological Station, Nanaimo, B.C. Rept. No. 1968–1. File No. N7–20–24. January.

Wickett, W. P. (1975). Mass transfer theory and the culture of fish eggs. *In* "Chemistry and Physics of Aqueous Gas Solutions." (Adams, W. A., ed.), pp. 417–434. Electro-chemical Soc. Princeton, USA.

Wilcock, D. N. (1971). Coarse bedload as a factor determining bedslope. *Bull. Int. Assoc. Scient. Hydrol.* **75**, 143–150.

Williams, D. D. and Hynes, H. B. N. (1974). The occurrence of benthos deep in the substratum of a stream. *Freshwat. Biol.* **4**, 233–256.

Williams, G. P. (1970). Flume width and water depth effects in sediment transport experiments. Prof. Pap. U.S. Geol. Surv. 562–H.

Williams, G. P. (1978). Bankful discharge of rivers. *Water Resour. Res.* **14**, 1141–1154.

Williams, P. F. and Rust, B. R. (1969). The sedimentology of a braided river. *J. sedim. Petrol.* **39**, 649–679.

Willoughby, L. G. (1974). Decomposition of litter in fresh water. *In* "Biology of Plant Litter Decomposition." (Dickinson, C. H. and Pugh, G. J. F., eds), vol 2, pp. 659–681. Academic Press, London and New York.

Wise, E. J. (1980). Seasonal distribution and life histories of Ephemeroptera in a Northumbrian river. *Freshwat. Biol.* **10**, 101–111.

Wolf, P. O. (1959). Discussion of "A study of the bankful discharges of rivers in England and Wales". by M. Nixon. *Proc. Inst. civ. Engrs.* **14**, 400–402.

Wolman, M. G. and Miller, J. C. (1960). Magnitude and frequency of forces in geomorphic processes. *J. Geol.* **68**, 54–74.

Wolman, M. G., Wahrhaftig, C. A., Cox, A. V. and Currey, R. R. (1972). Land use policy, sediment yield and water quality standards. *Geol. Soc. Am. Ann. Mtg. Asstr.* **4**, 710–711.

Yalin, M. S. and Karahan, E. (1979). Inception of sediment motion. *J. Hydraul. Div. Am. Soc. civ. Engrs.* **105**, (HY 11), 1433–1443.

Yang, C. T. (1973). Incipient motion and sediment transport. *J. Hydraul. Div. Am. Soc. civ. Engrs.* **99**, (HY 10), 1679–1704.

Yang, C. T. (1979). Unit power equations for total load. *J. Hydrol.* **40**, 123–138.

Yang, C. T. & Song, C. C. S. (1979). Theory of minimum rate energy dissipation. *J. Hydraul. Div. Am. Soc. civ. Engrs.* **105**, (HY 7). 769–784.

Yorke, T. H. (1976). Ten years of experience with automatic pumping sediment samplers. Proc. 3rd Fed. Inter-Agency Sed. Conf. 7–54, 57–64.

Zahar, A. R. (1951). The ecology and distribution of black flies (Simuliidae) in south east Scotland. *J. Anim. Ecol.* **20**, 33–62.

Appendix

Notation
All units in metric with S.I. system

A	cross-sectional area	n	Manning's number
C	mean suspended sediment concentration	r_0	shear stress associated with bed material
D	depth	β	permeability
K	permeability parameter	λ	porosity
L	dimension of length	υ	kinematic viscosity
P	wetted perimeter	ρ	density of water
Q	discharge	γ	specific weight of water
		σ	thickness of laminar sub-layer
R	$= \dfrac{A}{P}$ = hydraulic radius	ΔH	change in hydraulic head along flow path
S	energy slope	f	function
U	mean velocity		
U_*	shear velocity = $\sqrt{r_0/\rho}$	*Subscripts*	
V_s	settling velocity		
W	width	c	critical value
d	grain diameter	n	representative percentile
f	Darcy-Weisbach coefficient	s	sediment
g	acceleration due to gravity	$(a, b, c, f, k, m, z$ are intercept and exponent value
k_s	roughness associated with deposited sediment		

Rat Control

DAVID E. DAVIS

777 Picacho Lane, Santa Barbara, California, 93108 USA

WILLIAM B. JACKSON

Environmental Studies Center, Bowling Green State University,
43403 Bowling Green, Ohio, USA

I. Introduction

The control of rats has been an intractable problem for centuries. A reivew of the biology and ecology of rats may stimulate the application of scientific principles to their control and lead to more frequent reductions of human hazard and of damage from rats. Rats of the genus *Rattus* have successfully evolved the ability to live with man and to share some of his diseases and parasites. Rodents of a few other genera (e.g., *Mus, Acomys, Bandicota*) also have learned to live with man, but these genera will seldom be mentioned in this review, either because little is known about them or because their ecological requirements are different. The numerous other genera of rodents will not be discussed.

The purpose of this review is to evaluate the status of rat control from the perspective of population principles. In addition, procedures for the application of these principles to the development of the strategy and tactics for control are discussed. While the strategy of control is also discussed from the administrative viewpoint, this review does not present a manual of control techniques. It is axiomatic that an understanding of birth, death, and movement rates is required for the evaluation of control efforts and for planning effective control. The presentation of such basic information and principles should, therefore, help to construct control programmes based on the biology of the rat.

In approaching pest-control problems it is usual to present a cost-benefit ratio derived from the cost of a programme and the value of the benefits, both stated in monetary values. In rat control it is possible to obtain a reliable figure for the cost of a programme, but values for benefits are elusive. The possible benefits and threats of disease or damage can be identified; but data are scarce. The reasons for control include dislike or fear of rats, as potential disease carriers by causing loss or contamination of food, and damage of property.

This review includes considerations of changes in populations with and without control procedures and presents information on birth and death rates that may lead to improved control programmes. It also suggests how appropriate information about populations may be obtained but excludes details of poisons (Chitty 1954), traps, barriers (Brooks 1974) and disease. The role of rats in human disease is excluded as are most anecdotes about rats. Only references that indicate a relationship between the biology and some feature of rat control are included, hence many publications that record data without interpretation are omitted, primarily to avoid an encyclopaedic approach. For example, data from a small geographic area or for a small number of rats are ignored.

The genus *Rattus* belongs to the subfamily *Murinae* of the family *Muridae*. The family has 100 genera and almost 500 species throughout the world (Anderson and Jones, 1967). *Mus* has 24. The most important for this consideration are:

(1) *R. norvegicus*, called Norway, brown, barn, or wharf rat, is a terrestrial or burrowing rodent, weighs about 400 g, and has a tail shorter than the head plus body. It is a native of southern Siberia but has followed man throughout the world. (Legend claims that Linnaeus chose the name *Mus norvegicus* because at that time the Swedes hated the Norwegians!) It is probably the second-best known mammal in the world. Its anatomy and physiology have been studied experimentally (Cotchin and Roe, 1967; Farris and Griffiths 1962) and its behaviour (Barnett, 1963) and life history (Twigg, 1975) have been described.

(2) *R. rattus*, called roof, black, or alexandrine rat, is a somewhat arboreal rodent, weighs about 300 g, and has a tail longer than the head plus body. It is a native of southeast Asia (India, Malaysia, and other areas) and has followed man throughout the world, but generally has been most successful in subtropical places. It commonly lives ferally in suburban areas of Texas, California, Florida and other southern areas of USA.

(3) *R. exulans*, called Hawaiian or Polynesian rat, is a terrestrial and arboreal rodent, weighs about 100 g, and has a long tail. It is a native of southeast Asia, but man has transported it throughout the Pacific Islands. In some places it lives within habitations but usually lives nearby in crops, secondary growth or forest.

(4) *Rattus* spp. Many other species cause problems, but few have been studied in detail. Often taxonomic relationships are uncertain, so that some may be subspecies. Some names are: *R. r. mindanensis*, *R. argentiventer*. Although little has been published about the different forms, it is likely that the principles described here will be useful in planning control, even where local conditions are unusual. A rat belonging to a different genus, *Bandicota*, has been observed in great detail in India (Spillett, 1976).

II. POPULATION PRINCIPLES

For rat control the essential concepts concern population limitation and regulation. There has been confusion of these concepts for three decades, partly because of numerous synonyms or antonyms and local differences and partly because species differ in detail in the way they conform to such principles. At the risk of oversimplification, the following analogy indicates how the above terms apply. The speed of an automobile is *limited* by the physical condition of the highway (holes, ice, traffic) and *regulated* by the accelerator and brakes. For rats, the term *control* can include both population limitation and regulation.

A. LIMITATION

Numerous factors in the habitat, such as food, shelter, predators, disease, and competition, may limit population size. Measurement of these factors is difficult but not impossible, but it has rarely been attempted. One approach is to change the level of the limiting factors by a broad programme of sanitation or cleaning or by reconstruction of buildings. Because this approach changes all aspects of the habitat, the factor limiting the population at that time and place is included. Orgain and Schein (1953) measured the factors in the environment of rats in Baltimore, and their analysis indicated that shelter seemed in excess of needs and that food was limiting the population. Experimental removal of the shelter (fences, trash) resulted in a disturbance of the population but did not produce permanent reduction, showing that shelter was not the limiting factor. Reduction of food (garbage), however, caused a significant decline in numbers, indicating that food was the limiting factor at that time and place.

Unfortunately additional quantitative data from experimental manipulation of limiting factors are not available for rats, but enough is known about the principles involved to draw the following conclusions. Although rat populations are limited by some factor, the measurement of such factors is difficult and hence expensive. Experimental manipulation of individual factors either at different times or places, or serially, can reveal what factor is limiting. But the limiting factor changes with conditions and differs between places and time, so that when a factor is identified as limiting, another may have already supplanted it. Consequently, for suppression it is more practical to alter all factors that provide food or shelter so that the limiting factor is included.

B. REGULATION

The process of adjustment of numbers to a particular set of conditions is called regulation. By understanding how a population is regulated we then learn how to regulate it ourselves. The mechanisms that produce regulation are primarily birth and death rates with, in some situations, the involvement of dispersion. Before examining these parameters in detail the concept of density must be considered.

The key aspect of a population for understanding regulation is its relation to density (i.e. number per area). When a population is low (i.e. it is not dense relative to resources), its resources are not limiting and the population can increase. When a population is high (dense), its resources are less available and increase will be slow. When a population is very high and the resources are limiting, it will decrease in size. A population can exceed the available resources if immigrants move into the area (e.g. demolition of adjacent buildings), or if the resources are suddenly decreased (e.g. removal of food), or if a climatic change affects a resource (e.g. shelter).

1. *Density-independence*
Some factors have the same proportional effect at both high and low densities. Density-independent factors are unaffected by the number of animals present. Rainfall is generally accepted as a density-independent factor as it falls on both dense and sparse populations which cannot affect the level of precipitation. In some cases, however, density might affect the result of rainfall such as flooded burrows.

Density-independent factors include aspects of climate and soil, non-infectious disease, features of structural needs (e.g. broken pavement for shelter) and the quality of food. Most of these factors lose their independence in some circumstances, especially at the extremes of density. For example, holes in paving may be used as shelter. Rats do not make holes (except as a result of digging) and hence density does not affect the number of holes. As long as empty holes occur, the number of rats does not depend on the number of holes. When there is no more space the population is limited by the number of holes but is not regulated by them. (Some important complications resulting from territorial behaviour are ignored in this example.)

2. *Density-dependence*
Some factors have different proportional effects at high and at low densities which change as density changes. In considering the transmissible disease salmonella, a small percentage of rats may be infected at low densities, at medium densities a higher percentage may be infected due to increased frequency of transmission, and at high densities most may be infected due to rapid transmission complicated by a decrease in tolerance to the disease. Thus the percentage mortality increases with density. Such a factor can *regulate* a population because mortality increases as the population increases, thereby slowing down its rate of increase. Alternatively, as the population decreases mortality also decreases, thereby slowing its rate of decrease.

Density-dependent factors include infectious disease, parasitism, quantity of food and competition. While these factors can affect the population in a density-dependent manner, in nature they rarely do. Regulation can be achieved by one or several factors acting in a density-dependent manner. When salmonella regulates a population other parasites present may not be density-related. Thus when a factor is shown not to be dependent on density, even though it is classed as density-dependent, this result should not preclude the existence of density-dependence as a principle.

The relationship of natality and mortality to density may be positive or negative. If recruitment (births plus immigration) is constant, then the proportion dying from various causes may increase as density increases (e.g. the result of predation by cats). When the proportion dying equals recruitment (Fig. 1A) the population should theoretically

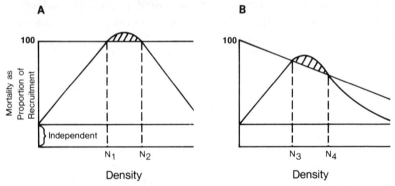

FIG. 1. (A) With recruitment constant, part of the population will die from density independent mortality factors. Another part will die (or emigrate) from density-dependent factors so that the population will become stationary at N_1. Due to delayed effects the mortality could exceed recruitment but would stabilize at N_2. (B) Because recruitment usually declines as density increases, the density-dependent mortality would follow different curves. N_3 and N_4 correspond to N_1 and N_2. In the cross-hatched area mortality exceeds recruitment.

stabilize, but in fact a delay in adjustment occurs so that the density continues to increase. The population size could adjust and remain at N_1 but it might, due to some other factors (e.g. behaviour), escape the relationship (e.g. increase faster than the cats) so that the proportion killed declines, thus allowing the density to increase. However, in nature recruitment will decline as a result of competition (Fig. 1B), so that the population will stabilize at a lower density (N_3) than in Fig. 1A. When escaping from stabilization, an inverse density dependence occurs, in which the population continues to increase until another factor takes over.

The relationship of proportional mortality has been shown as a straight line for expository reasons. In nature it is rarely linear and becomes concave indicating a reduced effect to the point of becoming negative. In other cases it may be sigmoid, reaching an asymptote. The actual shape is related to local conditions such as nature of habitat, kinds of predators, or competition with other species.

C. NUMBERS

When planning a control programme it is important to know the size of the population. The answer may be simple or complex depending on the type of problem to be solved. Absolute estimation may seem necessary but for wild rats is difficult to obtain, being expensive and arduous. Nevertheless, an estimate of the number of rats in Baltimore in 1952 was 115 000 or about one rat per nine persons (Brown *et al.*, 1955). The population of New York in 1949 was estimated at no more than 250 000 or one rat per 36 persons (Davis, 1950).

However, for many purposes a relative estimate is sufficient to know that a population is increasing or decreasing or whether more are present in one building than in another. The exact number is hardly ever required. The number per unit area can be estimated by a variety of procedures (see Davis and Winstead, 1980; Jackson, 1979b).

1. *Density*
Given a relative measure of a population, a rat control programme administrator can use it to plan the operation and to evaluate the results. While normally the population will be at the carrying capacity for that habitat the history of the population must also be considered. The population size must be related to previous control measures, since time is required for it to increase or to decrease to match the habitat and resources.

2. *Composition*
In addition to size, it is necessary to know as much as possible about the following population parameters. For some purposes a rough classification is satisfactory; for others considerable detail is necessary.

Age. This is usually indicated in time units, e.g. months, but it may be better described as a development stage, such as young, sub-adult, or adult. Measurements of length and weight may be used as a relative index of age (longer is older) and age structure may be calibrated from recaptured individuals in relation to size. (Unfortunately variation is large, so that size is not useful for adult individuals.) A curve for the relation of size to age reaches a plateau, and each point has a large standard deviation (McDougall, 1946; Perry, 1945; Tamarin and Malecha, 1972; Davis and Hall, 1951). Another useful unit of age is reproductive stage, i.e. sexually immature or mature.

Sex composition. This would seem to be simple, though a third gender, neuter, should be used more often. Outside the breeding season sex may have no significance.

3. Disease

A difficult but important character to measure is the health of a population or the prevalence of disease, including internal and external parasites. Part of the problem is that such determinations are laborious to assess. Usually the animal must be caught and examined for collection of appropriate tissues, and the elaborate laboratory procedures which follow may take weeks. For example, the prevalence of salmonella in rats requires that material be sent to a specialized laboratory. For these reasons few studies of prevalence of pathogens in populations are made. In many populations some approximations have to be made. Measures of weight or emaciation (lack of fat) may indicate the health of the population, or some symptom may indicate that some unspecified disease is prevalent. Thus the composition in terms of health may often be stated.

D. CHANGES OF NUMBERS

The basic concepts for the studies of population include a consideration of their dynamics. Population dynamics includes natality, mortality and movement or migration. In recent years the term has been extended to include a study of the changes in population which, of course, result from the interaction of these three forces. Natality produces an increase in population size, while mortality obviously reduces the population. Movements may either increase or decrease the population depending on their net value. Often, in the study of population dynamics, movement out of the population (emigration) is counted as deaths and movement into the population (immigration) are counted as births. From the viewpoint of population increase it may make little difference whether the increase comes from additions by birth or immigration.

1. Types of change

For rat control knowledge of previous changes is essential for understanding what measures to use and how the population will respond to them. To understand the causes of change one must determine what type of change is occurring.

Random change. This is the simplest type. Under these circumstances the population increases or decreases in a haphazard manner showing no clear regularity in its changes. Hence it should be assumed that many different factors are acting on the population, none of them dominating or occurring in a regular pattern. The population is regulated by the possibility that these factors do not all result in a persistent decrease or increase thereby causing the population to fluctuate. However, random fluctuation cannot be considered as lacking causes but as having so many causes that no systematic pattern becomes obvious. Suitable analysis may reveal some of the causes.

Oscillation. This is another type of population change. The number of animals increases and decreases in a regular manner. Under these circumstances the oscillations may resemble the swings of a pendulum which can either: (a) maintain constant amplitude; (b) decrease in amplitude due to friction until the pendulum becomes stationary (damped); or (c) increase. In nature some extraneous factor intervenes before the population becomes stationary, and the fluctuations continue erratically. In theory the amplitude of the oscillations should decrease. Such fluctuations may be coupled to some external factor, such as seasonal changes in climate. For example, on an island in Venezuela drastic changes in the food supply occurred between the dry and wet season. The roof rat population oscillated from about 40 rats at the end of the rainy season to about 10 at the end of the dry season (Gomez, 1960). Changes within breeding seasons occur elsewhere but often are obstructed by other factors.

Asymptotic oscillation. A special case of oscillation may be called asymptotic. In a constant habitat the damping of the oscillation may be so great that the population increases (or decreases) to its stationary level after displacement and remains there. Such populations can be called asymptotic because they approach an asymptote. Actually all populations would be asymptotic if the habitat remained constant long enough.

Descriptions of change. Mathematical models are a useful tool to describe the changes of a population conditions. The logistic curve (Lotka, 1925) is simple and easily derived. It is deterministic rather than

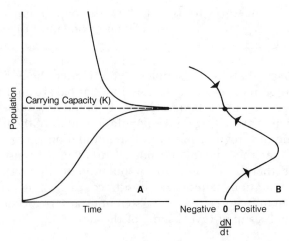

Fɪɢ. 2. (A) Logistic curve showing population increase to capacity (K) and also decline to capacity from a population above K (Lotka, 1925). (B) The rate of increase (dN/dt), plotted in arbitrary units, changes from close to zero at the start (0), to a maximum, returning to zero at the asymptote. If the population is above the asymptote the rate of change will be negative and the population will decline to K.

stochastic and thus fails to consider random events. The equation for the curve (Fig. 2A) is $N = \dfrac{K}{1 + e^{rt}}$ where N is number of rats, K is the number of rats at the asymptote, e is the base of natural logarithms, r is the intrinsic rate of increase, and t is time. The number of animals, when different from K, will change towards K. If N is greater than K, the rat population will decrease towards the asymptote; if N is smaller than K, the population will increase. K is synonymous with the number of rats at the carrying capacity for that time and place.

For analysis of changes, the first derivative, which is the rate of change of N with time, is perhaps more useful (Fig. 2B). When N is greater than K, the rate of change is negative and increases to zero at the asymptote. When N is less than K, the rate shows a sequence from low to high to zero.

The conventional use of the logistic model has been for growing populations, but for rat control the major interest is in a declining population. To produce a decline the value K should be reduced, K being the number of rats at the asymptote. An area has its own carrying capacity (K rats), but at a particular time it may have more or less rats which will tend to decline or increase towards K. Several cases exist for the reduction of K.

(1) K gradually declines over a long period of time. The actual

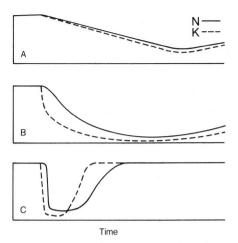

FIG. 3. Schematic relation of K and N when (A) K declines slowly and then increases. (B) K declines rapidly. N adjusts after delay, (C) K declines abruptly due to some pollutant and then increases. N follows after considerable lag.

number of rats also would decline, so that no perceptable difference between N and K exists (Fig. 3A).

(2) A change of habitat may rapidly reduce K so that N is greater than K. N then declines to the new asymptote (Fig. 3B). In practice the decline of K due to sanitation procedures would not be instantaneous, so the decline in N would reflect both a continuous decrease in K and the adjustment of N to K.

(3) No permanent change in K but a sudden and brief pollution that reduces K for days or weeks (Fig. 3C). Pollution here refers to poisons or toxins in the food or water, flooding, abrupt increase of predators, traps or pathogens and other temporary effects. The number of rats declines abruptly but promptly begins to increase and return to K.

Cases (2) and (3) are clearly two poles of a continuum. The change of habitat (Fig. 3B) might last only months, so that K would increase slowly over that time. The sudden pollution (Fig. 3C) might have effects that last for varying periods of time; for example radioactivity can persist for years. In most cases the pollution (poison) rapidly disappears, and the population quickly increases.

The regulation of N to K is executed by changes in birth and death rates or immigration and emigration rates, and a major regulation factor is competition (Davis, 1949). Because these rates are fundamental to understanding population regulation, their relation to competition in high density populations will be considered.

Competition has a special relation to K and N, but r should be considered first. The intrinsic rate of increase, r, equals b, the birth rate, minus d, the death rate. (For simplicity i, immigration, is included in b; e, emigration, is included in d.) Now, as can be seen in the equation, the value of r is the key to change, being constant for particular conditions but having different values under different conditions. As competition under different conditions affects both birth and death rates, r will also change (except in the case of a perfect balance). Competition decreases birth rate and increases death rate, so that r becomes less positive and eventually negative. Then the rate of increase (dN/dt) becomes negative, and the population declines (see Fig. 2B). Competiton exerts its effects through b and d by numerous mechanisms. However, the competition term $(-rN^2/K)$ is a ratio of N^2 to K. Hence if both K and N decline, the competition term could remain the same. In this case the number of rats (N) would decline but the competition term would not change.

This consideration of population principles sets the stage for a discussion of birth, death, and migration rates in rat populations. Many aspects of the habitat can affect the values of more than one rate (starvation reduces births and increases deaths) and competition affects all three rates. Nevertheless, each rate will be discussed separately and then unified in a discussion of populations.

III. Birth Rates

It is axiomatic that appraisal of the methods and efficacy of rat control requires information on birth rate, which in turn depends on knowledge of reproductive anatomy and physiology as well as of ecology. Because such information is not readily available in a form suitable for use in rat control, this section will describe those aspects needed to determine birth rates.

Although the rat is probably the second-best known mammal, much of the information about reproduction has been obtained from experiments designed to determine biochemical and endocrine mechanisms. Such information is important in the development of physiological principles but may not be useful in understanding the principles of rat control. Various text-books (see references) summarize the anatomy and physiology of the domesticated rat (always *Rattus norvegicus*) and merit perusal for increasing the understanding of birth rates. Only those aspects that pertain to control of wild populations will be mentioned or described here. Some aspects are outlined in general terms to complete the background information.

A. REPRODUCTIVE CHARACTERS

Some anatomical features are useful to determine the production of young and hence to measure the effect of methods that reduce the birth rate. In this section those aspects of development of gonads and anatomy of the testes and ovaries and of their ducts and of the pituitary that appear to be pertinent have been selected. Information on these topics is available in various texts. (Hebel and Stromberg, 1976; Moghissi and Hafez, 1972; Wynn, 1977).

The organs of the male include the testes, the ducts, the penis, and several glands. Rats lack secondary sex characters. In the immature rat the testes lie in the body cavity; at maturity they decend into an external sac. Evidence of presence of sperm can be obtained by examination of the testes macroscopically. When the tubules are visible, sperm are present.

The organs of the female include ovaries, oviduct, vagina, clitoris, and mammary glands. The ovary contain follicles, which increase in volume, burst for ovulation, and then transform into corpora lutea. A count of the corpora indicates the number of ovulations (Davis and Hall, 1950). A comparison of the count of corpora lutea with the count of embryos indicates how many ova failed to implant, and thus is a measure of reproductive efficiency. A count of embryos with some corrections for loss during development indicates the number that will be born. The mammae are small except during lactation. Pregnancy lasts about 25 days in wild rats (Miller, 1911) and lactation about 35 days.

B. PHYSIOLOGICAL CHARACTERS

The reproductive processes that determine the birth rate are thoroughly known from endocrinological experiments on the white rat. However, physiological measurements are rarely useful for rat control, except in a few cases such as those using chemosterilants. Therefore, this section will simply remind the reader that stimuli from the exterior pass to the hypothalamus, which sends neurotransmitters to the anterior pituitary and nerve stimuli to the posterior part. In the anterior part these "releasing hormones" stimulate the release into the circulatory system of hormones that control the production of sperm and of ova and the maintenance of the corpus luteum. The hormone (ACTH) that controls secretion of cortical hormones from the adrenal also comes from the pituitary.

1. *Processes*

Reproduction has various physiological stages or processes. Spermatogenesis, once started continues during all seasons (Davis, 1953b), even where a dry season occurs (Gomez, 1960). The stages of the oestrous cycle can be determined for wild rats as for domestic, but such information has little utility for control. Ovulation is spontaneous (e.g. does not require stimulation by the male). The age at first breeding influences the number of young (Asdell, 1941). The litter size is small at first, increases, and then declines (Feldman, 1925).

The blastocysts lie free in the lumen of the uterus for about 5–7 days before implantation. After implantation the embryo enlarges and can be counted with assurance starting 7–8 days after conception. Implantation is an intricate process and results in a uterine scar (Davis and Emlen, 1948) formed at the site of implantation. The presence of scars is proof of pregnancy; the number of scars is a good measure of the number of embryos for recent parturition. However, if the entire litter is resorbed before the eleventh day, no scars form (Conaway, 1955). Scars from previous pregnancies may not be determined with confidence except by tedious histological procedures (Davis and Hall, 1950), although bleaching the uterus permits making a good approximation. Count of scars provide data on number born.

After parturition a new oestrous cycle can start immediately, with ovulation occurring about 24 h post-partum. Thus a female can become pregnant at once even though lactating. The concomitance is rare in small rats but may reach one third in large rats (Davis and Hall, 1951). Lactation follows birth and lasts a month or more. A

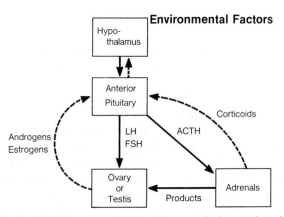

Fig. 4. Major feedback relations among hypothalamus, pituitary, adrenals and gonads. Solid lines indicate stimulation and dotted lines indicate inhibition. The hormones are given in generic groups to avoid involvement in details of chemistry.

lactating female can be identified by the absence of hair around the nipples as well as their large size.

Chemosterilants have a potential role in the control of rats by reducing the number of births and their action can occur on any of the processes described above. One type stops spermatogenesis and oogenesis (Davis, 1961); another type interferes with the oestrous cycle. When a steroid was used on wild rats at open dumps the number of sexually active females was reduced from 2·1% (62 females) to 0·6% (66 females) per male (Brooks and Bowerman, 1971). In another study an antagonist to oestrogen reduced the population in a large pen from 16 to 10, while reference populations increased (Gwynn, 1972). At present chemosterilants do not seem to be promising for control (Davis, 1961; Brooks and Bowerman, 1971), because they are expensive to apply and they reduce the population slowly. Compounds that are both sterilants (low dose) and toxicants (higher dose) may provide a more acceptable tool (Andrews et al., 1974; Jones, 1978).

2. Adrenal glands

The adrenals are not part of the reproduction apparatus but influence reproduction in so many ways that a discussion of their function is proper. Each gland is imbedded in fat and connective tissue just anterior to the kidney and consists of two parts. The medulla, part of the sympathetic nervous system, is surrounded by the cortex. The cortex has several layers that can be distinguished histologically and which serve different functions.

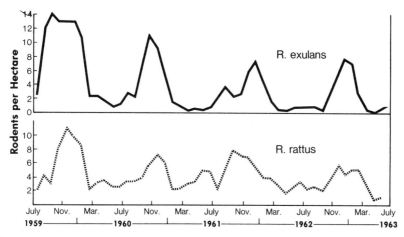

FIG. 5. Seasonal changes in numbers of two species of rats in Hawaiian sugarcane fields (Tomich, 1970).

The function of the adrenal in rats is well-known due to research on its numerous effects on diseases, but its principal action concerns the metabolism of carbohydrates and minerals. Removing the adrenal results in death unless specific replacement therapy is given. In addition the adrenal produces hormones that have specific effects on reproduction.

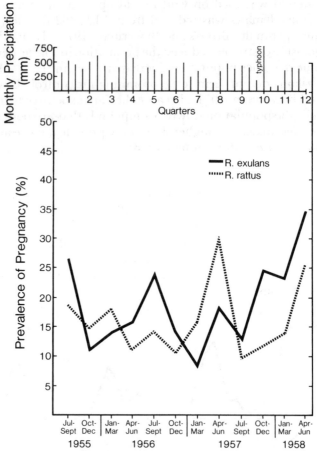

FIG. 6. Prevalence of pregnancy of mature female *R. rattus* and *R. exulans* on Ponape for three years (redrawn from Storer, 1962).

Activity of the adrenal follows hypothalamic stimulation of a hormone in the pituitary that initiates production of several hormones in the gland itself. Adrenocorticotrophin (ACTH) is produced in the hypothalamus, stored in the anterior pituitary, and released by a releasing factor. ACTH stimulates the cortex especially the zona fasciculata, to produce hormones that affect metabolism and directly affect reproductive functions. ACTH does not stimulate the medulla.

The hormones from the cortex are numerous and complex: (1) reproductive hormones, including estradiol, and androgens and (2) adrenal hormones, including cortisone, corticosterone, hydrocortisone. The adrenal is a complex chemical factory which, when stimulated, increases its production and may provide chemicals in excessive quantities. Some direct effects of cortical hormones on ovaries or testes may occur, but their important function is a feedback on pituitary hormones, especially ACTH. The cortical hormones inhibit the hypothalamus and pituitary hormones (follicle stimulating hormone, FSH and leutinizing hormone, LH) in various complex ways, thus interfering with reproduction (Fig. 4). Androgens produced by the adrenal may weakly mimic the action of testosterone.

The result of these complex influences is that ACTH can reduce or inhibit testicular and ovarian function. Events in nature such as disease, cold, and particularly aggressive behaviour may stimulate the pituitary to release ACTH, which in turn releases the cortical hormones. These events may result in the inhibition of LH and FSH and hence reproduction. These effects occur on the gonads and affect the whole sequence of reproduction from growth of gonads (maturation) through pregnancy to lactation (Fig. 4).

3. *Breeding Season*
Information obtained by examination of rat carcasses for the characters described above makes it possible to describe the extent of breeding in different months.

Rats may reproduce in any month but usually have one or even two breeding periods per year. Collections of large numbers of Norway (*Rattus norvegicus*) and roof rats (*R. rattus*) in many places in the United States, England, India, Africa, and Europe showed that peaks mostly occurred in the spring or at the beginning of the rainy season. No large collections have been made recently.

A small collection of roof rats (Daniel, 1972; Best, 1973) indicated a breeding season in September to March in New Zealand. The males had sperm throughout the year. In Queensland, *R. villosissimus* breeds in the spring; the proportion pregnant increased from 33·0% in June (winter) to 77·0% in September.

An intensive study of populations (*R. rattus* and *R. exulans*) in Hawaii (Tomich, 1970) showed that maximum numbers occurred in November, suggesting that a breeding season occurs between August to October (Fig. 5). Another study of reproduction of *R. rattus* and

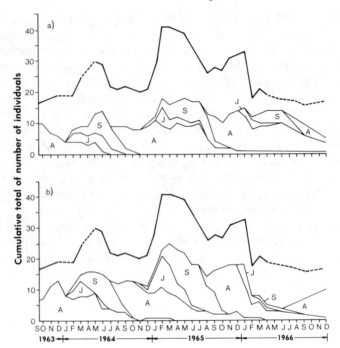

FIG. 7. Changes in numbers of *R. fuscipes* in a forest in New Zealand. The heavy line gives total of both sexes and the light lines indicate numbers by age classes for females (a) and males (b) (Wood, 1971). J = Juvenile; S = Subadult; A = Adult.

R. exulans in Hawaii (Tamarin and Malecha, 1972) confirmed that breeding occurred in August–September. In Ponape, *R. rattus* and *R. exulans* have maxima in various months but not consistently in different years (Fig. 6). In New Zealand, *R. fuscipes* which inhabits forests, changes in numbers from year to year and also within the year (Fig. 7). Due to its short life the age composition changes drastically within the year.

On city dumps at Nome, Alaska (64° 30 N) where temperatures reach −47°C, reproduction of *R. norvegicus* was high in summer but ceased in January. Males lacked sperm and the testes were abdominal (Schiller, 1956). This case is the only published evidence that reproduction can cease seasonally.

Unfortunately, no observational or experimental studies have examined the influence of individual environmental factors such as photoperiod, rain, temperature, etc., in regulation of breeding period in wild rats. The role of density will be discussed later under regulation.

In *Rattus exulans* populations in the New Guinea highlands, Dwyer (1975) found pregnancy rates higher in March–September than in

Table I. Reproduction rates of Norway rats (adapted from Storer, 1962)

Locality	Category	Total females	% Pregnant	Embryos/ pregnant female	Incidence of pregnancy
Baltimore warehouses 1946	Perf. vag.[1]	437	23·6	9·4	4·8
Baltimore farm 1946–7	Perf. vag.[1]	312	21·1	8·2	3·9
Baltimore population	Stationary	442	18·6	10·3	3·8
1947–9	Increasing	263	29·8	10·3	6·1
	Decreasing	162	30·9	9·9	6·3
Baltimore outdoors 1943–5	Perf. vag.[1]	1481	23·3	9·9	4·7
San Antonio, Texas farm	Over 130 mm	610	13·6	7·9	2·7
Norfolk	Adult	620	34·1	—	6·9
Urban	12–14 inches	6220	3·9	—	0·8
San Francisco	Over 14 inches	12962	20·7	8·9	4·2
Bombay	Over 100 g	27278	15·1	8·1	3·1
England	Adults	660	22·8	7·5	4·6
English farms	Rick	1180	30·0	8·9	6·0
	Non-rick	542	16·5	9·3	3·3
Germany	Over 163 mm	1824	17·7	—	3·6
Germany	Over 160 mm	620	31·4	8·7	6·4
Boston	Adult females	?	7·0	—	1·4
San Diego 1934	All females	945	21·2	—	—
Dothan, Ala. 1933–4	All females	1403	10·7	8·4	2·2
Savannah, Ga. 1932–3	All females	1842	13·7	8·4	2·8
Jacksonville 1934	18 inches	2015	14·3	8·9	2·9
Georgia 1947–9	Perf. vag.[1]	2202	18·9	9·2	3·8
Lactation rates					
Baltimore warehouses 1946	Perf. vag.[1]	434	34·2[2]	—	3·5[3]
Baltimore farm 1946–7	Perf. vag.[1]	307	22·0	—	2·2
Baltimore 1947–9	Stationary	439	43·0	—	4·3
(population)	Increasing	254	42·4	—	4·2
	Decreasing	162	39·9	—	4·0

[1] Perf. vag. —vaginal orifice was perforated.
[2] This column gives percentage lactating.
[3] This column gives incidence of lactation.

October–December for house and garden populations. For grassland populations breeding activity declined from March through August. After an inactive period breeding resumed in late December. The dry season began in late May and extended to mid-October, and the rainy season is an extended one with rainfall in excess of 2000 mm. Average litter size was 3·48 (118 females) and the first litters were smaller than subsequent ones, 7% of the embryos were resorbed.

Studies of *R. exulans* at Eniwetok Atoll (Marshall Islands) indicated a close parallel between rainfall and the prevalence of pregnancy (Temme, 1979), the major peak was April to November, with a secondary increase in April to June. Nearly 1560 sexually mature females were necropsied over a 15-year period. Breeding on islets of the atoll was not synchronous but may have been dependent on some local environmental factor.

In the Philippines, the breeding of rats (*R. argentiventer* and *R. r. mindanensis*) was initiated with the onset of the rainy season in June or July (Sumangil, 1965). On the island of Mindoro the breeding peak was reached in November to December, shortly before the rice harvest (Sanchez, 1977).

4. Birth rate

The number of young born and hence the true birth rate can rarely be determined for practical reasons. However, data on numbers of embryos can easily be obtained (Tables I, II, III), and used to calculate birth rates or to compare the number of embryos, prevalence of pregnancy, and prevalence of lactation in various populations. The extent of mortality of embryos has received little attention, possibly because large numbers of pregnant rats must be obtained and also because the age (size) of the embryo must be known. The only detailed examination (Perry, 1945) showed that about 57% of 113 litters lost some embryos and about 20% of the ova were lost. The loss was not affected by age of embryo (stage of gestation). In an examination of 526 embryos from *R. exulans* in the northern Marshall Islands only 11 (2·1%) were noted as resorbing (Temme, 1979). A recent study of ovulation rates in several species of native rats in Australia indicates a range from 3·5–9·4 corpora lutea (Breed, 1978). Lack of data on number of embryos or births prevents interpretation of these data to indicate intrauterine loss.

The number of young actually born is also difficult to obtain. In one study of a rural rat population in England, nests were examined at threshing time. In 85 nests, the average number of young was 7·31. At the same time 225 pregnant females averaged 8·65 embryos. The loss was 1·34 rats in the latter stages of gestation, at birth, and shortly after (Leslie *et al.*, 1952).

Determination of the *birth rate* is the objective of the collection of information described above. The birth rate (often called *b*) is the number of young produced per individual (not per female) in the

TABLE II. Reproduction rates of *roof rats* (*Rattus rattus*) (adapted from Storer, 1962)

Locality	Criteria	Total females	Vagina perforate	Pregnant	Pregnancy %	Embryos/ female	Annual incidence of pregnancy	Young/ Year/ female
	Weight (g)							
Ponape	90–269	769	625	98	15·7	3·8	3·2	12·2
Malaya	90–170+	1937	–	268	13·8	5·7	2·8	16·0
	All groups							
India	50+	13326	–	–	22·6	5·9	4·6	27·1
Belgaum	70+	11925	–	–	43·3	5·3	8·8	46·7
Egypt	65+	9788	–	–	33·3	–	6·7	–
	Body length (mm)							
Ponape	110–229	769	625	98	15·7	3·8	3·2	12·2
Guam	110–180+	121	112	16	14·2	3·4	–	–
San Antonio, Texas	>120	917	–	–	18·4	7·2	3·7	26·6
Georgia	110–239	–	4978	1288	25·9	5·6	5·2	29·0
Thomasville, Ga	>140	279	–	–	26·5	5·8	5·4	31·5
	Total length (mm)							
San Francisco, Calif.	>38	3633	–	–	26·5	7·4	5·2	38·5
San Diego, Calif.	>38	244	–	–	15·3	–	3·1	–
Honolulu	>38	473	–	–	36·2	6·3	6·4	40·4
	Other criteria							
Majuro	–	53	49	11	22·5	5·1	–	–
Tampa	Perforate vagina	–	275	–	30·0	6·4	6·1	39·0
Georgia	Perforate vagina	–	4978	–	25·4	5·6	5·1	28·5
London	Corpora lutea in ovary	598	–	–	31·6	6·5	6·4	41·6
London, (ships)	Corpora lutea in ovary	315	–	–	37·1	7·5	7·5	56·0
Cyprus	Corpora lutea in ovary	566	–	–	14·6	6·0	3·0	18·0
Cyprus,	Corpora lutea in ovary	572	–	–	12·9	6·1	2·6	15·8

TABLE II.—*cont.*

Locality	Criteria	Total females	Vagina per-forate	Preg-nant	Preg-nancy %	Embryos/ female	Annual incidence of preg-nancy	Young/ Year/ female
New Caledonia	Adult	—	16	6	37·5	5·0	—	—
Guam	Adult	37	—	13	35·2	4–8	—	—
Burma	Pregnant	—	—	15	—	3–8	—	—
Solomon Islands	Pregnant	—	—	—	—	4–11	—	—
New Zealand	Adult	—	280	—	—	6·1	4·8	29·3
Hawaii	—	—	172	19	—	5·1	2·3	11·8

population. Numerous qualifications are required to standardize the rate and to accommodate difficulties in collection of data. Although not strictly correct, the birth rate in rats is usually considered to be the average number of embryos (or scars) per individual in the population active enough to be trapped. This criterion of the population is purely pragmatic; it might be better to use the criterion of sexual maturity as the denominator. But unless the rats are captured, maturity cannot be determined. One can immediately think of deficiencies in this defini-tion, but the justification of this usage is that other definitions also produce difficulties in the collection of data. In any case any investigator has the right to state his own definition.

The number of embryos (births) are available for numerous places for several species (Tables I, II, III). A few collections have been made recently. In New Zealand *R. rattus* averaged 6·1 embryos per adult female (Daniel, 1972). In Hawaii *R. rattus* produced 5·1 embryos per female and *R. exulans* produced 4·0 (Tamarin and Malecha, 1972). In the Marshall Islands *R. exulans* produced 3·3 embryos per female in a sample of 1468 mature females (Temme, 1979).

The incidence (number of pregnancies per year) can be calculated from data on the prevalence of pregnancy. The incidence (I) equals the prevalence (P), times the duration (t) of collection of sample (t in days usually), divided by the length of time (d), in days, that embryos are detectable as swellings of the uterus:

$$I = \frac{Pt}{d} \text{ for example: } I = \frac{(0·158)\ (365)}{18} = 3·2.$$

To calculate incidence the duration of the breeding season must be known. For rats (except for Nome, Alaska) it lasts the entire year, how-ever, the collection must sample each part of the season where

TABLE III. Reproduction rates of Polynesian rats (*Rattus exulans*) (adapted from Storer, 1962)

Locality	Criteria	Total females	Orifica perforate	Pregnant	Pregnancy %	Embryos/ female	Incidence of pregnancy	Young/ year/ female
	Weight (g)							
Ponape	25–55	1148	1066	203	19·0	2·5	3·9	9·8
Malaya	20–45	140		39	27·9	4·5	5·7	25·7
New Zealand	20–99	33	24	10	41·6	4·7	8·4	39·5
	Body length (mm)							
Ponape	70–135	1148	1066	203	19·0	2·5	3·9	9·8
Majuro	100–135	207	189	68	36·0	3·0	—	—
	Other criteria							
Solomon Islands	—	—	—	—	—	2–8	—	—
New Caledonia	Adult	29	—	9	31·0	3·0	—	—
Burma	—	—	—	62	—	4·0		
Hawaii	—	—	35	7	—	4·0	4·3	17·2
Marshall Islands	—	1754	1468	161	11·0	3·3	—	—

TABLE IV. Relation of production of young rats to the size of females (Storer, 1962)

Species	Locality	Size	Number of females	Number of females	Incidence[1]	Young[2]
R. rattus	Ponape	90–129g	112	3·7	1·4	5·2
		130–169	223	3·6	3·9	14·0
		170–199	89	3·8	5·7	21·7
R. rattus	Georgia	110–129 mm	111	0	0	0
		130–159	1554	5·0	2·0	10·0
		160–189	2992	5·7	6·9	39·3
		190–239	320	6·3	6·7	42·2

[1] The number of times in a year that a female becomes pregnant.
[2] The number of young (embryos) produced by a female in a year.

maxima or minima occur. The best procedure is to collect samples at regular intervals all year, thus avoiding errors introduced by accidentally sampling at a maximum or minimum pregnancy season. An incidence could be calculated for a part of the year (say, maximum breeding) if any purpose were served.

From these data we find that the incidence varies with age (size) of female, location, and of course species. Some samples of data given in detail in Table IV illustrate some of these relations. Size is an index of age but other factors such as food or disease may influence the relation. The following trends occur with size (age); the number of embryos per female and the prevalence of pregnancy, which determines incidence, increase. Hence, the total young produced increases. These trends occur in various species in various places where enough data have been collected to show any relationships.

The potential increase in number resulting from births (assuming no immigration, emigration or mortality) is given by the equation for the finite birth rate $\lambda = \dfrac{N_t}{N_o}$ where N_t is the number at time t and N_o is the original number. Assuming that two adults produce about 20 young a year, then $\lambda = \dfrac{22}{2}$ or 11 times. This number is hypothetical because adults rarely live a year and they are not paired, however, the value can be used for comparisons by making the same assumptions for other populations. The finite rate (λ) can be changed to the infinitesimal (exponential rate) by letting $\lambda = e^b$. Thus:

$$N_t = N_o e^b \text{ and } b = 2\cdot4$$

The rate b has many mathematical virtues (Caughley, 1977), but here we will comment on a serious lack of information about rats. The young produced by a pair early in the year will reproduce before the end of the year, thereby increasing the population by an additional amount. At the moment data are not available to include this correction.

Although as shown above a birth rate can be calculated directly from births, the assumptions are so numerous that in practice it is not useful. Comparisons can be made between populations or among groups within a population by using the basic data for number of embryos, prevalence of pregnancy and prevalence of lactation.

IV. DEATH RATES

For effective control of rat populations an understanding of the rates and processes of mortality is essential. Definitions of terms will be presented later, but for the present such general phrases as high or low mortality are used. The level of mortality reveals nothing about the population level, even when the level of births is known. The same level of population can be maintained with a high mortality and a high reproduction as with a low mortality and a low reproduction.

This section describes mortality in rat populations from natural causes. In addition, it considers predation as a process to increase mortality by harvesting individuals. These effects will be combined in the next section with reproductive processes to show how the changes in numbers occur.

A. SOURCES OF MORTALITY

To know the relative importance of mortality it is important to know the major cause of death. In practice this is rarely achieved because dead rats are difficult to find and because diagnosis is practically impossible in a carcass found in nature. The best that can be done is to indicate some general experiences.

(a) *Habitat* conditions such as weather may cause mortality. A combination of a drought and a hurricane resulted in a decline in rat populations in Baltimore (Davis, 1956), but the relative role of deaths, births, and movements were not established. Many rats are killed on highways. Starvation due to shortage of food must be very rare, but it may accentuate the effects of pathogens.

(b) *Disease* among rats has been examined principally for its relation to human health. A catalogue could be prepared of the numerous human pathogens found in rats, but little or no knowledge of the effect of these pathogens on mortality of rats is available. Apparently viruses have not been examined in wild populations, but various bacteria have been studied, because they cause disease in humans. Plague (*Yersinia pestis*) is the most prominant and has caused severe mortality among rats (Pollitzer, 1954). Such mortalities are reported as a large number of carcasses found without details of rates or sex and age of the rats. Salmonella (various types) seems not to affect the mortality rate of rats. Its prevalence has been substantial, but an experimental epidemic failed to produce an increase in mortality (Davis and Jensen, 1952) even though salmonella spreads widely. Leptospira (various types), while less prevalent (Li and Davis, 1952), and rickettsia (especially *R. prowazeki*) lack effects on rats so far as known. Rats develop a solid immunity without symptoms. Numerous helminths exist in rats, without apparent harm. Only the nematode *Capillaria hepatica* has been studied in detail, and occurred in half to three-quarters of the rats and caused no disease (Davis, 1951; Farhang-Azad, 1976). Pathogenic fungi have not been detected in rats.

(c) The *stress syndrome*, resulting from stimulation of the pituitary-adrenal axis, has been implicated in a number of studies of wild populations (Christian, 1978). Stimulation of the hypothalamus, presumably by behavioural competition and other activity, results in activation of the pituitary to release ACTH, which stimulates the adrenal to produce numerous cortical hormones. These, in turn, decrease resistance by reducing the production of antibodies, interfering with phagocytosis, and altering some lesser mechanisms that affect immunity. Thus, under certain behavioural or habitat circumstances a non-specific decrease of resistance may occur that increases the fatality rate from pathogens that may be present in the rats. So far the difficult task of measuring such an effect has not been accomplished.

(d) *Predation* can produce deaths either by natural means (cats, *Yersinia*, etc.) or by planned programmes (traps and poisons). Whether or not it increases the rate of death is a complex topic. If sufficiently intense, predation can increase the death rate or change the age composition. Surprisingly little information is available on the effect of human predation on death rates, although many studies have shown that a population decreased after trapping. For example, three months of trapping *R. norvegicus* in Egypt resulted in a decline in captures from 0·26–0·13 rats per trap per night (Mahdi *et al.*, 1972). Presumably an increase in mortality caused the decline. Numerous studies report declines (e.g., Schein, 1950) but lack data on rates.

Perhaps it is unnecessary to care about a change in mortality rates resulting from control measures, the only concern being whether or not the population declined. However, for those who wish to understand changes in populations, a knowledge of the changes in birth, death, and movement rates is essential.

B. RATES

Mortality of rats is measured by rates and ratios, depending on the needs of the comparison. Much confusion exists in the literature, resulting from the failure to define terms and to a confusion of rates with ratios. The following discussion defines and uses terms in a rigorous manner to permit an understanding of the mortality process (Elandt-Johnson, 1975).

(a) *Definitions* of ratios (proportions) are: *mortality* equals the number that died divided by the total number in the population. Thus if 40 out of 100 rats die, the mortality ratio is 0·4. *Fatality* is the number dying divided by the number having a particular disease or condition. Thus, if 10 out of 20 rats with plague die, the fatality ratio from plague is 0·5. The *morbidity* is the number having a disease divided by the number in the cohort. Thus if 20 out of 100 rats have plague the morbidity is 0·2. These ratios occur at (not during) a particular time (e.g. a year or a day).

In contrast, rates occur during (not at) a particular time. The death rate (d) is the number that die during a particular time divided by the average population. Hence, if 40 rats die during a month out of a population of 100 at the start, d equals 40 divided by the average population over the month (which is about 80) or 0·5. When deaths occur during a period of time the probability of dying can be calculated, which is the number that die divided by the number in the cohort at the start. Thus if during a month 40 rats die out of 100, the probability of dying during the month is 0·40. Note that the numerical values for the probability of dying and for the mortality ratio are the same, but the concept differs.

The probability of dying (q) plus the probability of surviving (p) equals one ($q + p = 1$). Now $p = N_t/N_0 = e^{-dt}$ (Davis, 1960), where p is the probability of surviving, N_t is the number at time t, N_0 is the number at time zero, e is the base of natural logarithms, d is the death rate and t is time. Thus if p can be determined so can d.

An additional means of comparing mortalities is available in the life table. However, for rats, rarely are enough data available and rarely do the assumptions fit the data.

(b) The *collection* of data for calculations of mortality can be done by several methods. The best method is to find recently dead rats of known ages. The next best method is to recapture marked rats, thereby permitting calculations of survival. Lastly a large sample of rats whose ages can be determined, can be collected and mortality can be calculated on the assumption the sample is representative of the distribution of mortality.

(c) *Calculations* of mortality rates in rats are rare. A study of the population on a farm in Maryland indicated an annual probability of dying of 0·95. Two other studies gave similar results (see Davis, 1953b). A study in an Australian rain forest (Wood, 1971) of *R. fuscipes* indicated a probability of about 0·25 each month for adults and 0·40 for juveniles. Malayan rats, based on recaptures, had a monthly probability of dying of about 0·30 (Harrison, 1956).

(d) *Density dependence*, which is the relation of a mortality ratio to the density of the population, is crucial for understanding the action of mortality. As defined earlier, a relation is said to depend on density if it changes in proportion (either positively or negatively) as the population changes. If, for example, the mortality ratio changed from 0·4 to 0·6 as a population increased, one would say the relation was density-dependent. The importance is that positively density-dependent factors are the only ones that can regulate a population. If mortality from disease or predators (trapping or poisoning) remains constant as density of rats increases, the population will continue to increase. Hence for control the effect must increase proportionally to be successful. Unfortunately no data on this aspect of rat control are available. These remarks apply to a continuing relationship, not to the initial knockdown resulting from the introduction of a new disease or predator.

C. SUSTAINED YIELD

The desire of game managers, foresters, fishermen and others is to have a predictable yield that sustains itself for years. Such a yield is the last thing a rat control programme wants, but unfortunately it often occurs. The yield, of course, is the number of rats killed by the control procedures which together with natural mortality add up to the death rate; and when this equals but does not exceed the birth rate, a population will persist. How much predation (trapping or poisoning) should be increased can be determined from the following calculations.

The relation of the probability of survival to the death rate is:

$$p = e^{-d} \text{ where } p \text{ is probability of survival,}$$
$$e \text{ is the base of natural logarithm,}$$
$$d \text{ is the death rate.}$$

The death rate can be increased from d_1 to d_2 with traps or poisons. The ratio of the survival can be calcualted thus:

$$p_1 = e^{-d2} \text{ and } p_2 = e^{-d2}$$

Then: $\ln p_1 = -d_1$ and $\ln p_2 = -d_2$ and the ratio is:

$\dfrac{\ln p_1}{\ln p_2} = \dfrac{d_1}{d_2}$. If we assume that an increased predation will

increase d_1 to d_2, then the probability of survival will change logarithmically as in Table V. An increase in predation (traps, etc.) has a great effect when survival is low but less when survival is high. Obviously when survival is good, a greater effort must be exerted to improve mortality from trapping, etc.

Table V. Change in probability of survival when number of predators (d) are doubled. In this case $d_2 = 2d_1$ and hence $2 \ln p_1 = \ln p_2$.

Original (p_1)	Changed (p_2)
0·0100	0·0001
0·0320	0·0010
0·1000	0·0100
0·2230	0·0500
0·3150	0·1000
0·5000	0·2500
0·7050	0·5000
0·9450	0·9000
1·0000	1·0000

D. BIOLOGICAL CONTROL

The introduction of a pathogen or predator is often proposed as a panacea for rat control. Indeed several decades ago a prize of $25 000 was offered by the Philippine government for a demonstration of rat control by a pathogen. The offer was not collected. In theory biological control should work, but the biological control agent must kill (or sterilize) at a rate higher than the birth rate. In practice, although introduced predators can kill many rats, they shift to other prey when catching rats becomes difficult.

The mongoose has been introduced many times over the past two centuries on small and large islands; but after killing rats for a while, the mongoose shifted to other prey and did not affect the mortality rate of rats (Pimentel, 1955). The introduction of weasels (*Mustela siberica*) on some Pacific Islands seemed promising (Uchida, 1969) but requires more testing and evaluation of the environmental hazard (e.g. destruction of seabird nests). Success might be achieved by the introduction of hordes of mongooses onto an island, but this method is similar to using a trap.

Cats have also been introduced as biological control agents, but effects are limited. In an urban area the existing cats caught many young rats and altered the age composition but not the size of the population (Jackson, 1951). Cats on a farm, maintained by provision of extra cat food, reduced the rat population during the winter, but in the spring shifted to other prey, and the rats increased. The experience with salmonella has been similar (Davis *et al.*, 1976), as its introduction into rats on a farm did not result in a reduction of the population (Davis and Jensen, 1952).

In summary, prospects are poor for a biological control agent to be useful on rats (Wodzicki, 1973). The temporary and spectacular success

of myxoma virus in rabbits leads to persistent demands for more research on biological control. But recent history of rabbits (Davis *et al.*,
1976) shows that the virus and rabbits are evolving a harmonious relation, and the rabbit problem remains. Such would probably happen in
biological control of rats.

V. Movement Rates

Various kinds of movements occur that need to be distinguished.
Some have little relevance to control programmes and others are rare.
Migration in the form of a to-and-fro movement for the purpose of
breeding does not occur. Natural dispersion via drainage systems and
around natural barriers has been augmented by human activities, e.g.
village food supplies, trade routes and vehicles, domestic animals,
and cultivated crops.

A. GEOGRAPHICAL DISPERSAL

Movement occurs over long periods of time and is responsible for the
distribution of Norway and roof rats throughout Europe and many
other regions.

The Polynesian rat moved with the Pacific peoples in their canoes
as they moved from island to island across the Pacific from Asia
thousands of years ago. The later introduction of Norway and especially
roof rats into this basin was facilitated by Western commerce and especially by the operations of World War II (Storer, 1962; Atkinson,
1977). The history of spread of rats through New Zealand (Atkinson,
1973) shows that the Norway rat spread throughout the islands before
1800; but the subsequent spread of the ship (roof) rat was facilitated
by ships, trains, and road vehicles.

Dispersal of the Norway rat through the state of Georgia and the
subsequent restriction of the roof rat was documented by Ecke (1954,
1955). Similarly, the roof rat has typically infested the cities of the
United Kingdom and Middle East, but in the last decade Norway rats
have gradually displaced roof rats (Bentley, 1964; Jackson, 1979b),
again indicating that dispersal and displacement are continuous.

B. LOCAL DISPERSAL

When some change in habitat occurs rats must move. After a building
is demolished, an alley cleaned up, a warehouse emptied of grain, or a
dump closed, the rats move at least over short distances. Presumably

their mortality is high, but it has not been measured. Probably each rat travels until it finds a suitable habitat. Catastrophic changes in rice cultivation caused dispersal in California (Brooks and Barnes, 1972).

Seasonal movements are a special case resulting from habitat changes that occur regularly. Seasonal changes of habitat conditions (Gomez, 1960; Drummond, 1963) or harvesting of crops (Errington, 1935) cause movements to other areas. In the autumn with a decrease in habitat support, movements of rodents into barns, houses, and warehouses often cause concern.

If movements occur simultaneously in large populations over large areas, they may be called emigrations. Reports of emigrations have perhaps been exaggerated; none has been documented.

C. HOME RANGE

This concept emphasizes lack of movement. Many animals, including rats, live for most of their lives within a circumscribed area called a home range, which if defended is called a territory. Methods of marking rats for such studies are numerous (Taylor and Quy, 1973), and documentation of the existence of home range is ample (Davis, 1953a). Movements of Norway rats in urban Baltimore rarely exceed 50 m and on a farm were even less. In Cyprus, Australia, and Germany movement was also limited. Within buildings in India, movements were short (Prakash and Rana, 1973). On some occasions, however, regular long-distance movements occur. In a farming habitat Norway rats made regular trips of 1·5 and 3·3 km to a barn (Taylor and Quy, 1973).

Intensive studies of rats on Ponape (Storer, 1962) provided data on home range under Pacific Island conditions. The data from this and other studies may be compared with the statistic called the standard diameter, which contains 68% of the recaptures (Table VI) (Harrison, 1958). Note that the diameter for males was, as usual, larger than that for females. For Polynesian rats habitat differences may exist; the diameter in the coconut plantations was smaller than that for forest or grassland habitats.

In Hawaii, the movements of R. rattus and R. exulans are complicated by the harvesting of sugar cane at intervals of more than a year. After the dry vegetation is burned and the cane cut, the field is ploughed or ratooned for a second crop. The rats die in large numbers, mostly killed by the heavy machinery that collapses burrow systems, but a few retreat to the gulleys (Tomich, 1970; Nass et al., 1971a), where some are able to establish themselves among the permanent residents. When the new cane in adjacent fields provides cover, some

TABLE VI.   Ranges of Pacific Island and Malayan rats expressed in standard diameters
(adapted from Storer, 1962)

Species	Source	Sex	Number of individuals	Number of trappings	Standard diameter
Polynesian rat	Grassland	male	46	558	48
		female	71	685	34
	Coconut plantation	male	47	272	31
		female	49	310	28
	Rain forest	male	8	54	43
		female	12	91	27
	Grassland		36	294	26
Roof rat	All areas	male	14	96	83
		female	10	79	65
R. argentiventer	Grassland		14	58	80
R. jalorensis	Grassland		27	175	102

rats expand their home into the new habitat, eventually deserting the gulleys (Nass *et al.*, 1971b). The size of these ranges is about the same as found for the species in other places.

VI. POPULATIONS

In the section on principles, the nature of change of numbers was described, emphasizing the concept of equilibrium capacity but recognizing that in nature conditions rarely remained constant long enough for the population to achieve such an equilibrium. Here, since the decline of a population is the central theme, changes in numbers are described primarily from the viewpoint of an adjustment to a decreased capacity. The changes in numbers result from changes in birth and death rates and sometimes movement rates. Unfortunately in natural populations measurements of rates are rare. No distinction is needed between natural (floods, predators, etc.) and human (ploughing, poisoning, etc.) factors, since the effect on population is not influenced by the distinction. First, we will describe changes in populations that result from multiple, often unmeasured, factors, second, we will describe changes that seem to result from some particular factor.

A. HISTORIES OF POPULATIONS

In many cases population changes over months or years have been measured, but the influences of particular factors were either not measured or were obscured by other factors. Nevertheless these histories often provide insights to the causal relations or to predictions of changes in different situations. When measured for several months some changes have no consistent direction (random), whereas others have consistent declines or increases. In a few cases the population remained stationary (Davis, 1953b).

1. *Random changes*

Numerous reports during the past century describe spectacular local changes in numbers (Mohr, 1947, Errington, 1935) that result from some change in local supply of food and/or shelter. Unusual conditions may result in increases or declines. In Kenya in 1961, heavy rain prevented the harvesting of grain crops and weeds. Several species of rodents, including *R. natalensis*, increased to high levels. In fields that were cleanly cultivated the damage was less than in fields that contained weeds (Taylor, 1968).

Seasonal changes. These occur when some climatic change (e.g. temperature, rainfall) causes changes in the habitat conditions. If the seasonal changes are similar from year to year and within a year (e.g. in Hawaii) then the population changes are similar and not really random. But if the seasonal changes vary (e.g. in California) then the population changes are random. A seasonal increase usually results from an increase in reproduction, although a decrease in mortality could be important. The Norway rat increases by reproduction and invasion on islands where there are colonies of terns' nests (Austin, 1948), and roof rats increase when nestlings provide food (North, 1946). A review of population changes (Davis, 1953b) showed that in the northern hemisphere some cases of seasonal change occurred where habitat changed, but in urban areas little evidence of change in population exists. In some cases the number in a population changes little but the age composition alters due to the seasonal production of young. In some tropical areas seasonal changes are apparent. In Hawaii (Fig. 5) changes were regular. In Venezuela a population of roof rats increased during the rainy season (Gomez, 1960). A population of Norway rats increased in winter in New Zealand forests (Beveridge and Daniel, 1965). An obvious cause of seasonal change is the seasonal provision of food that corn ricks once provided in England (Venables and Leslie, 1942), which allowed reproduction in the winter and a great increase in numbers. A reverse change occurs in Alaska where reproduction stops in the winter (Schiller, 1956). An example of seasonal change in urban areas is the number of bandicoot rats per burrow in India, the numbers being low in February and July (Srivastava, 1968).

Changes over long intervals of time occur (McDougall, 1946) but are difficult to document and analyse. An analysis of records of rats killed in various cities in England shows considerable fluctuations but no systematic change that might be due to a predator–prey relation or to a seasonal change in habitat (Matheson, 1943). Serious outbreaks in the Philippines may result from the coincidence of weather patterns and planting schedules in sites adjacent to marshlands or second-growth forests (Libay and Fall, 1976). Densities of one rat m^{-2} have been indicated.

Movements. These occur when the habitat changes. Rats may move to other areas and, although survival is poor (Calhoun, 1948), some rats may establish residence. In Alaska, rats move in winter from dumps to houses and warehouses (Schiller, 1956). Rats in corn ricks had to move when they were seasonally demolished (Venables and Leslie, 1942). In Hawaii, rats move into growing sugar cane in the winter (Twigg, 1975). Rats leave fields flooded in spring (Chapman, 1938) and they may move from residential areas to a land-fill (Schroder and Hulse, 1979).

2. *Declining populations*

Descriptions of declines of rat numbers are spectacular. Sometimes rather crude but useful indices are presented. In certain counties in California in 1963–4, Norway rats became abundant as measured by those caught in muskrat traps and pounds of bait sold (Brooks and Barnes, 1972) and then abruptly declined in 1964. In New York State, a concerted programme in several cities from 1969–73 resulted in a reduction of "premises with rats" from 24·4%–3·9%. Changes in habitat (trash, junk cars, etc.) as well as the use of poisons were responsible (Brooks, 1974).

The current federal assistance to urban rat control programmes in 68 communities, where emphasis is placed on improved sanitation as well as use of rodenticides, has resulted in about 75% of the nearly 52 000 blocks in the designated "target areas" being converted to maintenance status, since the prevalence of premise rat infestation had dropped to 2% or less. Thus almost 6 million people could live in an "essentially rat free" environment (MMWR, 1979).

3. *Increasing populations*

The data for increases of rats are more numerous and detailed than those for declining populations. The landmark paper (Emlen *et al.*, 1948) established that after a decrease by removal, the population will return to the asymptote (see Davis, 1953b, 1977). This observation has been verified in several situations. In English villages and in Malta, populations after reduction returned to the original level in about a year and in some cases temporarily went above that level (Barnett *et al.*, 1951). In sewers the population returned to the original level 6–12 months after reduction (Bentley *et al.*, 1959), and in mines the population also returned towards the asymptote (Twigg, 1961). Population increase is sigmoid (Barnett and Bathard, 1953) when rats increase after either a seasonal or a contrived decline.

After introduction into a favourable habitat rats increase towards an asymptote. Rats invaded Lord Howe Island (Holland, 1952) and also the Solomon Islands (Johnson, 1946) and increased to a stationary level. Even populations of albino rats can become established in favourable habitats. A colony of 2000 was present before 1935 on a dump in Montana (Minckler and Pease, 1938). Data on the rate of increase of introductions are absent, but the curve of increase was probably sigmoid.

B. FACTORS

The histories of populations mentioned illustrate rates of change and something about the factors influencing the change. We now intend to discuss the environmental factors that influence changes in birth, mortality, and movement rates (Davis, 1952). In some situations the environment is fairly stable, and change is therefore reasonably predictable; in other situations the environment is unstable and predictions less precise. Previously we looked at some of these cases as examples of density-related, seasonal, physiological or behaviour-related components.

1. *Habitat*
The habitat is defined here to include all aspects of the physical environment including food. These aspects usually act in a density-independent manner in regulating a population. Weather conditions alone may be sufficiently severe to affect populations. In Baltimore in 1955, the rainfall in July was 89 mm and in August it was 450 mm, mostly during a hurricane. During these months, the average decline (possibly from excessive mortality) in population in 18 blocks was 34%, in contrast to average gains in three other years of 34%, 5%, and 3% respectively. The increase in September–November of 1955 was 40%, far greater than in other years (Davis, 1956). In Queensland

abundant rainfall was followed by plagues of rats (Crombie, 1944) and in Californian rice fields, heavy rain was responsible for the increase of rats (Brooks and Barnes, 1972). Presumably in both these cases the food increased.

Destruction of habitat regularly affects a population. In sugar cane fields burning of the cane did not result in deaths, but the machinery used to harvest the cane killed 73% of 35 rats that were tagged with radios. Other rats were later killed by mongooses (Nass et al., 1971a).

In numerous situations crops provide both food and shelter. R. conatus in Australia thrives in crops (Plomley, 1972); coconuts provide habitat for R. rattus (Wodzicki, 1972) and sweet potatoes are a cause for increase (Hiraiwa and Sumikawa, 1951). In California two introduced plants (blackberry and ivy) provide an ideal habitat for roof rats (Dutson, 1973, 1974). A special case demonstrating periodicity is the increase coincident with the flowering of bamboo, which occurs synchronously over large areas at intervals of several years. The rats increase until the bamboo dies (Hiraiwa and Sumikawa, 1951; Tanaka, 1957).

A special type of habitat manipulation is the use of poison, which can be considered as pollution of the environment, which temporarily lowers its carrying capacity, usually by making the food supply lethal but sometimes by indirectly reducing the supply of oxygen. In the use of poisons the rate of application is intended to kill large numbers before the rats learn to avoid the food or develop a resistance to the poison.

Following the use of poisons the rate of recovery in numbers follows the same pattern following reduction by trapping. A single application that pollutes the food supply for a short time reduces the population, which then increases (Schein, 1950). A persistent pollutant may keep the population at a low level as long as the pollution (poison) is maintained (Myllymaki, 1969; Drummond et al., 1977).

2. Predation

The process of predation can occur whether by cats, traps, bacteria, or other mortality factors that can, but usually do not, respond by increasing in numbers or frequency of killing (see Introduction). Natural predators rarely capture enough rats to alter the population (Kirkpatrick and Conaway, 1947; Uchida, 1969). The introduction of the mongoose has been a failure, as has the use of salmonella bacteria (Davis, et al., 1976).

Cats can capture many rats. Analysis of the frequency of rat remains in faeces showed that 6·7% contained rat hair. Most of the rats were young. Calculations based on food of wild cats in the laboratory indicate that a cat might catch 28 young rats a year. In a large urban area 70 cats lived amongst a population of 1000 rats. The rats produced about 9600 young during the year, and the cats caused about 20% of the total natural mortality. The population did not change (Jackson, 1951). In another study three cats were held on a farm by providing them with food. The population of rats declined in 1955 and remained at a low level. Few females were lactating in February–April when in previous years a fifth of the adult female rats were lactating. In June, however, the faeces of the cats suddenly showed many feathers from young pigeons, and small rats appeared in the traps: the cats had switched to pigeons (Davis, 1957). A different version of the predator relation was that the introduction of cats onto a farm did not decrease the number of rats but, if rats were absent, prevented their establishment (Elton, 1953).

Trapping is a form of predation that can be intensive if desired. A thoroughly supervised programme in Baltimore in 1942–4 showed that 50–95% of the rats in a city block could be captured in a few months. In one block all rats were captured (Emlen et al., 1948). Analysis of the size and sex of rats in relation to the day of capture in each block showed that large rats were captured before small ones but that no difference occurred between the sexes (Davis and Emlen, 1956). Numerous campaigns such as those of the India Plague Commission record numbers of rats captured but fail to separate the trapping, poisoning, and sanitation factors. In an attempt to control rats in Rangoon over 700 000 rats of four species were captured each year from 1934–7 and in 1938, 569 000 were captured (Harrison and Woodville, 1948), but no claim was made that the population declined.

Because most trapping programmes have had little effect on the rat numbers, it seemed desirable to examine the results from the viewpoint of a sustained yield. If removal by predators (traps) and other causes matches recruitment, then the number trapped remains constant. An experiment was organized to calculate the rate of increase at the inflection point of logistic equation and to remove the requisite number monthly. The results (Fig. 8) showed that the number could be kept near the inflection point for seven months. However, the monthly quota necessary to keep the population at the inflection point was twice the number predicted from the logistic calculations. After seven months the populations declined; in the winter a few rats were removed monthly for anatomical samples. These results show that the rate of increase at the inflection point is not a good prediction of a sustained yield. Some mechanisms must increase the production of young when adults are removed. These data provide some understanding of

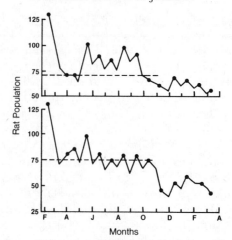

FIG. 8. In two urban areas whose populations had grown logistically the number was reduced to the inflection point (dotted line) and maintained close to it by removal of rats (redrawn from Davis and Christian, 1957).

the effects of removal by traps or poisons. It seems possible that trapping develops a sustained yield system with augmented recruitment matched by removal.

3. *Competition*

Knowledge of competition among members of the same species has been accumulated over half a century, so that for many species the principles of social rank and territory at the behavioural and physiological level are known. For rats, information from several sources suggested that competition was a factor in regulating populations (Davis, 1949). For example, the declining rate of increase (dN/dt) as the population approached the asymptote (Emlen *et al.*, 1948) could be explained by competition. Earlier work (Jennison, 1921; Eaton and Stirrett, 1928) suggested that reproduction was inhibited in crowded populations.

These suggestions were confirmed by results on Norway rats (Davis, 1949; Calhoun, 1949). However, a physiological mechanism was not recognized until the brilliant hypothesis of pituitary-adrenal feedback was proposed (Christian, 1950). The first tests on this hypothesis were performed on wild rats, but many other investigators have used other species both in the laboratory and in the field (see reviews by Christian, 1978, 1980). Other authors contributed to the testing of the hypothesis (Barnett, 1969, Ewer, 1971).

The hypothesis states that as numbers (and interactions) increase, the

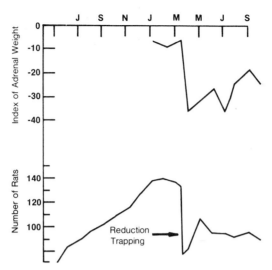

FIG. 9. The curve of rat number against time shows the combined population history of the populations in Fig. 8 first increased and then were reduced. The index of adrenal weight shows that the weight of the adrenals declined after the reduction in numbers (redrawn from Christian and Davis, 1955).

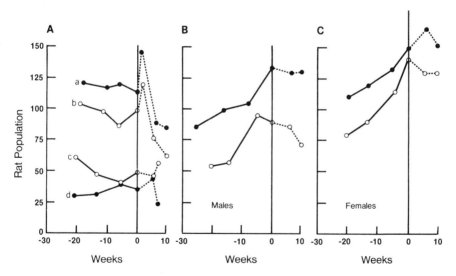

FIG. 10. Changes in numbers of wild Norway rats in urban areas after introduction of alien rats into the existing populations. (A) Males were added to areas with either high or low numbers. (B) From each of two areas that were increasing in numbers, some males were removed and replaced by alien males. (C) Same programme for females (redrawn from Davis and Christian, 1956).

hypothalamus activates the hormones of the pituitary, some of which, in turn, stimulate the adrenal cortex while others affect the gonads. To test these ideas in rats, measurements of adrenals and other organs were obtained from low, increasing and high populations. The adrenals of both sexes increased as the population increased (Christian and Davis, 1956). The next step (Christian and Davis, 1955) was to reduce a rat population and compare the size of the adrenals (Fig. 9). Reduction of rats by 42% resulted in a decline of 26% in the weight of the adrenals. In crowded populations adrenal hypertrophy should affect both reproduction and mortality according to research on other species in the laboratory (Christian, 1978). However, data on these aspects were difficult or impossible to obtain for wild rats.

Social competition occurs when a population is high or when strange rats are introduced. Mortality, at least of introduced rats, is high (Calhoun, 1948) and large males avoid each other (Davis, 1955). To test the effects of social strife, rats were introduced into areas where the populations were stationary and where populations were increasing (Fig. 10). Some of the stationary populations had high numbers and others low. In the high but not the low populations the numbers declined for ten weeks. In the increasing populations the numbers declined when either alien males or females replaced natives. Evidence also indicated that reproduction declined. Analysis of the weights of adrenals was inconclusive due in part, perhaps, to the difficulty of obtaining adequate samples at suitable times after the introductions.

Recent studies of rats in pens and on a land-fill (Andrews et al., 1972) confirm these general relations. The adrenals of rats in the land-fill increased in size as the population increased. In the pens the adrenals did not increase, but reproduction declined. Such evidence comes from many species of mammals and for rats in particular. The mechanisms differ in degree in various circumstances, in that mortality is sometimes increased (Tamarin and Malecha, 1971), and reproduction is sometimes decreased, and often both act to bring the population near the carrying capacity for the area.

The discussion so far has considered intraspecific competition but in some cases interspecific competition occurs among rats. A frequent observation, especially during the rat proofing programmes in southern cities of the USA in the 1930s, was that when Norway rats were eliminated roof rats increased, presumably because they were difficult to capture in the high parts of a building. The later spread of Norway rats throughout Georgia apparently drove out roof rats (Ecke, 1954). In New Zealand, Norway rats have driven the native R. exulans from most of the country (Watson, 1961). However, the roof rat also competes with *R. exulans* (Taylor, 1975). In most parts of the world several species occupy separate habitats and do not alter the overall population by competition.

VII. Control

Control of rats has been practised for centuries but usually in a haphazard manner that removes only a few offending individuals. The plans and techniques are simple and short-term, and the problem soon recurs. For economically justifiable results a more comprehensive approach is necessary (National Research Council, 1980).

A. OBJECTIVES

The obvious purpose is to reduce damage caused by rats and the associated hazard of disease, but simply killing large numbers of rats, especially in a brief campaign, will not achieve this purpose. The control programme should be designed to lower the carrying capacity of an area and thereby permanently reduce damage and hazards. In emergency situations, or following harvest, temporary action may be justified. However, such operations should be followed by plans for permanent control.

B. REASONS

While the direct damage that rats cause by destroying field crops and stored foods, contaminating food and clothing with urine, faeces, and hair, and damaging structures and fibre by gnawing is well known, the actual amounts are poorly documented (Jackson, 1977). A recent world-wide survey in tropical and subtropical regions further emphasizes the widespread prevalence of damage and the lack of reliable data from field studies (Hopf *et al.*, 1976). "Rat plagues", often fostered by excessive rains that force abandonment of unharvested crops and promote new plant growth, can devastate large areas (e.g. Taylor, 1968). In Gujurat State, India, three million hectares of groundnuts were recently saved from such a plague of rats by the widespread use of zinc phosphide baits and aluminium phosphide burrow treatments (Shah, 1979).

Estimates of damage, such as $10-rat-year, commonly appear in reports and are easily transposed (via the myth of 1 rat per person) to an annual US loss to rats in excess of $2 billion. Certainly a very large loss does occur worldwide, and, in areas where inadequate storage and transport facilities exist, rodents and insects may consume much of the harvest. More importantly, any food saved from such pests is available to the human population without additional energy investments in crop production.

Solid waste provides rodent harbourage and thus is a major contributor in urban fires; perhaps a quarter of the fires of undetermined origin result from rodents' gnawing on wires, dragging in oily rags or smouldering cigarettes, or filling ducts or chimneys with nests (Walcott and Vincent, 1975). Additional losses occur when rats are caught in electric motors.

While some rodent-borne diseases are readily recognized (plague, typhus, rat-bite fever), we hear little of their actual occurrence in the USA. Other diseases are less well known (e.g. rickettsial pox, leptospirosis, salmonellosis) but are of greater importance to us. Some of these diseases may be acquired from rats by house pets and then relayed to humans. On the farm, rats may introduce a pathogen or participate in the maintenance of the disease cycle for domestic animals (e.g. leptospirosis, trichinosis). In England during a recent foot and mouth disease outbreak, concern was expressed over the movement of rats across quarantine lines and their possible role as mechanical carriers, but the relationship was never documented.

Rat bites represent more visible evidence of a public health problem, with some 10 bites per million persons in the USA metropolitan areas reported annually, but unreported incidents probably number several times that. Children are usually involved, but the elderly or incapacitated may also be ravaged.

The low prevalence of disease has caused the USA Public Health Service to consider the rat as an aesthetic rather than a public health (disease) problem. Thus current grants to cities for their rodent control programmes are justified on the basis of their improving the environment and thereby reducing the rats, which have become symbols of urban deterioration.

General eradication of rats and mice is not feasible, but populations can be managed. Priority must be given to living and food processing and storage areas where contact with rats is least desirable. In noninhabited sites numbers of rats may be maintained at low (and tolerable) levels by relatively simple management techniques.

C. AGRICULTURE

In temperate climates rats are most notably serious pests of grain storage in field shocks, ricks, or silos. In more tropical regions field and forest infestations also cause concern. In Hawaii, three species (*R.*

norvegicus, R. rattus, R. exulans) are found in sugar-cane fields. In continental areas many other species (*Rattus* spp. as well as species from other genera) may be involved.

The farmer usually has the responsibility for dealing with his own rodent problems although government offices may provide advice and even assistance. His control efforts are likely to consist of placement of rodenticide baits only when enough rats or damages are seen.

In some countries, such as Britain, government control operators make regular service calls on farmers; and their persistent efforts may have facilitated the selection of anticoagulant-resistant rat populations. Elsewhere larger growers may develop rat management programmes through their trade organization (e.g. Hawaiian Sugar Planters' Association).

Hawaiian sugar-cane fields are often bordered by deep erosion gulleys which constitute harbourage for emigrants following the harvest and denuding of the fields. Later, individuals feed in the young cane fields but retreat to the edge areas for cover. By maintaining perimeter bait stations, the number of transgressing rats can be reduced. Once animals have shifted their entire home range into the spreading cane, aerial application of rodenticide (e.g. zinc phosphide) is the only control measure available (Nass *et al.*, 1971a, b).

In a similar way rats invade young rice fields initially using the bunds (dikes) as highways and then as burrow sites. If bait stations can be placed and maintained prior to "booting" (development of reproductive stalk) of the rice, depredations on the maturing rice are very much reduced. Barrios in the Philippines were organized around this programme, but the human interest was difficult to maintain. When the rice was young and few rats could be seen, control was thought unnecessary. However, when the rice had "headed" and damage was evident, the rats would no longer feed at bait stations and may have established themselves wholly within the field itself (Fall, 1977; Sumagil, 1965).

The roof rat is a pest of coconut palms in the Pacific Basin, often causing serious losses. The Polynesian rat, the "native" rat of the Pacific islands, even in the absence of the roof rat, typically does not colonize the coconut crowns (Storer, 1962). However, in the Tokelau islands, the Polynesian rat is the serious pest on coconuts (Wodzicki, 1969). The reason for such differences is not understood.

Cacao pods, gnawed by rats, are worthless, and such depredations have forced some plantations out of commercial production. Rats from habitats without cacao would not feed on pods in the laboratory, while rats from cacao plantations readily did so. When adults from the two groups were placed together, learning transfer did not occur. If young, recently-weaned rats had been present, learning by imitation might have taken place (Strecker and Jackson, 1962; Jackson, 1979a). Improved sanitation within the plantation (promptly picking ripe pods, removing damaged or fallen pods) could possibly slow the information transfer process.

In Malaysia, the traditional cultural practice for oil palms is to place pruned fronds in frequent neat piles within the plantation between tree rows. While desirable composting occurs, this refuse provides abundant harbourage for rats that feed heavily on ripening fruits. Regular baiting with rodenticide baits is considered preferable to disrupting this established pattern since the control programme can be conducted economically.

The "Green Revolution" while providing many benefits to tropical agriculture, has not been without its problems and critics. To the extent that rodent depredations on "miracle" rice constitute an economic threat to invested capital so the farmer must take remedial action. If he plants an off (dry) season crop when irrigation water is available, rats that formerly died in the annual drought period congregate in this new source of succulent food and control is even more difficult and necessary.

One component of these new schemes is intercropping, so that by early planting three or four successive crops can be grown on the same plot in a year. The inter-relationship of such practices to rodent populations is unknown. Would such cultural practices, if widespread, support a significant increase in rodents and thus require a carefully prescribed management regime?

VIII. ADMINISTRATION

The Health Department normally has prime responsibility for urban rodent control. In a few cities, the Public Works Department has operational direction. In the USA, only in Chicago does the programme have Cabinet status.

Because many agencies of city government are involved, administration and coordination of an integrated management programme is difficult. Cooperation and assignment of resources across administrative boundaries is required. For example, the Health Department may have responsibility for dealing with rat bites and potential diseases, health education, and enforcement of premise sanitation (garbage storage) regulations. The Housing Department will be responsible for enforcement of building codes and maintenance of structures. Public Works may be responsible for maintenance and repair of sewers and streets as well as the collection and disposal of refuse. A Department of Social Service may have responsibility for family counselling and welfare matters. Rarely is the programme integrated, so that it can deal with the total needs of the residents.

When voluntary compliance fails, resort to the courts is usually so complex, costly and time-consuming that little can be accomplished within the life span of a rat. Both residents and programme staff become frustrated.

Federal funds (about $13 million annually) have been available during the past decade to assist urban rat control programmes in the USA. To qualify a city must evaluate its rat infestation and refuse storage conditions, proceed through a formal application process, and then demonstrate each year that improved sanitation and reduced rodent infestation justify continued funding. Nearly a hundred cities or areas have been assisted, and some of them have so improved their environments that assistance was terminated. About 75% of the nearly 52000 blocks involved in current programmes are maintaining their present status (i.e. with 2% or less premise rat infestation). This affects nearly 6 million people (MMWR, August 10, 1979).

These Federal programmes, however, are directed only at target areas (usually inner city) with relatively high rat-infestation rates. The municipality is left to its own resources to deal with problems over the rest (usually the larger part) of the city. The government's programme is almost always limited to responding to complaints and it may refer the problem to a commercial pest control operator. While householder persistance may be rewarded, usually only a spot treatment occurs; the infestations next door or the causative conditions are rarely dealt with.

Seldom is there more than nominal political (and therefore financial) support for rat control programmes, unless cases of rat bite or disease are publicized. Often the medically-oriented departmental personnel do not appreciate the importance and effectiveness of an environmentally-oriented management programme (Marsh and Jackson, 1978).

Obviously a well-coordinated programme should maximize benefits for all participants. When the housing and fire inspectors, the social worker, the health educator, the sanitarian, the public works engineers, and other workers all recognize their mutual dependence, the total environment can be improved. When the courts—especially a separate housing court—can be involved, progress can be speeded up. When the private pest control operator is made a part of the total effort, his expertise can also be used.

Such an integrated approach requires administrative and field personnel with a great range of expertise, skills, and understanding. Administrative "territories" would be challenged. Because of such presumed threats, an effort such as this is difficult to implement in the public sector.

Yet the rat is unconcerned about such administrative arrangements; it profits from the protective cover of snow and the abundance of garbage made available when streets and alleys are made impassable by blizzards. Rat survival improves whenever increased food and shelter are made available. When the outdoor and indoor environments are improved, pests generally decrease; and the quality of human life is enhanced.

Schools are often regarded as an important programme adjunct, but it is difficult, though certainly possible to educate parents through their children and such educational efforts may need to be regarded as long-range strategies. Most effective, though most time-consuming, is the formation of block clubs or local groups which have political influence with the ward councilman and mayor and within their association (Kohuth and Marsh, 1976; Marsh and Jackson 1978).

Most discussions on urban rodents focus on inner city areas, but rats know no socioeconomic limits. The increase in dogs (and other pets) and the feeding of wild birds has fostered suburban rat infestations. Gardens with unharvested fruits and vegetables also provide a year-round food supply. Additionally the general use of food-disposal units in sinks provides a floating cafeteria in the sewer system, and rats are known to invade homes via the sanitary sewer.

Europeans have long talked of "rat-free" cities, a concept made notable by Telle (1968). Both in northern Germany and in Britain systematic village- or city-wide control programmes are undertaken by municipal authorities (Drummond, 1970). An initial intensive poisoning is followed by a maintenance programme with anticoagulants. Bait stations are established along rodent dispersion routes (e.g. ditches and streams, railroads, beaches, and shorelines) and perimeter farms, immigrating rats hopefully go no farther. With already high levels of environmental sanitation and good community organization, these programmes have enjoyed considerable success. However, the larger the city, the more difficult it is to maintain a rat-free environment.

A parallel programme has been instituted successfully in small villages in Gujarat State, India (Chaturvedi, 1975; Madsen, 1975) where the emphasis is on rodent-proof grain storage in each house. Periodic use of toxicants is required and 85–95% of the homes are rat-free. The programme is carried out by the villagers themselves with advice, training, and technical assistance from the regional Project Officer.

In the USA no similar programme exists, although the concept of "maintenance" blocks in the federal urban rat control programme is similar. Our difficulty is that, apart from these federally-assisted programmes, we usually do not view urban pest management as an area concern. Until we do, our efforts will be ineffective and inefficient.

IX. Management Techniques

The traditional approach to rodent control by most people (city programmes, pest control operators, householders) is to put out some rat poison. Since 1950, the anticoagulants have been readily available but their indiscriminant use has caused selection of resistant animals in Europe, North America, and Australia (Jackson and Ashton, 1979). Concern for an integrated approach, inherent in the USA Public Health Service programmes of the 1940s, is newly expressed.

Basic to the management of any species is the alteration of the environment to the detriment of the pest, as was adequately illustrated in Baltimore three decades ago. Removal of food (garbage) and shelter (backyard privies, broken concrete, junk) drastically reduced the rat population without resort to rodenticides (Fig. 3B). This basic tenet holds today.

In the home this involves repair of foundation walls, windows, and doors, removal of junk from the basement and yard, and proper storage of garbage. Adding grates to drains, capping vent pipes, and protecting from entry along utility wires or overhanging vegetation (especially in roof rat-infested areas) is helpful. In the commercial scene the same procedures are magnified by scale. In addition, there is a need to prevent importing rodents with arriving shipments.

If rodents are already a problem, once a decision to deal with them is made, instant results are desirable. Lethal techniques are especially valuable when an emergency is created by administrative decision and/ or a rat bite or disease problem.

Traps that kill or catch animals alive and reset themselves are excellent tools for small, restricted infestations and/or use by individual householders (Temme, 1979). For an extended programme, however, the manpower requirements are usually too great to make their use practical.

Toxicants of two types are available: quick acting, single-dose rodenticides give overnight kill. Zinc phosphide, the best known of these, is available in several different formulations. Bait shyness, occurring after consumption of a sublethal dose, frequently results, and reduces the efficacy of this rodenticide unless higher quality baits are properly placed. Other more toxic and hazardous chemicals are available only for use by certified applicators.

The other type of toxicant, the anticoagulants, being multiple-dose rodenticides, require repeated feedings over 3–5 days to be lethal. Death results from internal haemorrhage; but because the kill is delayed and rats continue to feed, some people find these poisons undesirable. These chemicals are readily available and widely used. Several new chemicals (second-generation anticoagulants), that are effective against anti-coagulant-resistant rats and mice have recently been or are about to be introduced in the USA; they are already available elsewhere. (Kaukeinen, 1979; Marsh, 1977; Lund, 1977).

These chemicals, in a more concentrated form, may be placed or blown into structural voids or places of rodent activity. The dust adheres to the fur, and the toxicant is ingested during grooming (DDT was the safest and most effective of these for mouse control, but its use in the USA is no longer permitted.)

Fumigants, except for carbon monoxide exhaust from internal combustion engines, are often available only to the specially certified applicator. Forcing such gases into outdoor burrows or into stored grains can be very effective in killing both insect and rodent pests. In the USA, calcium cyanide dust, the best of the burrow fumigants, has been withdrawn from the market because of low sales and assumed potential risk but its availability elsewhere has not changed. Aluminium phosphide tablets are available in many countries and are easily used to treat burrows and stored grain.

Chemosterilants, perhaps theoretically attractive, have not proved to be practical, partly because female rodents may mate with many males. Sex hormone derivatives have to be ingested regularly to be effective (Marsh and Howard, 1973). One compound (a chloro-hydrin), that permanently sterilizes the male at low doses and kills at higher doses, is being considered for use in the USA; it is already used elsewhere in the world (Jones, 1978).

Hazards from use of rodenticides are small. Less than 1% of reported human exposures in the USA to toxic substances are from rodenticides and most of the deaths are related to suicide (National Poison Center Network, 1978). Dangerous situations do occur when rodent baits are mixed into food channels, and multiple deaths can result but fortunately these are rare.

Non-lethal approaches are finding increasing favour, especially for use in situations where rodenticides are restricted because of possible food contamination or hazard to other species. Glue boards catch and hold rodents that attempt to cross them. Chemical repellents have not been used with any general success. Electromagnetic devices, supposedly able to suppress an amazing variety of pests, were removed from the USA market by the Environmental Protection Agency due to lack of effect (Smith, 1979). On the other hand, ultrasonic devices have shown some promise of effectiveness, especially when coordinated with sanitation and appropriate use of traps and baits (Jackson, 1980).

In the autumn, with the advent of cooler temperatures, reduced field foods and increased stored foods, rodents may move to indoor environments. Utilization of traps and toxic baits are especially effective at this time of year. Severe depression of the population either through natural winter mortality or man-made predation prior to the spring reproductive recrudescence has a long-term depressant impact on the population. Moderate depression (to the mid-point on the logistic curve), characteristic of many control programmes, tends to stimulate reproduction and population recovery. Thus many efforts at rat "control" are in fact "rat farming".

A. MEASUREMENT

One component of modern society is the demand for validation, but operational programmes cannot divert scarce resources to measure rodent populations precisely before and after control efforts. Often alternative procedures, such as developing indices for rat activity, are possible and can be carried out with limited manpower (Brown et al., 1955).

One research study in Baltimore was able to correlate a 60% reduction in rats with improvement in refuse storage and housing (Davis and Fales, 1950). In federally assisted rat control programmes, semi-annual evaluations of rodent activity and refuse storage practices are required. Only when the infestation rate is 2% or less can the block be placed in "maintenance" status, which requires less commitment of manpower.

Margulis (1977) attempted a time-series analysis of rat complaints in Newark to determine if they might be used for evaluation or prediction. He concluded "existing programmes .... serve more to identify the agency's role than to diminish the size of the rat population". There is no basis, however, for assuming general application of this concept.

For the pest control operator (PCO), his "call-back" rate is the best indication of his effectiveness. Each additional service stop between his scheduled (often monthly) visits represents an attack on his profits and therefore stimulates greater attention to details in his regular programme. Where the PCO can stimulate customer initiative or incorporate environmental management (repair of structures, better sanitation, improved storage) into his efforts, he can often extend his effectiveness and increase his profitable return.

Regulatory agencies provide another evaluation procedure. US Food and Drug Administration inspectors in a single year seized or condemned nearly 5M lb of contaminated food (Dykstra, 1966). In extreme cases food storage or processing operations may be closed by government inspectors because of actual or potential rodent contamination of products. Recent claims of exportation of contaminated grain by the US stirred federal agencies into enforcing clean food standards (Risser, 1975).

It is ultimately the user of food products that determines standards. Rodent hairs or droppings in raw materials or finished goods, even if cooked and sterile, are considered contaminants and therefore undesirable. However, small hairs may be virtually impossible to remove from food during processing. Visual or electronic inspection for droppings or parts of rodents in food products cannot be 100% effective. Consequently, the best approach is to prevent rodent infestations in storage, transport, and processing facilities.

X. Conclusions

The premise adopted in this review is that understanding of population principles applied to rats will promote better control. In addition, knowledge of the principles involved will indicate what information should be collected to assist in its understanding. It is true that more data are desirable and that elaboration of principles will continue. However, we now know enough about rats to be confident that the understanding of the following principles will be helpful.

(1) Numbers of rats are limited by some factor in the environment. Numerous examples show that the population stops increasing and achieves some level until changes in the environment alter conditions.

(2) The attainment of a level or of a rate of increase is regulated by combination of density-independent and density-dependent factors that affect the numbers. Only the density-dependent factors can truly regulate.

(3) The numbers of rats may be estimated with satisfactory accuracy for control purposes. Changes in numbers with time may be random or may follow a sigmoid curve.

(4) A knowledge of birth rates is essential for timing control efforts

and for understanding the increases after a reduction in numbers. Various anatomical and physiological characteristics permit assessment of breeding condition and numbers. The determination of the extent of breeding season is particularly useful.

(5) Information on death rates allows prediction of the effect of additional causes of mortality and of the importance of timing of efforts. By such information a condition of sustained yield can be prevented, and reasons for failure of various schemes for biological control can be seen.

(6) Movements have contradictory effects in control. Since rats have a small home range intensive control in a small area will be very effective. But since rats that are displaced by habitat change may travel far the chance of reinfection, though small, is real.

(7) Case histories of populations illustrate the range of variation in population change and in habitat factors. From these the possible results of a control plan can be evaluated. The complexity of relations among the aspects of habitat, the variety of predation, and the pervasiveness of competition are apparent.

(8) Finally, some of the fundamental features of administration suggest a range of problems to be considered in planning and justifying a control programme.

References

Anderson, S. and Jones, J. K. (1967). "Recent Mammals of the World". Ronald Press Co., New York.

Andrews, R. V., Belknap, R. W. and Keeman, E. J. (1974). Demographic and endocrine responses of Norway rats to antifertility control. *J. Wildl. Manag.* **38**, 868–874.

Andrews, R. V., Belknap, R. W. and Southard, J., Lorincz, M. and Hass, S. (1972). Physiological, demographic and pathological changes in wild Norway Rat population over an annual cycle. *Comp. Biochem. Physiol.* **41A**, 149–165.

Asdell, S. A. (1941). The influence of age and rate of breeding upon the ability of the female rat to reproduce and raise young. *Cornell Univ. Agr. Exp. Sta. Mem.* **238**, 3–28.

Atkinson, I. A. E. (1973). Spread of the ship rat (*Rattus r. rattus* L.) in New Zealand. *J. Roy. Soc. N.Z.* **3(3)**, 457–472.

Atkinson, I. A. E. (1977). A reassessment of factors, particularly *Rattus rattus* L. that influenced the decline of endemic forest birds in the Hawaiian Islands. *Pacific Sci.* **31(2)**, 109–133.

Austin, O. L. (1948). Predation by the common Rat (*Rattus norvegicus*) in the Cape Cod colonies of nesting terns. *Bird-Banding* **19(2)**, 60–65.

Barnett, S. A. (1963). "The Rat. A Study in Behaviour" Aldine, Chicago.

Barnett, S. A. (1969). Grouping and dispersive behaviour among wild rats. *In* "Aggressive Behavior". (Garottini, S. and Sigg, E. B., eds), pp. 3–14. Exerpta Medica Amsterdam.

Barnett, S. A. and Bathard, A. H. (1953). Population dynamics of sewer rats. *J. Hygiene* **51(4)**, 483–491.

Barnett, S. A., Bathard, A. H. and Spencer, M. N. (1951). Rat populations and control in two English Villages. *Ann. App. Biol.* **38(2)**, 444–463.

Bentley, E. W. (1964). A further loss of ground by *Rattus rattus* L. in the United Kingdom during 1956–61. *J. Anim. Ecol.* **33**, 371–373.

Bentley, E. W., Bathard, A. H. and Riley, J. D. (1959). The rates of recovery of sewer rat populations after poisoning. *J. Hyg.* **57(3)**, 291–298.

Best, L. W. (1973). Breeding season and fertility of the roof rat, *Rattus rattus rattus*, in two forest areas of New Zealand. *New Zealand. J. Sci.* **16**, 161–170.

Beveridge, A. D. and Daniel, M. J. (1965). Observations on a high population of brown rats (*Rattus norvegicus* Berkenhout 1767) on Mokoia Island, Lake Rotorua. *N.Z. Jour. Sci.* **8(2)**, 174–189.

Breed, W. G. (1978). Ovulation rates and estrous cycle lengths in several species of Australian native rats (*Rattus* spp) from various habitats. *Aust. J. Zool.* **26**, 475–480.

Brooks, J. E. (1973). A review of commensal rodents and their control. *Critical Rev. Envir. Control* **3(4)**, 405–453.

Brooks, J. E., and Barnes, A. M. (1972). An outbreak and decline of Norway rat populations in California rice fields. *Calif. Vector News.* **19(2)**, 1–14.

Brooks, J. E., and Bowerman, A. M. (1971). Estrogenic steriod used to inhibit reproduction in wild Norway rats. *J. Wildl. Manag.* **35(3)**, 444–449.

Brown, R. Z., Sallow, W., Davis, D. E. and Cochran, W. G. (1955). The rat population of Baltimore 1952. *Amer. J. Hyg.* **61(1)**, 89–102.

Calhoun, J. B. (1948). Mortality and movement of brown rats in artificially supersaturated populations. *J. Wildl. Manag.* **12(2)**, 167–172.

Calhoun, J. B. (1949). A method for self-control of population growth among mammals living in the wild. *Science,* **109** (2831), 333–335.

Caughley, G. (1977). "Analysis of Vertebrate Populations." John Wiley, Chichester.

Chapman, F. B. (1938). Exodus of Norway rats from flooded areas. *J. Mamm.* **19**, 376–377.

Chaturvedi, G. C. (1975). Organization of rodent control in rural areas. *Proc. All India Rodent Sem.*, 297–303.

Christian, J. J. (1950). The adreno-pituitary system and population cycles in mammals. *J. Mamm.* **31**, 247–256.

Christian, J. J. (1978). Neurobehavioral endocrine regulation of small mammal population. *In* "Populations of Small Mammals under Natural Conditions" (Snyder D. D., ed.). Spec. Pub. 5, Pymatuning La. Ecol. Univ. Pittsburgh.

Christian, J. J. (1980). Endocrine factors in population regulation. Plattsburgh Symposium.

Christian, J. J. and Davis, D. E. (1955). Reduction of adrenal weight in rodents by reducing population size. *Trans. North Amer. Conf.* **20**, 177–189.

Christian, J. J. and Davis, D. E. (1956). The relationship between adrenal weights and population status of urban Norway rats. *J. Mamm.* **37(4)**, 475–486.

Chitty, D. (1954). "The Control of Rats and Mice" Vols I and II, Rats. Clarendon Press, Oxford.

Conaway, C. H. (1955). Embryo reabsorption and placental scar formation in the rat. *J. Mamm.* **36(4)**, 516–532.

Cotchin, E. and Roe, J. C. (1967). "Pathology of Laboratory Rats and Mice". F. A. Davis Co., Philadelphia.

Crombie, A. C. (1944). Rat plagues in Western Queensland. *Nature, Lond.* **154**, 803–804.

Daniel, M. J. (1972). Bionomics of the ship rat (*Rattus r. rattus*) in a New Zealand indigenous forest. *New Zeal. J. Sci.* **15(3)** 313–341.

Davis, D. E. (1949). The role of intraspecific competition in game management. *Trans. North Amer. Wildl. Conf.* **14**, 225–231.

Davis, D. E. (1950). The rat population of New York, 1949. *Amer. J. Hyg.* **52(2)**, 147–152.

Davis, D. E. (1951). The relation between the level of population and the prevalence of Leptospira, Salmonella and Capillaria in Norway Rats. *Ecol.* **32(3)**, 465–468.

Davis, D. E. (1952). A perspective on rat control. *Pub. Health Rep.* **67(9)**, 888–893.

Davis, D. E. (1953a). Analysis of home range from recapture data. *J. Mamm.* **34(3)**, 352–358.

Davis, D. E. (1953b). The characteristics of rat populations. *Quart. Rev. Biol.* **28**, 373–401.

Davis, D. E. (1955). Social interaction of rats as indicated by trapping procedures. *Behaviour* **8(4)**, 335–345.

Davis, D. E. (1956). The frequency of reduction of rat populations by weather. *Ecol.* **37(2)**, 385–387.

Davis, D. E. (1957). The use of food as a buffer in a predator-prey system. *J. Mammal.* **38(4)**, 466–472.

Davis, D. E. (1960). A chart for estimation of life expectancy. *J. Wildl. Manag.* **24(3)**, 334–348.

Davis, D. E. (1961). Principles of population control by gametocides. *Trans. North Amer. Wildl. Nat. Res. Conf.* **26**, 160–166.

Davis, D. E. (1977). Advances in rodent control. *Zeit. angew. Zool.* **64**, 193–211.

Davis, D. E. and Christian, J. J. (1956). Changes in Norway rat populations induced by introduction of rats. *J. Wildl. Manag.* **20(4)**, 378–383.

Davis, D. E. and Christian, J. J. (1957). Population consequences of a sustained yield program for Norway rats. *Ecol.* **39(2)**, 217–222.

Davis, D. E. and Emlen, J. T. (1948). The placental scar as a measure of fertility in rats. *J. Wildl. Manag.* **12(2)**, 162–166.

Davis, D. E. and Emlen, J. T. (1956). Differential trapability of rats according to size and sex. *J. Wildl. Manag.* **20(3)**, 326–327.

Davis, D. E. and Fales, W. T. (1950). The rat population of Baltimore, 1949. *Amer. J. Hyg.* **52(2)**, 143–146.

Davis, D. E. and Hall, O. (1950). Polyovuly and anovular follicles in the wild Norway Rat. *Anat. Rec.* **107**, 187–192.

Davis, D. E. and Hall, O. (1951). The seasonal reproductive condition of female Norway (Brown) rats in Baltimore, Md. *Physiol. Zool.* **24(1)**, 9–20.

Davis, D. E. and Jensen, W. L. (1952). Mortality in an induced epidemic. *Trans. North Amer. Wildl. Conf.* **17**, 151–160.

Davis, D. E. and Winstead, R. L. (1980). Estimating the numbers of wildlife populations. *In* "Wildlife Techniques Manual" (Schemnitz, S., ed.), 4th ed., pp. 221–245. Wildlife Society, Washington DC.

Davis, D. E., Myers, K. and Hay, J. B. (1976). Biological control among vertebrates. *In* "Theory and Practice of Biological Control" (Huffaker, C. B. and Messenger, P. S., eds). Academic Press, New York and London.

Drummond, D. C. (1963). Social behaviour and movements of rats. *Ann. Appl. Biol.* **51** (1963), 343–345.

Drummond, D. C. (1970). Variation in rodent populations in response to control measures. *Symp. Zool. Soc. London* **26**, 351–367.

Drummond, D. C., Taylor, E. J. and Band, M. (1977). Urban rat control: further experimental studies at Folkestone. *Envir. Health* **85**, 265–267.

Dutson, V. J. (1973). Use of the Himalayan blackberry, *Rubus discolor*, by the roof rat, *Rattus rattus*, in California. *Calif. Vector News* **20(8)**, 59–68.

Dutson, V. J. (1974). The association of the roof rat (*Rattus rattus*) with the Himalayan blackberry (*Rubus discolor*) and Algerian ivy (*Hedera canariensis*) in California. *Proc. Vert Pest Conf.* **5**, 41–48.

Dwyer, P. D. (1975). Observations on the breeding biology of some New Guinea murid rodents. *Aust. Wildlife Res.* **3**, 33–45.

Dykstra, W. W. (1966). The economic importance of rats. *In* "Rodents as factors in disease and economic loss". *Proc Asia-Pacific Interchange*, 47–52.

Eaton, P. and Stirrett, C. S. (1928). Reproduction rate in wild rats. *Sci.* **67**, 555–556.

Ecke, D. H. (1954). An invasion of Norway Rats in southwest Georgia. *J. Mammal.* **35**, 521–525.

Ecke, D. (1955). Analysis of populations of the roof rat in southwest Georgia. *U.S. Public Health Service Monogr.* **27**, 1–20.

Elandt-Johnson, R. C. (1975). Definitions of rates: some remarks on their use and misuse. *Amer. J. Epidemiol.* **102**, 267–271.

Elton, C. (1953). The use of cats in farm rat control. *Brit. J. Anim. Behav.* **1**, 151–155.

Emlen, J. T., Stokes, A. W. and Winsor, C. P. (1948). The rate of recovery of decimated populations of brown rats in nature, *Ecol.* **29(2)**, 133–145.

Errington, P. L. (1935). Wintering of field living Norway rats in Wisconsin. *Ecol.* **16**, 122–123.

Ewer, R. F. (1971). The biology and behaviour of free-living population of Black rats (*Rattus rattus*). *Animal Behav. Monog.* **4(3)**, 126–174.

Fall, M. W. (1977). Rodents in tropical rice. *Rodent Research Center Tech. Bull* **36**, 1–39 Univ. Philippines at Los Baños.

Farhang-Azad, A. (1976). Ecology of Norway rat populations and *Capillaria hepatica*. PhD dissertation, The Johns Hopkins Univ.

Farris, L. J. and Griffiths, J. Q. (1962). "The Rat in Laboratory Investigation". Hafner, New York.

Feldman, H. (1925). Fertility of the rat, *Mus norvegicus*. *Proc. Nat. Acad. Sci.* **11**, 718–721.

Gomez, J. C. (1960). Correlation of a population of roof rats in Venezuela with seasonal changes in habitat. *Amer. Mid. Nat.* **63(1)**, 177–193.

Gwynn, G. W. (1972). Effects of a chemosterilant on fecundity of wild Norway rats. *J. Wildl. Manag.* **36**, 550–556.

Harrison, J. L. (1956). Survival rates of Malayan rats. *Bull. Raffles Mus. Singapore* **27**, 1–26.

Harrison, J. L. (1958). Range of movement of some Malaysian rats. *J. Mamm.* **39(2)**, 190–206.

Harrison, J. L. and Woodville, H. C. (1948). An attempt to control house rats in Rangoon. *Trans. Roy. Soc. Trop. Med. Hyg.* **42(3)**, 247–258.

Hebel, R. and Stromberg, M. W. (1976). "Anatomy of the Laboratory Rat". Williams and Williams, Baltimore.

Hiraiwa, Y. K. and Sumikawa, S. (1951). Rapid and marked proliferation of Norway rats grown semi-wild on the small island, Tojima, Ehime Prefecture. *Sci. Bull.* **13(1–4)**, 406–414.

Holland, C. W. (1952). A plea for the native birds. *Queensland Nat.* **14(4)**, 80–82.

Hopf, H. S., Morley, G. E. J. and Humphries, J. R. O. (eds) (1976). Rodent damage to growing crops and to farm and village storage in tropical and subtropical regions. Center for Overseas Pest Research.

Jackson, W. B. (1951). Food habits of Baltimore, Maryland, cats in relation to rat populations. *J. Mamm.* **32(4)**, 458–461.

Jackson, W. B. (1977). Evaluation of rodent depredations to crops and stored products. *EPPO Bull.* **7(2)**, 439–458.

Jackson W. B. (1979a). Comments on cacao feeding by rats. *Rodent Newsletter*, Central Arid Zone Research Institute Jodhpur **3(3)**, 17.

Jackson, W. B. (1979b). Assignment Report, Regional Training Centre for Rodent Biology and Control, Iraq. WHO EM/VBC/IS.

Jackson, W. B. (1980). Ultrasonics protect egg farm. *Pest Control* **48(8)**, 28–38.

Jackson, W. B. and Ashton, A. D. (1979). Present distribution of anticoagulant resistance in the United States. *In* "Vitamin K metabolism and Vitamin K-dependent proteins" (Suttie, J. W., ed.), pp. 392–397. Univ. Park Press, Baltimore.

Jennison, G. (1921). Rat repression by sexual selection. *J. Roy. San. Inst.* **41**, 358–363.

Johnson, D. H. (1946). The rat population of a newly established military base in the Solomon Islands. *Naval Med. Bull.* **46(10)**, 1627–1632.

Jones, A. R. (1978). The antifertility action of α chlorohydrin in the male. *Life Sciences* **23**, 1625–1646.

Kaukeinen, D. E. (1979). Experimental rodenticide (Talon) passes lab tests; moving to field trials in pest control industry. *Pest Control.* **47(1)**, 19–23.

Kirkpatrick, C. M. and Conaway, C. H. (1947). The winter foods of some Indiana Owls. *Amer. Mid. Nat.* **38(3)**, 755–766.

Kohuth, B. J. and Marsh, B. T. (1976). Neighborhood Improvement Guide, City of Cleveland.

Leslie, P. H., Venables, U. M. and Venables, L. S. U. (1952). The fertility and population structure of the brown rat (*Rattus norvegicus*) in corn ricks and some other habitats. *Proc. Zool. Soc. London* **122**, 187–238.

Li, H. Y. and Davis, D. E. (1952). The prevalence of carriers of Leptospira and Salmonella in Norway Rats of Baltimore. *Amer. J. Hyg.* **56(1)**, 90–100.

Libay, R. L. and Fall, M. W. (1976). Observations on an exceptionally dense population of wild rats in marshland. *Kalihasan (Philippine J. Biol)* **5**, 207–212.

Lotka, A. J. (1925). "Elements of Physical Biology". Williams and Wilkins, Baltimore.

Lund, M. (1977). New rodenticides against anticoagulant-resistant rats and mice. *EPPO Bull.* **7(2)**, 503–508.

Madsen, C. R. (1975). Sidhpur rodent control and grain storage project. *Proc. All India Rodent Seminar*, 16–28.

Mahdi, A. H., Arafa, M. S. and Gad, A. M. (1972). Assessment of trapping with HCH (Gammexane) as anti-rat and flea measures in A.R.E. 1968. *J. Egypt. Pub. Health Assoc.* **47(2)**, 96–105.

Margulis, H. L. (1977). Rat fields, neighborhood sanitation and rat complaints in Newark, New Jersey. *Geograph. Rev* **67(2)**, 221–231.

Marsh, R. E. (1977). Bromadiolone, a new anticoagulant rodenticide. *EPPO Bull.* **7(2)**, 495–502.

Marsh, R. E. and Howard, W. E. (1973). Prospects of chemosterilant and genetic control of rodents. *Bull. WHO* **48**. 309–316.

Marsh, B. T. and Jackson, W. B. (1978). Environmental control of rats. *Pest Control* **46(8)**, 12–14, 16,37,38,43,54, **46(9)**, 26–29,40–40d,42.

Matheson, C. (1943). A study of rodent destruction in cities. *Roy. Sanit. Inst. J.* **63**, 38–50.

McDougall, W. A. (1946). An investigation of the rat pest problem in Queensland canefields: 4. Breeding and life histories. *Queensland J. Agric. Sci.* **3(1)**, 1–43.

Miller, N. (1911). Reproduction in the brown rat (*Mus norvegicus*). *Amer. Nat.* **45**, 623–635.

Minckler, J. and Pease, F. D. (1938). A colony of albino rats existing under feral conditions. *Science* **87**(2264), 460–461.

Moghissi, K. and Hafez, E. S. E. (1972). "Biology of Mammalian Fertilization and Implantation". C. C. Thomas, Springfield.

Mohr, C. E. (1947). Major fluctuations in some Illinois mammal populations. *Trans Ill. State Acad. Sci.* **40**, 197–204.

Morbidity, Mortality Weekly Report 1979. Urban rat control—United States, January–March 1979, August 10, 1979.

Myllymaki, A. (1969). An early approach to a rat-free town in Finland. *Ver. Wass. -Boden-Lufthyg. (Berlin-Dahlem)*, **32**, 162–166.

Nass, R. D., Hood, G. A. and Lindsey, G. D. (1971a). Fate of Polynesian rats in Hawaiian sugar cane fields during harvest. *J. Wildl. Manag.* **35**(2), 353–356.

Nass, R. D., Hood, G. A. and Lindsey, G. D. (1971b). Influence of gulch-baiting on rats in adjacent sugarcane field. *J. Wildl. Manag.* **35**(2), 357–360.

National Poison Center Network (1978). Annual report to the Rohm and Haas company, first year 1978. Unpublished.

National Research Council (1980). "Urban Pest Management". National Academy Press, Washington, DC.

North, M. E. W. (1946). Mait Island—a bird rock in the gulf of Aden. *Ibis* **88**, 478–501.

Orgain, H. and Schein, M. W. (1953). A preliminary analysis of the physical environment of the Norway rat. *Ecol.* **34**(3), 467–473.

Perry, J. S. (1945). The reproduction of the wild brown rat (*Rattus norvegicus* Erxleben). *Proc. Zool. Soc. London* **115**(1–2), 19–46.

Pimentel, D. (1955). The control of the mongoose in Puerto Rico. *Amer. J. Trop. Med. Hyg.* **4**(1), 147–151.

Plomley, N. J. B. (1972). Some notes on plagues of small mammals in Australia. *J. Nat. Hist.* **6**(4), 363–384.

Pollitzer, R. (1954). "Plague". World Health Organization.

Prakash, I., and Rana, B. D. (1973). A study of field populations of rodents in the Indian desert. *Zeit. Angew. Zool.* **60**(1), 31–41.

Risser, J. (1975). Assail corn from U.S. as 'rubbish' *Des Moines Register* **126**(33), 1, 8.

Sanchez, D. S. (1977). Population patterns of sympatric rodents: *Rattus r. Midanensis* Mearns and *R. argentiventer* Robinson and Kloss associated with the rice crop in oriental Mindoro, Philippines. MSc. Thesis. University of Philippines, Los Baños, Laguna.

Schein, M. W. (1950). Field test of the efficiency of the rodenticide compound W.A.R.F. 42. *Pub. Health Rep.* **65**(11), 368–372.

Schiller, E. L. (1956). Ecology and health of *Rattus* at Nome, Alaska. *J. Mamm.* **37**, 181–188.

Schroder, G. D. and Hulse, M. (1979). Survey of rodent populations associated with urban land fill. *Amer. J. Pub. Health* **69**(7), 713–715.

Shah, M. P. (1979). Population explosion of rodents in groundnut fields in Gujurat. *Rodent Newsletter* Central and Zone Institute **3**(3), 19–20.

Smith, R. J. (1979). Rodent repellers attract EPA strictures. *Science* **203**, 484–486.

Spillett, J. J. (1976). The ecology of the lesser bandicoot rat in Calcutta. Bombay Nat. Hist. Soc. and the Johns Hopkins University Center for Medical Research and Training, Calcutta 1–223.

Srivastava, A. S. (1968). Dynamics of population of field rats in relation to the alive burrows. *Mammalia* **32(2)**, 164–171.

Storer, T. I. (ed.) (1962). "Pacific Island Rat Ecology" B. P. Bishop Mus. Bulletin 225.

Strecker, R. L. and Jackson, W. B. (1962). Cacao plantings. *In* "Pacific Island Rat Ecology" (Storer, T. I., ed.), pp. 208–213. B. P. Bokoa Museum, Bulletin 225.

Sumangil, J. P. (1965). Handbook on the ecology of rice field rats and their control by chemical means. Bureau of Plant Industry, Manila.

Tamarin, R. H. and Malecha, S. R. (1971). The population biology of Hawaiian rodents: demographic parameters. *Ecol.* **52(3)**, 383–394.

Tamarin, R. H. and Malecha, S. R. (1972). Reproductive parameters in *Rattus rattus* and *Rattus exulans* of Hawaii, 1968 to 1970. *J. Mamm.* **53(3)**, 513–528.

Tanaka, R. (1957). An ecological review of small-mammal outbreaks with special reference to their association with the flowering of bamboo grasses. *Kochi Women's Univ. Bull.* **5(1)**, 20–30.

Taylor, K. D. (1968). An outbreak of rats in the agricultural areas of Kenya in 1962. *East African Forestry J.* **34**, 66–77.

Taylor, K. D. and Quy, R. J. (1973). Marking systems for the study of rat movements. *Mammal, Rev.* **3(2)**, 30–34.

Taylor, R. H. (1975). What limits Kiore (*Rattus exulans*) distribution in New Zealand? *New Zeal. J. Zool.* **2(4)**, 473–477.

Telle, H. J. (1968). Zur Problematik des Begriffes "rattenfrei". *Anz. für Schädlingskune* **41(8)**, 119–125.

Temme, M. (1979). Polynesian rat (*Rattus exulans*) populations in the northern Marshall Islands. PhD dissertation, Bowling Green State Univ.

Tomich, P. O. (1970). Movement patterns of field rodents in Hawaii. *Pacific Sci.* **24(2)**, 195–234.

Twigg, G. (1975). "The Brown Rat". David and Charles, London.

Twigg, G. I. (1961). Infestations of the brown rat (*Rattus norvegicus*) in drift mines of the British Isles. *J. Hyg. Cambridge* **59(3)**, 271–284.

Uchida, T. A. (1969). Rat-control procedures on the Pacific Islands, with specific reference to the efficiency of biological control agents. I. Appraisal of the monitor lizard, *Varanus indicus* (Daudid), as a rat-control agent on Ifaluk, Western Caroline Islands. *J. Fac. Agr. Kyushu. Univ.* **15(3)**, 311–330.

Venables, L. S. V., and Leslie, P. H. (1942). The rat and mouse populations of corn ricks. *J. Anim. Ecol.* **11(1)**, 44–68.

Walcott, Robert M. and Vincent, B. W. (1975). The relationship of solid waste storage practices in the inner city to the incidence of rat infestation and fires. EPA/530/SW/150.

Watson, J. S. (1961). Rats in New Zealand: A problem of inter-specific competition. *Proc. 9th Pacific Sci. Cong.* **19**. 15–17.

Wodzicki, K. (1969). Preliminary report on damage to coconuts and on the ecology of the Polynesian rat (*Rattus exulans*) in the Tobelau Islands. *Proc. N.Z. Ecol. Soc.* **16**, 7–12.

Wodzicki, K. (1972). Effect of rat damage on coconut production on Nukunonu Atoll, Tokelau Islands. *Oleagineux* **27(6)**, 309–314.

Wodzicki, K. (1973). Prospects for biological control of rodent populations. *Bull. World Health Org.* **48**, 461–467.

Wood, D. H. (1971). The ecology of *Rattus fuscipes* and *Melomys cervinipes* (Rodentia: Muridae) in a southeast Queensland rain forest. *Aust. J. Zool.* **19**, 371–392.

Wynn, M. (ed.) (1977). "Biology of the Uterus" Plenum Press, New York.

Biosphere Reserves

R. GOODIER

Nature Conservancy Council, Edinburgh EH9 2AS, Scotland

J. N. R. JEFFERS

Institute of Terrestrial Ecology, Grange-over-Sands, Cumbria LA11 6JU, England

I. INTRODUCTION

The last decade has seen the birth of several initiatives concerned to protect the natural or semi-natural environment of areas judged to be of international importance. This concern is in marked contrast to the earlier pattern in nature conservation in which the selection of national parks or nature reserves was made strictly on the basis of a framework describing the national resources and where any international element was generally accidental, depending on the coincidence of areas on either side of a frontier. To a very limited extent, the attempt has been made in some countries to give recognition to the particular supra-national interest of certain special localities; thus Great Britain's Nature Conservation Review (Ratcliffe, 1977) recognizes certain areas as being of more than purely national significance in terms of the contribution which might be made by these areas to conserving features of the natural environment which are particularly significant, or unique, at least within western Europe. The international significance of certain sites of importance for wildfowl has been recognized under the Ramsar convention but, here again, the sites concerned are primarily those chosen by reference to a national selection but which have subsequently been considered to be of international significance.

The Man and Biosphere (MAB) Programme, initiated by the General Conference of the United Nations Educational, Scientific and Cultural Organization (UNESCO) at its sixteenth session in 1970, aimed to "develop within the natural and social sciences a basis for the rational use and conservation of the resources of the biosphere and for the improvement of the relationship between man and the environment; to predict the consequences of today's actions on tomorrow's world and thereby to increase man's ability to manage efficiently the natural resources of the biosphere". Among the 14 major international project themes adopted within the MAB programme, Project 8 is concerned with "The conservation of natural areas and the genetic material they contain" and, within this project, the main emphasis has been on the conservation of representative examples of all the major types of ecosystems of the world in a series of biosphere reserves.

The World Conservation Strategy published by the International Union for the Conservation of Nature (IUCN) (1980) also emphasizes the importance of conserving genetic diversity and sees it as a priority that site preservation programmes should protect representative samples of ecosystem types, unique ecosystems and habitats of threatened species, and also the wild relations of economically valuable and other useful plants and animals and their habitats. It lays stress on the need to coordinate national protected area systems with international ones, particularly with that established under the biosphere reserve programme of MAB Project 8 so that a complete network of protected representative samples of ecosystems may be established as soon as possible.

II. THE FUNCTION OF BIOSPHERE RESERVES

In 1973, UNESCO, the United Nations Food and Agriculture Organization (FAO) and IUCN convened an expert panel on MAB Project 8 to discuss the classification and inventory of protected areas, the criteria for their selection and the use to be made of them (UNESCO, 1973). Among the main recommendations of this panel was the suggestion that a classification system should be prepared to include all the biomes and biotic subdivisions of the world, possibly along the lines indicated by Dasmann (1973), and that the primary objective of the international programme should be "to ensure that adequate examples of all important and representative biotic sub-divisions are protected". The panel also indicated that any proposed protected area should be "representative of a widespread biome or a main subdivision, and therefore valuable as a standard against which to judge the result of human use or modification elsewhere in the biome; alternatively, uniqueness may make an area worthy of conservation as the only example of its kind".

Subsequently, a task force convened by UNESCO devised criteria and guidelines for the selection of biosphere reserves (UNESCO, 1974). The task force proposed that biosphere reserves should be characterized as follows:

"(1) Biosphere reserves will be protected areas of land and coastal environments. Together, they will constitute a world-wide network linked by international understanding on purposes, standards and exchange of scientific information.

(2) The network of biosphere reserves will include significant examples of biomes throughout the world.

(3) Each biosphere reserve will include one or more of the following categories:

(a) Representative examples of natural biomes.

(b) Unique communities or areas with unusual natural features of exceptional interest. It is recognized that representative areas may also contain unique features, e.g. one population of a globally rare species; their representativeness and uniqueness may both be characteristic of an area.

(c) Examples of harmonious landscapes resulting from traditional patterns of land use.

(d) Examples of modified or degraded ecosystems capable of being restored to more natural conditions.

(4) Each biosphere reserve should be large enough to be an effective conservation unit, and to accommodate different uses without conflict.

(5) Biosphere reserves should provide opportunity for ecological research, education and training. They will have particular value as benchmarks or standards of measurement of long-term changes in the biosphere as a whole. Their existence may be vital to other projects in the MAB programme.

(6) A biosphere reserve must have adequate long-term legal protection.

(7) In some cases, biosphere reserves will coincide with, or incorporate, existing or proposed protected areas, such as National Parks, sanctuaries or nature reserves."

The above review, and extracts from the 1973 and 1974 UNESCO MAB Reports on Project 8, indicate how the aim of MAB Project 8 became rapidly focused during these two years on the main theme of the establishment of a world network of biosphere reserves, this term appearing first in the 1974 Report. In the process, the concept of this world network of protected areas became more complex, evolving from one based on the selection of mainly natural representative and unique areas to one which included "cultural landscapes" and modified or degraded areas capable of restoration. The 1973 Report (UNESCO MAB Report No. 12) touched on the problem of the conceptual framework for the selection of representative areas and it is to this issue that we now turn our attention.

III. The Selection of Biosphere Reserves

From the earliest discussions within the MAB 8 Project, it was understood that the main framework from which selection of protected areas would be made was a classification of the principal ecological systems of the world. However, it was recognized by the expert panel in 1973 that an important hindrance to the initiation of the selection process was the absence of any general-purpose classification of these systems covering the whole world, and this lack was re-affirmed by the UNESCO MAB Task Force Report in 1974 which stressed the need for the development of a generally acceptable classification of the world's biomes and for surveys and inventories of biotic communities to determine their nature and extent. As we shall see later, however, the extent to which the biotic province concept has conditioned the selection of biosphere reserves so far established is a moot point.

The first steps towards producing a classification of biotic communities of the world specifically designed as a framework for the selection of biosphere reserves were taken in a brief paper by Dasmann (1972) on behalf of IUCN. In this paper, Dasmann concluded that the biome system of Clements and Shelford (1939) was the most useful starting point for the definition of the world's major ecological divisions. According to Clements and Shelford (1939), the biome is an area characterized by a prevailing regional climax vegetation and its associated animal life. Dasmann's (1972) paper was further developed in the IUCN Occasional Paper No. 9 of 1974, "Biotic Provinces of the World" (Dasmann, 1974). Both these papers recognized the need to elaborate the scheme based on major biomes (e.g. tundra biome, forest biome, etc.) to take account of major biogeographically determined continental subdivisions, such as the differences within the lowland tropical rain forest biome in the Neotropical, African and Oriental situations.

Within the major regional or subregional biomes, it was considered that a further level of subdivision was necessary to arrive at biotic provinces. These provinces represented the limit of subdivision into natural areas that it was practical to achieve in terms of a world system, though, even here, it was recognized that there might be areas of such heterogeneity within these biotic provinces as to require special distinction as, for example, isolated mountain areas. While the physiognomy of the prevailing climax vegetation (biome type) is the first basis for recognition of a biotic province, the presence of a distinctive flora or fauna can be used to establish its boundaries. Dasmann (1973) showed that there were strong similarities between the North American biotic provinces defined by vegetation and the subdivisions of North America recognized in zoological work on small mammals and birds. From detailed study of the records of bird and mammal fauna, Dasmann was able to revise the biotic provinces of North America proposed in the 1972 paper to produce a more satisfactory framework for biosphere reserve selection.

The most recent attempt at "a classification of the biogeographic provinces of the world" was prepared for the UNESCO MAB Programme by Udvardy (1975) under the auspices of IUCN. This classification further elaborates the system developed by Dasmann, but introduces some changes in terminology; thus, in the hierarchical system, the largest units are the eight main biogeographical realms (covering Dasmann's seven biogeographical regions) and the realms are subdivided into biogeographical provinces (Dasmann's biotic provinces). Each province is characterized by a major biome type or biome complex according to a classification of biome types which includes 14 categories. Thus, in Udvardy's classification, each biogeographical province is characterized by code numbers indicating first the realm, second the province within that realm, and third the dominant biome type.

The most systematic adoption of the biogeographical province system in the selection of biosphere reserves has taken place in the United States, where it has been conveniently summarized in a paper by Franklin (1977). Franklin reports on a system of "biotic provinces" based on those prepared by Udvardy, but with additional subdivisions determined by the US MAB 8 Committee. Thus, while Udvardy recognizes 16 biogeographical province subdivisions within the USA, the US MAB 8 Committee arrived at a selection of biosphere reserves in 22 provinces or major subdivisions within continental USA (including Alaska).

Characteristic of the American approach has been the equal prominence given to the conservation and research functions of biosphere reserves. As well as the essential selection of representative ecosystem sites for conservation in each biotic province, the existence of potential for major ecological research and monitoring programmes was regarded as critical. In the early discussions of the US MAB 8 Committee, it became clear that individual reserved land areas capable of fulfilling both these functions equally well would be rare. Also, while within the international context biosphere reserves can include ecosystems substantially modified by man, priority was given in the USA to the selection of natural or near-natural areas (Franklin, 1976). The two categories of land in the USA which were most likely to provide candidates for the biosphere reserve series were, first, the National Parks and Wilderness areas, and, second, the experimental reservations established by various agencies such as the US Forest Service, US Department of Agriculture and the Energy Resources and Development Agency.

It was found that, within each biotic province, examples of these two categories could often be complementary in providing for the requirements of biosphere reserves. Thus, the National Parks and Wilderness areas generally contained fine examples of natural or near-natural ecosystems, but had not been the subject of much research or monitoring, while the experimental reservations were the site of much past and on-going research but did not contain sufficiently large natural areas to be considered as biosphere reserves on their own. The US MAB 8 Committee, therefore, decided that the best solution was a matching of selected large natural reserves with nearby "research-rich" and experimentally oriented reservations. This biosphere reserve "cluster" concept will be discussed further in the sections of the paper dealing with biosphere reserve organization, and with research and monitoring.

In the American situation, described above, we have a large nation within which lie many complete biotic provinces defined according to a system of biotic province classification which was largely developed in relation to that country. Elsewhere, the situation is frequently very different in that the biotic provinces usually overlap national boundaries and each nation has a rather different concept of how the biotic provinces within their territory should be defined and their boundaries drawn. Very often, the emphasis placed on the role of biosphere reserves contrasts, for a number of very good reasons, with that recognized in the USA. The countries bordering the Mediterranean are good examples of such a contrasting approach and their joint deliberations regarding the establishment of biosphere reserves in the Mediterranean area are recorded in the UNESCO MAB Report No. 45 (UNESCO, 1979a), entitled "Workshop on Biosphere Reserves in the Mediterranean region". This workshop meeting had been called primarily to adapt the concept of biosphere reserves to the special needs of the Mediterranean region (*sensu lat*, including several biogeographical provinces), and to improve the classification of biogeographical provinces within the region as a framework for future biosphere reserve selection. So far as the concept of biosphere reserves most appropriate for the region was concerned, the workshop came to the conclusion that, because the Mediterranean region is so diverse and is bounded by so many different biotic provinces, it was important that examples of the interfaces between provinces should be included within biosphere reserves as well as examples of typical ecosystems within the provinces at the centre of their distribution. From the biogeographical point of view, six principal transitions between the Mediterranean and adjoining regions were identified. It was stressed by the Workshop that it was at these great inter-regional (i.e. inter-biotic province) interfaces that most interactions, transitions and conflicts occurred, and, generally, the most active process of change and the most serious ecological imbalances deserving of attention within the context of the MAB programme. The Workshop considered that the inventory of biogeographical provinces devised by Udvardy (1975), while useful in considering the distribution of biosphere reserves at a global level, was too general to provide a satisfactory framework for the selection of biosphere reserves in the Mediterranean region. However, the reported discussion illustrates the considerable difficulty in arriving at a classification judged satisfactory to the local participants which, at the same time, can be related to the international framework. The classification proposed for the Mediterranean region is both polyphyletic and complex. In the first place, the definition of the region itself bears only a rather remote relationship to the biogeographical provinces recognized by Udvardy, its boundary being based "in Europe on climate and vegetation; in North Africa and south of the Near East on an arbitrary boundary corresponding with the isohyet of 100 mm average annual precipitation; in Northern Turkey, the limits of the Pontic region; in the east an arbitrary borderline with the Irano-Turanian region". With the Region, the first division is into three main classes of ecosystem: first, natural (mainly forest) ecosystems; second, ecosystems representing degradation stages of forest and steppe ecosystems; and third, the main types of agro-ecosystems (defined by principal crops or grazing animals). A preliminary working classification was presented in tabular form for each of these three categories and the Workshop considered that an ideal network of biological reserves should contain between them adequate areas representing every division in the tables. This requirement is difficult to interpret in the case of the natural ecosystems, but comprises at least 18 categories for the degraded ecosystems and 33 categories for agro-ecosystems. A single reserve of substantial size could encompass several of these categories, but probably not a large number if the area of each is to be "adequate".

The Mediterranean Workshop also stressed the important point that the biosphere reserve system should contain representatives of azonal ecosystems such as saltmarshes, riverine forests and wetlands. Another significant issue raised by the Workshop was the need for relating the biosphere reserve system to the system of the biogenetic reserves which the Council of Europe has proposed should be set up in member states. This issue will be discussed at greater length in the section of the paper (Section V) dealing with the organization of biosphere reserves and their relationship to other protected areas.

In Great Britain there are, according to Udvardy, only two biotic provinces, but 13 biosphere reserves have been established. The reason for this apparent anomaly is related to the basis of selection of nature reserve areas in Great Britain. As described in the Nature Conservation Review (Ratcliffe, 1977), the selection of nature reserves is essentially based on the conservation of representative habitats which have not, hitherto, been related to any system of biogeographical provinces. The habitats themselves are polyphyletic, comprising categories that are basically vegetation formation types, such as woodlands, grasslands etc., together with geographically determined coastlands, edaphically determined wetlands and topographically determined uplands. Essentially, nature reserve sites have been chosen as being the nationally "best" representatives of each habitat, with some account taken of major geographical variation, though some sites do contain a combination of habitats. As a result, the best examples of major ecosystems within the main biotic provinces are contained in several habitat-orientated sites—and these may be considered a site cluster—though here the cluster concept is conceived in a rather different way from that developed in the USA. Essentially, the diversity of approach to biosphere reserve selection is the result of two different perspectives (IUCN, 1979); one is derived from the distribution of plants and animals and leads to the definition of regions that are biogeographically distinct; the other is derived from the structural formations of vegetation and the associated animals—biomes. To a certain extent, these two perspectives are complementary—either approach developed alone is liable to lead to an incomplete system. Thus, having identified biogeographical provinces, it is necessary to ensure that reserves are established which contain the main biomes that occur in it; conversely, starting from the identification of biomes, it is necessary to ensure that biogeographical variation is adequately represented. Thus, for most countries, the selection and establishment of biosphere reserves is a two-pronged process, both *ad hoc* and systematic elements being present. Initially, nations have submitted for approval areas which would tend to qualify under a range of criteria and within most conceivable classificatory frameworks of biogeographical provinces. Then, more gradually, consideration is being given to filling gaps in the system, revealed as the result of national or international discussions. We now turn to examine the progress that has been made so far in the establishment of the international network of biosphere reserves.

IV. The Establishment of Biosphere Reserves

By October 1979, 177 biosphere reserves had been accepted by UNESCO into the Man and the Biosphere scheme. Table I summarizes the numbers of reserves proposed by each country that has so far participated in the biosphere reserve programme. The out-

TABLE I.   Alphabetical list of countries (by continent) showing number of biosphere reserves, and total areas

Continent	Country	Number of biosphere reserves	Total area of reserves (ha)
Europe	Austria	4	27 600
	Bulgaria	17	25 900
	Byelorussian SSR	1	76 200
	Czechoslovakia	3	168 900
	Denmark (Greenland)	1	70 000 000
	France	3	21 500
	German Democratic Republic	2	3 500
	Hungary	4	105 900
	Italy	3	63 700
	Norway	1	1 555 000
	Poland	4	25 600
	Romania	3	41 200
	Spain	3	101 000
	Switzerland	1	16 900
	United Kingdom	13	43 600
	Yugoslavia	2	350 000
		65	72 626 500

TABLE I.—*cont.*

Continent	Country	Number of biosphere reserves	Total area of reserves (ha)
Asia	China, People's Republic of	3	425 600
	Indonesia	4	367 000
	Iran	9	2 609 700
	Mauritius	1	3 600
	Pakistan	1	31 350
	Philippines	1	23 500
	Sri Lanka	2	9 300
	Thailand	3	26 100
	USSR	6	671 900
		30	4 168 050
Africa	Cameroun	1	170 000
	Central African Republic	2	1 640 200
	Congo, People's Republic of	1	111 000
	Ivory Coast	1	330 000
	Kenya	3	791 360
	Nigeria	1	460
	Senegal	1	750
	Sudan	2	7 751 000
	Tunisia	4	32 000
	Uganda	1	220 000
	Zaire	2	283 000
		19	11 329 770
North America	Canada	2	58 150
	United States of America	33	8 627 900
		35	8 686 050
Central America	Honduras	1	250 000
	Mexico	3	358 200
		4	608 200
South America	Bolivia	2	300 000
	Chile	4	1 571 750
	Columbia	3	2 514 400
	Peru	3	2 506 700
	Uruguay	1	200 000
		13	7 092 850
Australia	Australia	11	4 737 300
	TOTAL	177	109 248 720

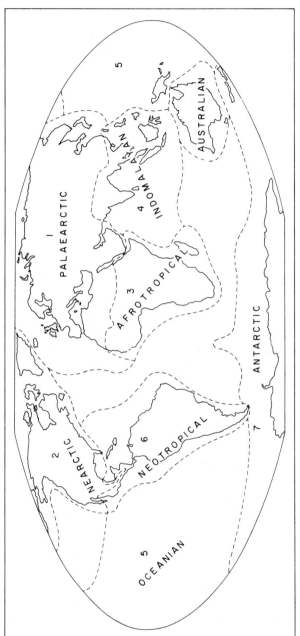

Fig. 1. Biogeographical realms of the world (adapted from Udvardy, 1975)

standing contributor is clearly the United States, while Bulgaria and the United Kingdom have proposed 17 and 13 Reserves, respectively, though these areas are necessarily rather small. Denmark, by contributing Greenland to the scheme, has the honour of proposing the biosphere reserve with the greatest area. Many developing countries are well represented on this list.

TABLE II.   Representation of biosphere reserves in biogeographical realms

Realm	Total	Number of provinces		Number of biosphere reserves	Area of reserves
		Represented	Not represented		
Nearctic	22	14	8	35	78 118 770
Palaearctic	44	20	24	84	6 362 670
Afrotropical	29	11	18	16	11 301 340
Indomalayan	27	10	17	12	458 580
Oceanian	7	1	6	2	20 300
Australian	13	9	4	10	4 724 970
Antarctic	4	1	3	1	12 343
Neotropical	47	18	29	17	8 240 030

Table II summarizes the numbers and areas of biosphere reserves by the biogeographical realms illustrated in Fig. 1. The Palaearctic realm has the largest number of reserves, but the Nearctic realm has the largest area of reserves. Interestingly, the Neotropical realm has the largest number of biogeographical provinces not yet represented, closely followed by the Palaearctic, Afrotropical and Indomalayan realms.

Table III shows the representation of biosphere reserves within biome types, and emphasizes the relatively large number of reserves of mixed mountain and highland systems. Temperate broadleaf forest and subpolar deciduous forest, and evergreen sclerophyllous forest, scrub or woodland biomes are also well represented. However, temperate needleleaf forest, tropical grassland and savanna, and river and lake system biomes are represented by only one reserve each.

The frequency distribution of biosphere reserves by total area is given in Table IV. Twenty of the existing reserves have areas less than 1000 ha. The median area is 18 770 ha, with upper and lower quartiles of 70 000 ha and 3076 ha, respectively. Of the 177 reserves, 30 have total areas greater than 330 000 ha.

Similarly, the frequency distributions of biosphere reserves by minimum and maximum altitudes are shown in Tables V and VI, respectively. Minimum altitudes range from − 18 m to 3000 m, with a median

TABLE III.   Representation of biosphere reserves within Biome Types

Biome type	Number of reserves	Total area of reserves (ha)
1. Tropical humid forest (including mangroves)	11	3 065 110
2. Subtropical and temperate rainforest	6	2 524 500
3. Temperate needleleaf forest	1	782 000
4. Tropical dry or deciduous forest (including monsoon)	13	4 447 540
5. Temperate broadleaf forest and subpolar deciduous forest	36	1 253 560
6. Evergreen sclerophyllous forest, scrub or woodland	18	1 462 500
7. Warm desert and semi-desert	15	11 935 010
8. Cold winter desert and semi-desert	5	354 800
9. Tundra and barren arctic desert	7	77 403 480
10. Tropical grassland and savanna	2	1 561 950
11. Temperate grassland	9	462 210
12. Mixed mountain and highland systems	54	7 584 700
13. Mixed island systems	10	431 920
14. River and lake systems	1	700 00

TABLE IV.   Numbers of biosphere reserves by total area

Total area (ha)	Number of biosphere reserves
Less than 1000	21
1000–4999	34
5000–9999	18
10 000–49 999	31
50 000–99 999	16
100 000–499 999	37
500 000–999 999	10
1 000 000–9 999 999	10
10 000 000–100 000 000	1

TABLE V.   Numbers of biosphere reserves by minimum altitude

Minimum altitude in metres	Number of biosphere reserves
Less than 100	59
100–499	62
500–999	22
1000–1499	18
1500–1999	8
2000–2499	4
2500–2999	3
3000 or over	1

TABLE VI.   Numbers of biosphere reserves by maximum altitude

Maximum altitude in metres	Number of biosphere reserves
Less than 500	62
500–999	22
1000–1499	14
1500–1999	25
2000–2499	17
2500–2999	11
3000–3499	9
3500–3999	3
4000–4499	6
4500–4999	0
5000–5499	2
5500–5999	3
6000–6499	2
6500 and over	1

TABLE VII.   Numbers of biosphere reserves by annual precipitation

Annual precipitation in mm	Number of biosphere reserves
Less than 250	9
250–499	11
500–749	15
750–999	8
1000–1249	8
1250–1499	4
1500–1749	2
1750–1999	5
2000–2249	1
2250–2499	3
2500–2749	5
2750 and over	1

TABLE VIII.   Numbers of biosphere reserves by mean temperature

Mean annual temperature in °C	Number of biosphere reserves
Less than −15	1
−15 to −11	0
−10 to −6	0
−5 to −1	0
0 to 4	3
5 to 9	12
10 to 14	10
15 to 19	13
20 to 24	8
25 to 29	9
30 and over	1

of 228 m and upper and lower quartiles of 800 m and 0 m. Sixteen of the reserves have minimum altitudes greater than 1450 m. Maximum altitudes range from 0–6768 m, and 58 of the reserves have maximum altitudes less than 500 m. The median maximum altitude is 1292 m, with upper and lower quartiles of 2400 m and 300 m. Ten of the reserves have maximum altitudes greater than 4300 m.

Tables VI and VII summarize the frequency distributions for annual precipitation and for mean annual temperature, respectively. Mean annual precipitation ranges from 65–2750 mm, with a median of 778 mm and upper and lower quartiles of 1300 mm and 450 mm respectively. Nine of the reserves have mean annual precipitations greater than 2250 mm. Similarly, mean annual temperature ranges from $-13\cdot5$–$32°C$, with a median of $15°C$ and upper and lower quartiles of $21°C$ and $7\cdot9°C$.

Finally, the numbers of biosphere reserves with various kinds of legal status, with research, monitoring and educational or training programmes, with various forms of land use and human impact, and with various types of ownership are given in Tables IX–XII. Nearly 40% of the biosphere reserves are National Parks, while 51% of the reserves are strict nature reserves. Experimental reserves and other legally protected areas comprise about 11 and 14% of the total number, respectively.

Over three-quarters of the declared biosphere reserves have a research programme and nearly 20% have some form of monitoring programme. Only 8% have an environmental education and training programme, and about 13% of the reserves have an existing management plan.

Of the 177 biosphere reserves (Table XIII), 29% contain some kinds of permanent human settlement, and tourism currently has an impact on 43%. Forestry and agriculture are each practised in nearly 15% of reserves, but areas of grazing exist in 25%. In contrast, only about 4% of reserves have some form of engineering, for example mining or road works, within their boundaries.

Finally, 85% of biospheres are owned by the nation, country or state. Individual states or counties own a further 11%, and local communities own about 10%. About 25% are believed to involve some kind of private ownership in at least part of the reserve.

V. The Organization of Biosphere Reserves

A. THE ZONING AND GROUPING OF RESERVES

In Section III, dealing with the selection of biosphere reserves, we

TABLE IX. Numbers of biosphere reserves with legal status

Legal status	Number of biosphere reserves
National Parks	69
Strict Nature Reserves	92
Experimental Reserves	19
Historical/Archaeological sites	4
Other legally protected areas	23

TABLE X. Numbers of biosphere reserves with research, monitoring and educational /training programme

Activity	Number of biosphere reserves
Research programme	138
Monitoring	36
Environmental education and training	15
Existing management plan	25

TABLE XI. Numbers of biosphere reserves by types of land use and human impact

Land use and human impact	Number of biosphere reserves
Permanent human settlement	51
Forestry currently being practised	24
Agriculture currently practised	28
Areas of grazing	45
Tourism currently having an impact	75
Some form of engineering, e.g. mining, road works,	7

TABLE XIII. Numbers of biosphere reserves by types of ownership

Type of ownership	Number of biosphere reserves
National/country/state	151
Individual states/counties	19
Local communities, etc	17
Privately owned	44

touched briefly on the problem of incorporating, within the series of biosphere reserves, the two main functions for which the reserves are established, namely the conservation of representative ecosystems, both natural and man-altered, and the facilities for carrying out long-term research and monitoring on ecosystems. It is the emphasis on the use of the biosphere reserves for improving man's understanding of the biosphere and, deriving from this understanding, an increasingly wise use of the biosphere, which distinguishes biosphere reserves from reserves which have a more strictly preservation, or even recreation, function. This is not to say that biosphere reserves will exclude the need for other types of reserves—the latter are essential if the full range of biotic diversity is to be conserved, and if the biosphere reserves are not to become too isolated.

According to a recent report on "The Biosphere Reserve and its relation to other protected areas" (IUCN, 1979), there is a current consensus that a biosphere reserve should ideally contain the following four elements:

(a) *A core area* of "the main ecosystem or ecosystems to be protected large enough to be sustaining, which should, if possible, represent the ecosystem in a climax state, including natural seral stages leading to the climax. If the climax community no longer exists, it should contain the sub-climax communities in as natural a state as possible. In this case, part of these sub-climax communities should be set aside to proceed by natural succession towards the climax".

(b) *A buffer area* surrounding the core area to "screen it from surrounding forms of land use", the buffer area comprising the same ecological systems as the core area. The buffer area can also be used as an area for research involving active intervention from which comparisons can be made with the undisturbed core area. In some circumstances, carefully controlled traditional activities including timber production, hunting, fishing and grazing are permissible.

(c) *A reclamation or restorative area*—it is suggested that, "in some instances, areas may be added which contain degraded ecosystems. These can serve two purposes. They can be used for experiments in rehabilitation; and small portions can be retained in their degraded state as demonstration areas to show the damage that can be done by unwise land use and to illustrate the extent of recovery".

(d) *A stable cultural area*—"managed to protect and study ongoing cultures and land use practices which are in harmony with the environment. Local residents and their activities are to continue, but new techniques may be strictly controlled".

It has been suggested (IUCN, 1979) that, ideally, all the components

TABLE XIII. Alphabetical listing of biosphere reserves (at November 1979)

Aggtelek Biosphere Reserve	Hungary
Aleutian Islands Wildlife Refuge	United States of America
Alibotouch Reserve	Bulgaria
Arasbaran Wildlife Refuge and Protected Area	Iran
Arjan National Park-International Reserve	Iran
Ayers Rock-Mount Olga National Park	Australia
Babia Gora National Park	Poland
Baevi Doupki Reserve	Bulgaria
Bamingui-Bangoran Conservation Area	Central African Republic
Banados del Este	Uruguay
Basse-Lobaye Forest	Central African Republic
Beaver Creek Experimental Watershed	United States of America
Beinn Eighe National Nature Reserve	United Kingdom
Berezina Reserve	Byelorussian SSR
Bialowieza National Park	Poland
Big Bend National Park	United States of America
Bistrichko Branichte Reserve	Bulgaria
Boatine Reserve	Bulgaria
Braunton Burrows National Nature Reserve	United Kingdom
Caerlaverock National Nature Reserve	United Kingdom
Cairnsmore of Fleet National Nature Reserve	United Kingdom
Camargue National Reserve	France
Cascade Head Experimental Forest	United States of America
Caucasian Reserve	USSR
Central Plains Experimental Range	United States of America
Central-Chernozem Reserve	USSR
Changbai Nature Reserve	People's Republic of China
Channel Islands National Monument	United States of America
Cibodas Reserve	Indonesia
Cinturon Andino	Columbia
Circeo Foret Domaniale	Italy
Claish Moss National Nature Reserve	United Kingdom
Collemeluccio-Montedimezzo	Italy
Coram Experimental Forest	United States of America
Coweeta Hydrologic Laboratory	United States of America
Croajingolong National Park	Australia
Danggali Conservation Park	Australia
Desert Experimental Range	United States of America
Dinder National Park	Sudan
Dinghu Native Reserve	People's Republic of China
Djebel Bou-Hedma National Park	Tunisia
Djebel Chambi National Park	Tunisia
Djendema Reserve	Bulgaria
Doupkata Rserve	Bulgaria
Dyfi National Nature Reserve	United Kingdom
Everglades National Park	United States of America
Fango Foret Dominiale	France
Fitzgerald River National Park	Australia

TABLE XIII.—*cont.*

Fraser Experimental Forest	United States of America
Fray Jorge National Park	Chile
Geno National Park	Iran
Glacier National Park	United States of America
Gossenkollesee	Austria
Grazalema Reserve	Spain
Great Smoky Mountain National Park	United States of America
Gurgler Kamm	Austria
H. J. Andrews Experimental Forest	United States of America
Hara National Park	Iran
Hauy Tak Teak Reserve	Thailand
Hortobagy National Park	Hungary
Huascaran Reserve	Peru
Hubbard Brook Experimental Forest	United States of America
Hurulu Forest Reserve	Sri Lanka
Ichkeul National Park	Tunisia
Isle of Rhum National Nature Reserve	United Kingdom
Jornada Experimental Range	United States of America
Juan Fernandez National Park	Chile
Kamtchia Reserve	Bulgaria
Kavir National Park	Iran
Kiskunsag Biosphere Reserve	Hungary
Komodo Island Game Reserve	Indonesia
Konza Prairie Research Natural Area	United States of America
Kosciusko National Park	Australia
Koupena Reserve	Bulgaria
Krivoklatsko Reserve	Czechoslovakia
La Luki Forest Reserve	Zaire
Laguna San Rafael National Park	Chile
Lake Ferto Biosphere Reserve	Hungary
Lake Rezaiyeh National Park	Iran
Lal Sohara National Park	Pakistan
Les Iles Zembra et Zembretta National Park	Tunisia
Lobau Reserve	Austria
Loch Druidibeg National Nature Reserve	United Kingdom
Lore Kalamanta Game Reserve	Indonesia
Luknajno Lake Reserve	Poland
Luquillo Experimental Forest	United States of America
Macchabee/Bel Ombre Nature Reserve	Mauritius
Macquarie Island Nature Reserve	Australia
Mae Sa-Kog Reserve	Thailand
Malindi-Watamu Marine Biosphere Reserve	Kenya
Mantaritza Reserve	Bulgaria
Manu Reserve	Peru
Mapimi Reserve	Mexico
Maritchini Ezera Reserve	Bulgaria
Miankaleh Wildlife Refuge	Iran
Michigan University Biological Station	United States of America
Michilia Reserve	Mexico
Miramare Marine Park	Italy

TABLE XIII.—*cont.*

Mohammad Reza Shah National Park	Iran
Mont St. Hilaire	Canada
Montes Azules	Mexico
Montseny Natural Park	Spain
Moor House-Upper Teesdale Biosphere Reserve	United Kingdom
Mount Kenya	Kenya
Mount Kulal	Kenya
Mount McKinley National Park	United States of America
Murray Valley	Australia
Neusiedler See-Osterreichischer Teil	Austria
Niwot Ridge Biosphere Reserve	United States of America
Noatak National Arctic Range	United States of America
Noroeste Reserve	Peru
North Norfolk Coast Biosphere Reserve	United Kingdom
Northeast Greenland National Park	Denmark
Northeast Svalbard Nature Reserve	Norway
Odzala National Park	People's Republic of China
Olympic National Park	United States of America
Omo Reserve	Nigeria
Ordesa-Vinamala Reserve	Spain
Organ Pipe Cactus National Monument	United States of America
Ouzounbodjak Reserve	Bulgaria
Parangalitza Reserve	Bulgaria
Pietrosu Mare Reserve	Romania
Pilon-Lajas National Park	Bolivia
Prince Regent River Nature Reserve	Australia
Prioksko-Terrasni Reserve	USSR
Puerto Galera Reserve	Philippines
Radom National Park	Sudan
Repetek Reserve	USSR
Retezat National Park	Romania
Rio Platano	Honduras
Rocky Mountain National Park	United States of America
Rosca-Letea Reserve	Rumania
Ruwenzori National Park	Uganda
Sakaerat Environmental Research Station	Thailand
Samba Dia Foret Classee	Senegal
San Dimas Experimental Forest	United States of America
San Joaquin Experimental Range	United States of America
Sary-Chelek Reserve	USSR
Sequoia-Kings Canyon National Parks	United States of America
Sierra Nevada de Santa Marta	Columbia
Sikhote Alin Reserve	USSR
Silver Flowe-Merrick Kells Biosphere Reserve	United Kingdom
Sinharaja Forest Reserve	Sri Lanka
Slovak Karst Reserve	Czechoslovakia
Slowinski National Park	Poland
Southwest National Park	Australia
Srebarna Reserve	Bulgaria
St Kilda National Nature Reserve	United Kingdom

TABLE XIII.—*cont.*

Stanislaus-Tuolumne Experimental Forest	United States
Steckby-Loedderitz Forest Nature Reserve	German Democratic Republic
Steneto National Park	Bulgaria
Swiss National Park	Switzerland
Tai National Park	Ivory Coast
Taoarp Atoll	France
Tanjung Puting Nature Reserve	Indonesia
Tara River Basin	Yugoslavia
Taynish National Nature Reserve	United Kingdom
Tcherventa Stena Reserve	Bulgaria
Tchouprene Reserve	Bulgaria
Three Sisters Wilderness	United States of America
Torres del Paine National Park	Chile
Touran Wildlife Refuge and Protected Area	Iran
Trebon Basin Reserve	Czechoslovakia
Tsaritchina Reserve	Bulgaria
Tuparro Territoria Faunistico	Columbia
Ulla Ulla Biological Reserve	Bolivia
Unnamed Conservation Park of S. Australia	Australia
Valley Vessertal Nature Reserve	German Democratic Republic
Velebit Mountain	Yugoslavia
Virgin Islands National Park	United States of America
Virginia Coast Reserve	United States of America
Waterton Lakes National Park	Canada
Waza National Park	Cameroun
Wolung Nature Reserve	People's Republic of China
Yangambi Floristic Reserve	Zaire
Yathong Nature Reserve	Australia
Yellowstone National Park	United States of America

making up the reserve should be contiguous, and that, where it is not possible to contain all the four types of area, the reserve should contain at least a core area protected by a buffer zone. The zonation and grouping of biosphere reserves to accommodate different management approaches has received a good deal of attention because, as there are so few parts of the world where the ideal situation occurs, there is generally need to modify the ideal prescriptions listed above. Quite often, it is not possible to obtain sufficiently large areas to encompass both the core zone and a zone where manipulative research can take place, or there are other reasons for selecting distinct areas containing the same biome within the one biogeographical province for these two purposes. One approach to these problems is the adoption of the biosphere reserve cluster principle developed first in the USA, but likely to be quite broadly applied elsewhere. As mentioned earlier, the biosphere reserve cluster approach arose because of a disinclination to accommodate manipulative research areas in the largely natural ecosystems of the core and buffer zones and the wish to capitalize on the long history of research on some of the generally smaller experimental reserves which are often situated nearby and contain the same biomes. Among the most noteworthy examples of this approach in the biosphere reserves so far designated are the Southern Appalachian biosphere reserve cluster, and the paired reserves in the Pacific North West of the USA.

The southern part of the Eastern Forest biogeographical province of North America is represented by a biosphere reserve cluster containing, as the core area, the Great Smoky Mountain National Park, and, associated with it, the Coweeta Experimental Reserve of the US Forest Service and the Oak Ridge Reservation of the Energy Research and Development Administration (ERDA) (Johnson *et al.*, 1977). The Great Smoky Mountain National Park covers 213057 ha and consists primarily of eastern deciduous hardwoods with large stands of hemlock and pine. Up to the 1930s, about 60% of this area was heavily logged or grazed, but the remainder was largely undisturbed. The whole area is now strictly protected so far as is possible against adverse human influence and the areas damaged in the forest in the past are being allowed to revert to a more natural condition. The Coweeta Hydrological Laboratory, approximately 40 km to the south of the Great Smoky Mountain National Park, covers 2185 ha of mixed deciduous hardwood, and has a 35-year history of intensive manipulative and observational research, largely concerned with the hydrology of the several experimental catchments within it, but also with studies of nutrient element circulation. It was a research site for the International Biological Programme (IBP). The Oak Ridge Reservation, occupying about 16000 ha, and lying about 40 km north-west of the Smoky Mountain National Park, contains the Environmental Sciences Laboratory of ERDA and has been the site of much research into the deciduous forest biome, particularly during the course of the International Biological Programme. It is an area which was for a long time maintained as farmland with scattered forest but, since the Reservation was established about 40 years ago, has reverted to the early stages of forest succession. Thus, taken together, these three areas provide valuable facilities for studying many aspects of both climax and successional stages of the Eastern Forest, and, at Coweeta, the consequences of various types of modification of the forest cover on its hydrology and biology.

The second example of implementation of the "cluster" concept in the USA is the association, in the Sierra-Cascade biogeographical province, of the Three Sisters Wilderness core area of 80900 ha with the nearby H. J. Andrews Experimental Forest of 6050 ha. The Three

Sisters Wilderness is an undisturbed natural area of *Tsuga heter*\
phylla, *T. mertensiana* and *Abies amabilis* forest, and alpine vegetation ои
the High Cascades. The H. J. Andrews Experimental Forest nearby is
an area dominated by *Tsuga heterophylla* and *Abies amabilis*, which,
though unlogged until 1950, is now subject to controlled clear cut
logging of areas from 4–16 ha in size. The effects of logging on watershed characteristics have been studied as part of the IBP research programme which was carried out on the site.

A rather different approach to the organization of biosphere reserves
in the USSR has been described by Chichikin (unpublished data) in
relation to the Berezina Reserve in Byelorussia. This reserve has a total
area of 76 201 ha and has been divided into three zones. The strictly protected core zone of 34 000 ha occupies the south-eastern part of the
territory. Of the remainder, 34 700 ha is an area available for scientifically oriented management directed towards the conservation of
historically formed ecosystems and preventing damaging effects of increased human influence. The area includes controlled forestry
operations and measures controlling wild animal populations. Within
the reserve, there is a small zone amounting to about 2 600 ha in which
agriculture is still carried out in a traditional manner, without radical
improvement or the use of new techniques. Maintenance of fertility is
through organic fertilizers, and the use of chemical herbicides and
pesticides is not allowed. Even within this area, only 25% consists of
arable land, hay meadows and pastures, most of the remainder being
forest. A fourth zone of 4700 ha is allocated to residential use and
contains scientific and educational facilities, including excursion routes.
Part of this zone corresponds to a rehabilitation area where landscape
improvement is being carried out.

The reserve cluster concept developed in the USA and the zoning
concepts promoted by IUCN (1979) exemplify various configurations
developed to facilitate management and research needs of biosphere
reserves. Some initial consideration has also been given to the very
difficult problem relating to size and configuration of reserve areas
required to conserve the range of biota contained in the reserve. This
is, of course, a critical issue with fundamental implications for the long-
term viability of all reserve establishment programmes, not only biosphere reserves. Di Castri and Loope (1977) draw attention to the
relevance of MacArthur and Wilson (1967) island biogeographic theory
to the exploration of the problem and express the hope that some of
the research on biosphere reserves might be directed to obtaining a
better understanding of how much land is required to conserve a
measurable proportion of the world genetic diversity even though it is

not intended that the biosphere reserve series should be so comprehensive as to fulfil the conservation function adequately by itself. The relevance of island biogeographic studies to the conservation of genetic diversity by reserve establishment has been reviewed by Diamond (1975) whose general conclusions on the size, shape and distribution of protected areas have been incorporated in the World Conservation Strategy (IUCN, 1980).

B. THE RELATION TO OTHER PROTECTED AREAS

We have already mentioned that the system of biosphere reserves is not intended as a substitute for other reserve establishment initiatives, but should supplement them and be supplemented by them in a mutually supporting way. In practice, it can be seen that the biosphere reserves so far established are already linked to a variety of other national reserve series such as national parks, nature reserves, wilderness areas and experimental forests. It is difficult to use these categories precisely because there is no internationally applied terminology for such areas. We have seen, however, that the biosphere reserve areas in the USA include all the four categories mentioned above and all the UK biosphere reserves are either National Nature Reserves or Sites of Special Scientific Interest under special protection.

Selected biosphere reserves will also feature as World Heritage Sites although so far only three, Yellowstone National Park, Everglades National Park and Bialowieza National Park, have received this designation. While biosphere reserves focus on natural ecosystem conservation and research, and are linked to other MAB project areas and internationally coordinated by the intergovernmental structure of MAB, World Heritage Sites are established under the World Heritage Convention with a different set of criteria in which emphasis is on their universal and outstanding value rather than representativeness and potential for research. As well as the three sites mentioned above, some of the others in the World Heritage List qualify as biosphere reserves although they are not yet included formally in the network. The inclusion of biosphere reserves fulfilling World Heritage criteria in the World Heritage List should be vigorously pursued because technical assistance may be available under the World Heritage Fund and protection may be enhanced under the World Heritage Convention.

C. THE COUNCIL OF EUROPE BIOGENETIC RESERVES

A major independent regional initiative, which is nevertheless closely related to the aims of the MAB biosphere reserve programme, is the Council of Europe's proposal for the establishment of a network of biogenetic reserves. These reserves have the aim of conserving "representative examples of European flora, fauna and natural areas". The establishment of the European network of biogenetic reservoirs will be according to a framework laid down by a committee of the Council which lists types of habitat, biocenoses and ecosystems to be given priority, taking into account national and international inventories, studies in the field of nature conservation carried out by the Council of Europe and the relationship with UNESCO biosphere reserves and other international designations. Initially, priority is being given to the establishment of reserves in heathlands, maquis and wetlands because of their decline in the face of development pressures and also, reflecting the phytosociological basis of much European thinking on this matter, the "loci typici" on the vegetation map being prepared under Council of Europe auspices. While no biogenetic reserves have as yet been established, 49 areas have already been proposed by Italy and 19 by France for inclusion in the series. A special symposium on wetlands convened in 1978 identified 67 wetlands for inclusion in the biogenetic reserve network and 11 for establishment as biosphere reserves.

VI. RESEARCH AND MONITORING ON BIOSPHERE RESERVES

It has been recognized from the first that one of the most important functions of biosphere reserves is to provide opportunities for ecological research, particularly in relation to studies of man's interaction with the environment. Biosphere reserves are regarded as having particular value as benchmarks or standards for measuring long-term changes in the biosphere as a whole, and are consequently seen as being particularly good sites for establishing monitoring programmes.

A. THE ZONING AND GROUPING OF RESERVES

Franklin (1977) lists several kinds of research and monitoring activities for which the biosphere reserves may be used:

(a) Long-term baseline studies of environmental and biological features.

(b) Research to help develop management policies for the reserves.

(c) Experimental or manipulative research (outside strictly preserved areas), particularly on the ecological effects of human activities.

(d) Environmental monitoring.

(e) Study sites for various MAB research projects.

Johnson and Bratton (1978) have distinguished between the research and monitoring which is done within the reserves (core zone) to examine the changes which take place in natural systems over a long period of time and the research done in nearby areas (which may be the manipulative zone of the reserve) where man-made modifications to the environment, including various types of land use can be compared with the conditions within the reserve core zone.

Obviously, the five categories of Franklin and the two main research functions distinguished by Johnson and Bratton are not completely independent of each other, and other kinds of research could be added related to the specialized situations on particular reserves. Some biosphere reserves, such as the Repetek Reserve in the USSR (Sokolov and Gounine, 1979) and the Coweeta Experimental Forest in the USA already mentioned, have a long research history resulting from being the site of a field station for many years. In some reserves, such as the H. J. Andrews reserve in the USA and Moor House in the UK (Heal and Perkins, 1978), much valuable baseline information has been accumulated as a consequence of the sites being used for International Biological Programme projects prior to their establishment as biosphere reserves.

In these "research-rich" reserves, there is already a good deal of baseline information on environmental and biological features from which to develop management policies for the reserve and programmes of monitoring directed at identifying and quantifying long-term changes in the ecological systems. However, in many biosphere reserves, such information is scanty and the improvement of it is a priority need.

B. STANDARDS FOR DIFFERENT LEVELS OF RESEARCH AND MONITORING

It has been recognized that the scientific information available on the biosphere reserves within the international series is variable and that the resources available in different countries to acquire such information are equally disparate. For this reason, it was proposed, at an international conference on the problem of monitoring of biosphere reserves held in the USA in 1978 (US National Committee for MAB, 1979), that a stepped programme should be adopted so that all countries could participate at a level commensurate with their resources.

The report of the conference divided the scientific work relating to the needs of biosphere reserves into four main categories—chemical, biological, geophysical and anthropological. Within each of these categories, three broad levels of work were recognized. At level 1, it was proposed that a common base of parameters should be measured in all biosphere reserves which relate broadly to global, regional and local trends in the environment. At this level, basic information is needed on features such as climatic conditions, major plant and animal communities, land use, physical features and land tenure. Emphasis was placed on simplicity and reliability of information so that studies made at different sites and times could be compared with confidence. In the biological category, studies at this level amounted to little more than an inventory, a simple but necessary precursor for proceeding to the second level. Level 2 investigation continues to build up the information base, but also begins to utilize the data establishment in level 1 to assess key processes and identify additional parameters. For example, an important part of long-term monitoring is the determination of rates and direction of change in biological processes compared with other areas. This determination usually requires more complex measurements than are obtained at level 1 and, in some cases, may be best served by measurements of biological productivity so that effective intercomparisons can be made. Level 3 investigations, being the most sophisticated and technologically intensive, will perhaps be implemented on only a few specially-selected biosphere reserves in the first instance. Projects are primarily based on hypotheses developed through levels 1 and 2 and, according to the USA MAB study, "are intended to provide intensive sophisticated study of key aspects of the system". The objectives of the three levels of monitoring within the geophysical, chemical, biological and anthropological categories, as described in the USA study on long-term monitoring on biosphere reserves, are summarized in Fig. 2.

Johnson and Bratton (1978) propose a procedural sequence of prediction—monitoring—assessment for the organization of biological monitoring work in biosphere reserves. The prediction phase of this procedure identifies specific hypotheses of change, by bringing together existing information and integrating biological and environmental monitoring which has already taken place, so as to enable the key issues to be selected from among the numerous possibilities for monitoring. The second phase of monitoring is then directed to testing the predictive hypothesis formulated in the first stage. The monitoring then "becomes purposeful and only a means to an end". The third phase, or assessment, is defined as the process of interpretation of the data obtained, leading to the testing of the accuracy of the predictions and enabling the generation of better hypotheses and improving prediction, thus allowing the succeeding monitoring stage to have greater accuracy and precision. Johnson and Bratton illustrate their paper with four examples from the Great Smoky Mountain National

	Level 1	Level 2	Level 3
GEOPHYSICAL	Development of basic information necessary to characterize Biosphere Reserve sites and, more important, establish reference materials important to research and monitoring projects in other categories.	Provision of additional data needed to assess the dominant processes which characterize a site or area. These data are intended to reveal gross trends and provide a more detailed characterization of the site for comparative analysis with other sites against which to measure perceived anthropogenic change for special correlative studies.	Establishment of a detailed data base which separates random physical changes from predictable occurrences, analyses of local, regional or global change. To isolate physical aspects of the site which may be considered for their value as indicators.
CHEMICAL	pH and conductivity measurements in rainfall and surface waters. This minimum level monitoring programme is recommended for remote areas and other locations where equipment and personnel are not readily available to carry out a wider range of chemical monitoring.	Measurement of selected cations and anions in atmospheric deposition, surface water, accumulation in animals, soil, litter and vegetation, and atmospheric gases and particulates. Information from this level surveys and establishes baselines in parameters with the intention of identifying those of possible global or regional significance.	Monitoring of trace metals and organics in all main compartments of the environment.

BIOLOGICAL	Monitoring biological characteristics of ecosystems with minimum use of equipment, minimal cost and limited trained personnel to provide survey information which facilitates subsequent expansion of monitoring effects to level 2 (mainly inventory surveys).	Collection of quantitative data necessary to develop long-term management plans, to study population dynamics and begin to integrate data contributing to the understanding of community structures and dynamics.	Integration of population models into community models; integration of biological data with those of geophysical, chemical and demographic monitoring effects. This will lead to the development of management systems and impact evaluations for the reserves.
ANTHROP-OLOGICAL	Documentation of direct human-related effects on natural ecosystems in Biosphere Reserves, by preparation of land use maps, information on human and livestock populations, etc.	Quantification of economic uses of Biosphere Reserves and aspects of human populations which can be anticipated to have a dominant effect on the site/ quantity economic production and yield of ecosystems and establish relations between human use and biological productivity.	Quantification of system inputs such as energy fertilizer and outputs such as emigration, yield/area for selected economic factors, etc, contributing to development of a management plan for the area.

FIG. 2. Objectives of long-term geophysical, chemical, biological and anthropological monitoring on Biosphere Reserves (adapted from US National Committee for MAB 1979)

Park relating to the monitoring of wild boar grazing effects, changes in the vegetation during forest succession, the effects of acid precipitation on stream biota, and the effects of possible climatic amelioration on transition zones of high altitude vegetation.

Proposals for the development of a pollutant monitoring system for biosphere reserves has been described in more detail by Wiersma *et al.* (1978), also in relation to a pilot project carried out in the Great Smoky Mountain National Park. In their paper, Wiersma *et al.* described the general criteria to be used in the selection of specific sampling sites within each biosphere reserve, taking account of such factors as topography, soils, vegetation types, sampling site size and sample point selection. They discuss the determination of the number of samples required to allow for the required level of precision and the variability of the particular parameter on the individual sampling sites and describe a pre-sampling programme carried out in the Great Smoky Mountain National Park to determine the minimum level of pollutants detectable, the variability of samples and the value of some new sampling techniques. The media sampled were soil, unincorporated litter, vegetation, water and air. Wiersma *et al.* concluded that limits for detection of pollutants in their pilot exercise were adequate for vegetation and soil, were marginally adequate for trace elements in the water samples, and could be made adequate for the air samples. The number of samples required to satisfy the desired aim of distinguishing a difference of plus or minus 10% at the 95% confidence level was not considered unreasonable, even though field sample variability was generally high. The pilot project revealed unexpectedly high lead levels in forest litter at the high altitude sites sampled, apparently derived from non-natural sources.

Several of the biosphere reserves in the USSR are sites with a long research history. Up to the present time, the main emphasis has been on the monitoring of vegetation change by means of a system of permanent observation sites. This monitoring has taken place at the Caucasus, Prioksko-Terrasny and Sikhote Alin Biosphere Reserves (Krinitsky, 1977) and also at the Repetek Biosphere Reserve (Sokolov and Gounine, 1979).

A draft "programme of scientific research in Biosphere Reserves", prepared by the USSR Academy of Sciences in 1977, identified six main categories of desirable scientific work:

(a) Inventory of natural and man-made ecosystems, involving, for example, ecosystem mapping using aerial photographs and detailed large-scale mapping of key plots of dominant and rare ecosystems.

(b) Study of the structural and functional organization of primary and derived ecosystems including studies on the abiotic environment, soils, biological productivity eco-physiological adaptation or organisms, food chains and effects of organisms on the environment leading to modelling of food chain and ecosystems.

(c) Study of the historical development of ecosystems to form a basis for measures for restoring ecosystems to a more natural condidion and for forecasting changes.

(d) Research into problems linked to identification of environmental indicators of pollution levels, etc, including studies on the ecological manifestation of various pollutant levels in microorganisms and in animal and plant tissues.

(e) Monitoring the condition of the biosphere, including the following main categories of parameters: firstly geographical and geophysical, secondly, chemical, including levels of pollutants in air, atmospheric deposition, water and soil, and thirdly, biological, including populations of certain species of animals and plants characteristic of ecosystems, ratio of primary to secondary production, ratio of the animal biological production to the general biomass, etc.

(f) Scientifically derived management methods for ecosystems, including determination of degree of interference necessary to maintain natural systems.

These proposals are comprehensive, and, even in countries such as the USSR where substantial scientific manpower resources are available, it seems certain that they will take many years to implement. However, examination of the contents of the above categories indicates that they can be fitted within the hierarchical long-term monitoring framework developed at the 1978 meeting in the USA. Thus, the first category of the USSR list fits within level 1 of the hierarchy, where inventory work predominates, whereas the subsequent categories contain both level 2 and the more highly sophisticated level 3 investigations.

Turning from a proposed research framework in the USSR to a compendium of current research being carried out in the four Polish biosphere reserves so far established (Bialowieza National Park, Babia Gora National Park, Slowinski National Park, Luknajno Lake Nature Reserve), one finds a pattern of research which is probably characteristic of a large number of biosphere reserves which have been selected from among long-established nature reserves. Thus, the 83 projects listed from these four biosphere reserves fall within the following broad categories:

| Botanical: | Phytosociological | 13 |
| | Ecosystem dynamics | 12 |

	Plant community management	5
	Grazing effects	2
	Plant species inventory	8
	Soils and methodology	3
Zoological:	Animal species inventory	6
	Bird ecology	12
	Other animal population studies	6
Pollution studies		3
Human impact studies		5
Other		8

A large proportion of these projects are level 1 inventory type studies, a necessary preliminary to further work, but it is hoped that the emphasis in these and other biosphere reserves will change towards a larger proportion of level 2 and 3 studies designed to investigate ecosystem processes more closely related to the needs of biosphere reserve management.

In the report presented by the MAB Secretariat to the Sixth Session of the International Co-ordinating Council, attention was drawn to the part being played by the biosphere reserves already established to MAB pilot projects relating to other international MAB themes. Inter-linkages with Project 1 (Tropical forests) are found in Basse Lobaye (Central African Republic), Omo Reserve (Nigeria), Tai Forest (Ivory Coast), Yangambi (Zaire), Mae Sa-Kog (Thailand), Sakaerat (Thailand), Puerto Galera (Philippines), and Rio de Platano (Honduras). Reserves linked to research on Project 3 (Grazing lands and arid zones) are: Mount Kulal (Kenya), La Michilia (Mexico), Mapimi (Mexico), Banados del Este (Uruguay), Waza National Park (Cameroun), Bou-Hedma (Tunisia), Beaver Creek (USA), and Repetek (USSR). In relation to Project 6 (Mountain systems), research is being carried out in the following biosphere reserves: Torres del Paine (Chile), Ulla Ulla (Bolivia), Swiss National Park, Cascade Head Experimental Forest (USA) and Caucasian Biosphere Reserve (USSR). Links with Project 5 (Fresh water ecosystems and coastal areas) are found in Neusiedlersee (Austria), Ichkeul (Tunisia), Lake Fertő (Hungary), Atoll de Taiaro (France), Malindi-Watamu (Kenya), Coweeta (USA), and Sary Chelek (USSR).

VII. Biosphere Reserve Information Systems

One of the purposes of biosphere reserves is to facilitate comparisons between ecological habitats and scientific studies on comparable reserves, and between reserves and similar areas outside the reserves

BIOSPHERE RESERVES UNITED KINGDOM

NAME:Beinn Eighe National Nature Reserve

APPROVAL BY MAB BUREAU: June, 1976

GEOGRAPHICAL LOCATION: N57035'; W05°26'; North Scotland, Ross and Cromarty. Falls within biogegraphical province 2.31.12 of Udvardy (1975).

ALTITUDE: 0–1053 m

AREA: 4800 ha

LEGAL PROTECTION: National Nature Reserve

LAND TENURE: 4200 ha are owned by the Nature Conservancy Council, the remaining 600 ha being managed under a Nature Reserve Agreement with the owners.

PHYSICAL FEATURES: It is part of the stable foreland of the Moine Thrust and is composed of a platform of Lewisian Gneiss, cut by igneous dykes, upon which are placed mountains of Torridonian Sandstone and Cambrian quartzite. These rocks have been faulted and subject to minor thrusts. Deep glacial erosion has carved the mountains into aretes and peaks with steep cliffs, corries and screes. Drainage is radial.

VEGETATION: The most notable feature of the vegetation is the remnants of native Scots pine (*Pinus sylvestris*) forest. Fragments of oak and birch wood also survive in places but the bulk of the lowland zone has been denuded of its former woodland and now presents a complex of wet heath and blanket bog communities. In the montane zone the presence of herb-rich dwarf shrub heath and *Calluna-Juniperus communis nana-Arctostaphylos* heath is noteworthy. There are small outcrops of dolomitic mudstone which support herb-rich grassland.

NOTEWORTHY FAUNA: A wide variety of animal habitats are present, ranging from sea and freshwater, to woodland, heath, bog and mountain slopes to mountain summits, screes and cliffs. The fauna is relatively sparse but representative of West Highland deer forest country.

ZONING: Managed National Nature Reserve.

MODIFICATION BY MAN: Clearance of the forest, often by burning, began centuries ago and has been carried out to assist grazing or to use the timber for iron smelting and boat building. The reserve area has, in the past, been grazed by sheep, cattle, goats and deer, though only the latter has been widespread and is the only type of grazing amenity permitted.

SCIENTIFIC RESEARCH POTENTIAL: The primary object of management on the reserve is to restore the woodland habitat to low ground and to maintain the montane habitats. Deer management over the unit as a whole is co-ordinated on agreed lines.

PRINCIPAL REFERENCE MATERIAL: The reserve is mentioned in scientific literature and in the Nature Conservancy Council trail and reserve publications.

STAFF: 2 wardens, 2 estate workers.

ADDRESS OF LOCAL ADMINISTRATION: The Regional Officer (West Scotland)
Nature Conservancy Council
Fraser Darling House
9 Culduthel Road
Inverness 1V2 4AG
UNITED KINGDOM

FIG. 3. Example of biosphere reserve proforma information contributing towards the Directory of Biosphere Reserves (UNESCO, 1979d) and the MAB Information:

which are subject to various forms of man-induced change or management. To enable biosphere reserves to be compared objectively a substantial amount of information, collected in a standard way, has to be obtained for them both internationally and nationally. The first comprehensive synopsis of a standard set of information relating to all the biosphere reserves so far established has recently been published by UNESCO (1979b), as part of their MAB Information System. This compilation presents, in a number of different ways, information derived from standard proformas submitted to UNESCO by the managing authorities of all the biosphere reserves (see Fig. 3, for example of standard proforma layout). Detailed information on MAB 8 projects relating to biosphere reserves, together with details of other MAB projects, is also listed in a recent UNESCO MAB Information System compilation (1979c). A "Directory of Biosphere Reserves", published by UNESCO in 1979, is essentially a compilation of the information proformas illustrated in Fig. 3.

In the USA, a coordinated attempt to acquire and organize an information base related to the USA Biosphere Reserves, the Information Synthesis Project, has been undertaken by the Oklahoma Biological Survey (Risser and Cornelison, 1979). This information base includes much more detail than it has proved possible to incorporate so far within the MAB Information System on Biosphere Reserves mentioned above. The Information Synthesis Project includes "1. a description and characterisation of the US biosphere reserves, 2. an evaluation and analysis of these sites as a total system in the context of the environmental characteristics of the USA, 3. a summary of extant research and monitoring projects at each reserve, 4. a bibliography of the information about these sites individually and collectively". Each site has been characterized according to a series of parameters selected to describe the environmental and biological attributes. Future development of the system will enable comparison between ecosystem level parameters and responses of the ecosystems to changes and manipulations caused by man.

VIII. THE PROTECTION AND MANAGEMENT OF BIOSPHERE RESERVES

If biosphere reserves are to fulfill their generally very long term purposes, they must be adequately protected against detrimental influences which diminish their value, whether these influences arise within the reserve itself, peripherally, or from some more distant source. The selection of biosphere reserves is being made, for the most, from among areas which satisfy certain minimum levels of safeguard. Thus, of the 177 biosphere reserves established by the end of 1979, 69 are protected as national parks and 19 as experimental reserves. The legal status of the remaining areas is mostly that of strict nature reserve. However, many of these areas are subject to actual or potential pressure for other uses, often increased recreational use (tourism has a major impact on 76 out of the 177 established reserves) and few are large enough to be immune from the effects in peripheral land use which may, for example, greatly influence the population of grazing animals for which the reserve is only part of their range. Another issue deserving urgent attention is the need to prevent, or at least minimize, long range contamination of these reserves by atmospheric pollutants, particularly in heavily industrialized countries. Because of the great variety of biosphere reserves, it is difficult to make generalizations which hold for all of them but, in general, the larger the reserve, the less vulnerable it is to peripheral influences because adequate buffer areas can be incorporated. For the smaller biosphere reserve core areas, where inadequate buffer zone land is available within the reserve itself, consideration should perhaps be given to stricter regulation of adjacent land use so as to reduce, for example, contamination from pesticide or fertilizer use in neighbouring agricultural or forest lands or the hazards of fire in dwarf shrub or steppe communities.

Many biosphere reserves incorporate other land uses, such as grazing by domestic stock or forestry operations, in their manipulative or buffer zones, if not in their core zones. However, the pressure towards introducing multipurpose use where it does not contribute to the value of the reserve should be resisted, as there are often areas available outside biosphere reserves where such uses can be accommodated more appropriately.

Legal aspects of biosphere reserve protection have been discussed by de Klemm (summarized in UNESCO, 1979a) who points out that there is no obligation for states to establish biosphere reserves and that biosphere reserves have no legal status as such. Furthermore, designation of a biosphere reserve by a government and subsequent approval by the MAB International Co-ordinating Council and UNESCO does not create, on the part of the State concerned, a legal obligation to protect the biosphere reserve. To create such a legal obligation would require a treaty. However, the fact that states have designated biosphere reserves and applied for their approval by the MAB Council implies that the states concerned have accepted an obligation to conserve the reserves. In honouring this obligation, states should take account of the need to provide adequate long term legal protection for the reserves. This will involve consideration of the general framework of conservation legislation, possibly the enactment of byelaws for particular reserves, and protection that is adequate both in law and in fact.

One of the functions of the international network of biosphere reserves is to promote international cooperation and understanding on issues of biosphere management. It is, therefore, to be hoped that the growing international understanding of the criteria for biosphere reserve selection and establishment will be followed by greater understanding of the management needs of biosphere reserve areas.

Very few biosphere reserves have such stable self-contained natural ecosystems that they can be left entirely alone so that only natural ecological forces control development. Even if the vegetation of the area is apparently stable it is often subject to influence by grazing animals whose range extends far beyond the reserve boundary and whose populations and behaviour, and hence influence on the reserve vegetation, may be determined by man-induced changes at a considerable distance from the reserve. In the early stages of the development it is perhaps neither possible nor desirable to aim for uniformity of approach to the very diverse problems of reserve management but in the longer term, a good case can be made for working towards more explicit and precise statements of management objectives and methods in management plans. These management plans should be published so as to promote both comment and understanding. Management should clearly be closely related to the long term conservation and research needs of the reserve and should itself be subject to a monitoring process so that the success, or failure, of management aims and methods adopted can be properly assessed. Lack of adequate monitoring of management is a common failing in many existing nature reserves so that much useful information, which could lead to a better management in future, is being lost.

IX. SUMMARY AND CONCLUSIONS

In this paper, we have described how Project 8 of the Man and Biosphere Programme, promoted by UNESCO, for the conservation of natural areas and the genetic material they contain, became largely directed to the establishment of an international series of biosphere reserves designed to conserve examples of the living ecosystems characteristic of all the main biogeographical provinces of the world. While its local application is still subject to argument, the broad framework for classification of the main biogeographical provinces of the world is generally accepted and has been used as a basis not only for selection of biological reserves but also for collection and presentation of information about the reserves of which 177 had been established by November 1979. This framework has enabled progress in the establishment of biosphere reserves to be analysed in relation to the representation of the biogeographical provinces of the world and major gaps to be identified.

In relation to the organization of reserves we have described the various solutions which have been implemented or proposed to enable biosphere reserves to accommodate their several functions and for an adequate level of protection to be achieved. The zoning of reserves, together with the association of strictly preserved and experimental reserves as in the reserve cluster concept developed in the USA, helps to ensure that the conservation and research function of biosphere reserves are compatible. The relationships of biosphere reserves to certain other international reserve proposals, notably to the Council of Europe biogenetic reserves, are described.

Ecosystem research and long term monitoring of environmental change are of vital importance for biosphere reserves if they are to achieve their aims. While, in the case of some long established reserves incorporated into the biosphere reserve network, there is a substantial baseline of environmental information, particularly where they were study sites for the International Biological Programme, such information is inadequate for the majority of reserves and the implementation of baseline studies followed by a monitoring programme is a high priority. Proposals for a graded monitoring system relatable to the level of resources available for monitoring are described.

Some of the problems of biosphere reserve management are outlined. These relate both to the intrinsic difficulty in ensuring the long term internal stability of isolated ecological systems and also to the protection of these ecosystems against detrimental external influences such as pollution. Management procedures for biosphere reserves need to be made explicit, and, for this reason, the publication of management plans is recommended, to promote the wider dissemination of knowledge gained from dealing with the practical problems of reserve management.

The international programme for the establishment and safeguarding of biosphere reserves is still in its initial stages and much remains to be done, both by way of reserve establishment, particularly in those biota such as the tropical rain forests which are being rapidly destroyed by man's activities, and in promoting ecological research and long term monitoring to ensure that these important resources are managed in a sustainable way. The establishment of the biosphere reserve system within the Man and Biosphere Programme has already promoted valuable international cooperation and deserves strong support to reach a successful conclusion, for it seeks not only to promote the conservation of representative natural and semi-natural ecosystems of the world as recommended in the World Conservation Strategy, but to use these areas for research that will enable man to use the biosphere more wisely. Perhaps this linking of the conservation of ecosystems with ecological research is the most important element to stress in the biosphere reserve concept. It is not enough to declare certain areas as biosphere reserves. It is not enough to promote research on the dynamics of natural and semi-natural ecosystems. Both these activities need to proceed in harmony and the MAB Programme for biosphere reserves provides just such an opportunity.

Acknowledgements

The authors are grateful to Dr B. von Droste of the Division of Ecological Sciences, UNESCO, for helpful comments on a draft of this paper and to Mrs Penelope Ward for her work on the MAB Information Service.

References

Di Castri, F. and Loope, L. (1977). Biosphere reserves, theory and practice. *Nature and Resources* XIII, 1. pp 2–7. UNESCO, Paris.

Clements, F. E. and Shelford, V. E. (1939). " Bioecology." John Wiley, New York.

Dasmann, R. F. (1972). Towards a system for classifying natural regions of the world and their representation by national parks and reserves. *Biological Conservation* **4**, 247–255.

Dasmann, R. F. (1973). A system for defining and classifying natural regions for purposes of conservation. A progress report. IUCN Occasional Paper No. 7, Morges.

Dasmann, R. F. (1974). Biotic provinces of the world. IUCN Occasional Paper No. 9, Morges.

Diamond, J. M. (1975). The island dilemma: lessons of modern biographic studies for the design of natural reserves. *Biological Conservation* **7**, 129–146.

Franklin, J. F. (1976). The conceptual basis for selection of United States biosphere reserves and features of established areas. USA/USSR Symposium, March 1976.

Franklin, J. F. (1977). The Biosphere Reserve programme in the United States, *Science* **195**, 262–267.

Heal, O. W. and Perkins, D. F. (eds). (1978). "Production Ecology of British Moors and Montane Grasslands." Springer, Berlin.

International Union for the Conservation of Nature. (1979). The Biosphere Reserve and its Relationship to other Protected Areas. IUCN, Morges.

International Union for the Conservation of Nature. (1980). World Conservation Strategy. IUCN, Gland.

Johnson, W. C. and Bratton, S. P. (1978). Biological monitoring in UNESCO biosphere reserves with special reference to the Great Smoky Mountains National Park. *Biological Conservation* **13**, 105–115.

Johnson, W. C., Olson, J. S. and Reichle, D. E. (1977). Management of experimental reserves and their relation to conservation reserves: the reserve cluster. *Nature and Resources* **XIII** 1, pp 8–14. UNESCO, Paris.

Krinitsky, V. (1977). Research at USSR state reserves and their role in monitoring biosphere changes. *Nature and Resources* **XIII**1, 15–17. UNESCO, Paris.

MacArthur, R. H. and Wilson, E. O. (1967). "The Theory of Island Biogeography." Princeton University Press, Princeton, New Jersey.

Ratcliffe, D. A. (1977). "A Nature Conservation Review." Cambridge University Press, Cambridge.

Risser, P. G. and Cornelison, K. D. (1979). Man and the Biosphere: US Information Synthesis. Project MAB—8. Biosphere Reserves. Oklahoma Biological Survey, Norman, Oklahoma.

Sokolov, V. and Gounine, P. (1979). The Repetek Biosphere Reserve. *Nature and Resources* **XV** 1, 15–18. UNESCO, Paris.

Udvardy, M. D. F. (1975). A classification of the biogeographical provinces of the World. IUCN Occasional Paper No. 18, Morges.

United Nations Educational, Scientific and Cultural Organization (1973). Expert Panel on Project 8: conservation of natural areas and of the genetic material they contain. MAB Report Series No. 12. UNESCO, Paris.

United Nations Educational, Scientific and Cultural Organization (1974). Task force on criteria and guidelines for the choice and establishment of biosphere reserves. MAB Report Series No. 22. UNESCO, Paris.

United Nations Educational, Scientific and Cultural Organization (1979a). Workshop on biosphere reserves in the Mediterranean region. MAB Report Series No. 45. UNESCO, Paris.

United Nations Educational, Scientific and Cultural Organization (1979b). MAB Information System: Biosphere Reserves. Compilation 1, May 1979. UNESCO, Paris.

United Nations Educational, Scientific and Cultural Organization (1979c). MAB Information System: Compilation 4, July 1979. UNESCO, Paris.

United Nations Educational, Scientific and Cultural Organization (1979d). Directory of Biosphere Reserves. UNESCO, Paris.

United States National Committee for Man and the Biosphere (1979). Long-term Ecological Monitoring in Biosphere Reserves. Washington D.C.

Wiersma, G. B., Brown, K. W. and Crockett, A. (1978). Development of a pollutant monitoring system for biosphere reserves and results of the Great Smoky Mountains Pilot Study. Fourth Joint Conference on Sensing of Environmental Pollutants. American Chemical Society; pp 451–456.

Subject Index

319

Cumulative List of Authors

Numbers in **bold** face indicate the volume number of the series.

Bennett, E. G. A., **5**, 350
Bradshaw, A. D., **4**, 142
Brooker, M. P., **6**, 91
Burdon, J. J. **5**, 145
Burton, W. G., **3**, 86
Cannell, R. Q., **2**, 1
Carter, N., **5**, 272
Cartling, P. A., **6**, 154
Caughley, G., **1**, 183
Crisp, D. J., **6**, 154
Corbett, J. R., **3**, 230
Davies, D. E., **6**, 221
Davies, D. R., **2**, 87
Davies, J. H. C., **3**, 1
Densem, J. W., **5**, 221
Dixon, A. F. G., **5**, 272
Dunn, J. A., **3**, 43
Edwards, R. W., **5**, 221
Evans, A. M., **3**, 1
Field, C. R., **4**, 63
Finch, S., **5**, 67
Cambell, R., **1**, 247
Gill, C. J., **2**, 129
Goodier, R., **6**, 279
Jackson, W., **6**, 221

James, C., **4**, 201
Jefferies, J. J. R., **6**, 279
Johnson, M. S., **4**, 142
Johnson, R., **5**, 350
Leakey, R. R. B., **6**, 57
Lowe, H. J. B., **5**, 350
Matthews, J. D., **1**, 49
McLean, I. F. G., **5**, 272
Mead-Briggs, A. R., **2**, 184
Milner, N. J., **6**, 154
Mortimer, A. M., **1**, 1
Murton, R. K., **1**, 89
Pirie, N. W., **4**, 2
Roberts, H. A., **6**, 1
Sagar, G. R., **1**, 1
Scott, P. R., **5**, 350
Scullian, J., **6**, 154
Shattock, R. C., **5**, 145
Steele, J. H., **4**, 103
Teng, P. S., **4**, 201
Thresh, J. M., **5**, 2
Watt, N. J., **5**, 272
Westwood, N. J., **1**, 89
Wright, D. A., **3**, 331
Wolfe, M. S., **5**, 350

Cumulative List of Chapter Titles

Numbers in **bold** face indicate the volume number of the series.

Adaptive biology of vegetatively regenerating weeds, **6**, 57.

Biosphere reserves, **6**, 279.
Birds as pests, **1**, 89.
Breeding phaseolus beans as grain legumes for Britain, **3**, 1.

Cereal aphids: a case study and review, **5**, 272.
Chemical attraction of plant feeding insects to plants, **5**, 67.

Design and management of reservoir margins for multiple use, **2**, 129.
Development of forest science, **1**, 49.

Effects of discharge on sediment dynamics and consequent effects on invertebrates and
   sediments in upland rivers, **6**, 154.

Disease in plant communities, **5**, 145.

Ecological principles for the restoration of disturbed and degraded land, **4**, 142.

European rabbits, the European rabbit flea myxomatosis, **2**, 184.

Fish from sewage, **5**, 221.
Future of pesticides and other methods of pest control, **3**, 331.